IFIP Advances in Information and Communication Technology

308

IFIP – The International Federation for Information Processing

IFIP was founded in 1960 under the auspices of UNESCO, following the First World Computer Congress held in Paris the previous year. An umbrella organization for societies working in information processing, IFIP's aim is two-fold: to support information processing within its member countries and to encourage technology transfer to developing nations. As its mission statement clearly states,

> *IFIP's mission is to be the leading, truly international, apolitical organization which encourages and assists in the development, exploitation and application of information technology for the benefit of all people.*

IFIP is a non-profitmaking organization, run almost solely by 2500 volunteers. It operates through a number of technical committees, which organize events and publications. IFIP's events range from an international congress to local seminars, but the most important are:

- The IFIP World Computer Congress, held every second year;
- Open conferences;
- Working conferences.

The flagship event is the IFIP World Computer Congress, at which both invited and contributed papers are presented. Contributed papers are rigorously refereed and the rejection rate is high.

As with the Congress, participation in the open conferences is open to all and papers may be invited or submitted. Again, submitted papers are stringently refereed.

The working conferences are structured differently. They are usually run by a working group and attendance is small and by invitation only. Their purpose is to create an atmosphere conducive to innovation and development. Refereeing is less rigorous and papers are subjected to extensive group discussion.

Publications arising from IFIP events vary. The papers presented at the IFIP World Computer Congress and at open conferences are published as conference proceedings, while the results of the working conferences are often published as collections of selected and edited papers.

Any national society whose primary activity is in information may apply to become a full member of IFIP, although full membership is restricted to one society per country. Full members are entitled to vote at the annual General Assembly, National societies preferring a less committed involvement may apply for associate or corresponding membership. Associate members enjoy the same benefits as full members, but without voting rights. Corresponding members are not represented in IFIP bodies. Affiliated membership is open to non-national societies, and individual and honorary membership schemes are also offered.

Jozef Wozniak   Jerzy Konorski
Ryszard Katulski   Andrzej R. Pach (Eds.)

Wireless and Mobile Networking

Second IFIP WG 6.8 Joint Conference, WMNC 2009
Gdańsk, Poland, September 9-11, 2009
Proceedings

 Springer

Volume Editors

Jozef Wozniak
Jerzy Konorski
Gdańsk University of Technology
Department of Computer Communications
11/12 Narutowicza Str., 80-233 Gdańsk, Poland
E-mail: {jowoz, jekon}@eti.pg.gda.pl

Ryszard Katulski
Gdańsk University of Technology
Department of Radiocommunication Systems and Networks
11/12 Narutowicza Str., 80-233 Gdańsk, Poland
E-mail: rjkat@eti.pg.gda.pl

Andrzej R. Pach
AGH University of Science and Technology
Department of Telecommunications
Al. Mickiewicza 30, 30-059 Krakow, Poland
E-mail: pach@kt.agh.edu.pl

CR Subject Classification (1998): C.2.1, C.2.5, C.1.3, C.3, I.5.4

ISSN   1868-4238
ISBN-10   3-642-03840-9 Springer Berlin Heidelberg New York
ISBN-13   978-3-642-26906-6 Springer Berlin Heidelberg New York

springer.com

© IFIP International Federation for Information Processing 2009
Softcover reprint of the hardcover 1st edition 2009

Typesetting: Camera-ready by author, data conversion by Scientific Publishing Services, Chennai, India
Printed on acid-free paper   SPIN: 12744121   06/3180   5 4 3 2 1 0

Preface

Recent spectacular achievements in wireless, mobile, and sensor networks have dramatically changed our lives in many ways. However, the rapid evolution of wireless systems not only promises increased functionality, reliability, availability, and security, as well as putting a wide variety of new services at the users' disposal – it also creates a number of design challenges that our research community is now facing. Scientists and engineers need to come up with, and promptly implement, novel wireless network architectures, while system operators and planners rethink their business models and attend to the growing expectations of their customer base.

To provide a suitable forum for discussion between researchers, practitioners, and industry representatives interested in new developments in the respective research area, IFIP WG 6.8 launched three separate series of conferences: MWCN (Mobile and Wireless Communications Networks), PWC (Personal Wireless Communications), and WSAN (Wireless Sensor and Actors Networks). In 2008, MWCN and PWC were merged into the IFIP Wireless and Mobile Networking Conference (WMNC 2008), held in Toulouse, France, from September 30 to October 2, 2008. MWNC 2008 and PWC 2008 topics were subsequently revised with a view to covering the whole spectrum of hot issues in wireless and mobile networking. As a result, IFIP WG 6.8 decided to add WSAN as another WMNC track. WMNC 2009 therefore subsumed three tracks, MWCN, PWC, and WSAN, each with its own steering committee, but with a common TPC of more than 100 members. It was held in Gdańsk, Poland, September 9-11, 2009, hosted by Gdańsk University of Technology.

A total of 65 regular papers (split quite evenly among the three tracks) were submitted by authors from 20 countries. All submissions were evaluated by the program committee members assisted by external reviewers – at least three per paper. Finally, 30 papers were accepted and included in this proceedings volume. They reflect the state of the art, current discussions, and development trends in wireless and mobile networks and services. A variety of internationally significant topics are covered, including wireless LANs, sensor, ad hoc, UMTS and cellular networks, mobility, localization, routing, quality of service, IMS, network management, energy efficiency, and security.

The PWC track continued to penetrate its traditional area of communication-layer protocol design and performance evaluation. Particular attention was paid to multimedia traffic accommodation, including low-layer considerations in respect to EDCA queues and bandwidth reservation for voice connections, network-layer multicast video streaming, and session establishment signaling. Optimization of existing WiMAX protocols, and coexistence of popular wireless technologies like WiFi and Bluetooth within a common unlicensed band were given both analytical and experimental treatment. Universal teletraffic studies and cross-layer design principles were applied to wireless network design, with a relatively large number of papers dealing with handover performance and engineering. The authors did not turn away from security-related issues in view of possible uncooperative behavior of wireless terminals; one that has recently become of particular importance is trust building and management.

The WSAN track reflected two distinct groups of problems, one connected with the organization of ad-hoc wireless sensor networks, and the other covering the theory and practical algorithms of sensor location. Monitoring of the current state of the environment, especially geared towards threat detection and out-of-the-ordinary values of various parameters, is the subject of keen interest of local governments and commercial operators alike. A monitoring system constitutes a wireless wide-area network with sensor terminals, each of which consists of a measurement part and a transmission part with access to some telecommunication services. Its main function is the reliable and real-time transfer of monitoring data to a management database. From the design perspective, self-organization and security also play a central role.

Finally, the MWCN track focused on three major areas: mobile and wireless application and services, 3G/4G networks and next generation systems, and architectures, protocols and technologies. One of the major goals of mobile networks is to provide multimedia broadcast/multicast service on a mass scale. However, this requires successful completion of the relevant standardization processes, techno-economic studies, and calls for new concepts of multicast schemes. Another hot topic is the convergence of telecommunications and web services based on the integration of context information, Service Oriented Architectures (SOAs) and the IP Multimedia Subsystem (IMS). One of the key issues in real-time service provision in mobile networks based on e.g., WiMAX technology, is to ensure suitable handover schemes that exploit cross-layer optimization. As mobile networks become increasingly heterogeneous, network operators look beyond GSM/UMTS and get interested in wireless LAN/MAN and 3GPP Long Term Evolution (LTE). This creates additional issues e.g., handovers between access networks and operators.

As conference organizers, we are extremely grateful to the steering committees, the TPC, and all the external reviewers, for the hard work they put into the preparation and distribution of the call for papers, as well as the timely delivery of paper reviews. It is their highly competent and friendly support that enabled us to put together the consistent and, in our opinion, high-quality final program of WMNC 2009. We would like to thank all the authors for their submissions and interest in our conference, as well as Springer for their cooperation during the preparation of this volume. We feel indebted to many organizations and individuals who made this conference possible, in particular, the International Federation for Information Processing, Gdańsk University of Technology, IEEE Poland Section, Lord Mayor of Gdańsk, Alcatel-Lucent, and Nokia Siemens Networks.

Lastly, we are grateful to the local staff, service providers, and student volunteers for the countless hours they spent helping with the organization of WMNC 2009.

September 2009

Józef Woźniak
Jerzy Konorski
Ryszard Katulski
Andrzej Pach

Organization

General Chair

Jozef Wozniak — Gdansk University of Technology, Poland

Track Chairs

14th Personal Wireless Communications Conference (PWC)

Jerzy Konorski — Gdansk University of Technology, Poland

11th Mobile and Wireless Communication Networks Conference (MWCN)

Andrzej Pach — AGH University of Science and Technology, Poland

3rd Wireless Sensors and Actor Networks Conference (WSAN)

Ryszard Katulski — Gdansk University of Technology, Poland

Steering Committees

PWC

Pedro Cuenca, Spain
Zoubir Mammeri, France
Robert Bestak, Czech Republic
Otto Spaniol, Germany
Jozef Wozniak, Poland
Ignas G. Niemegeers, The Netherlands
Sonia Heemstra De Groot, The Netherlands

MWCN

Pedro Cuenca, Spain
Georg Carle, Germany
Cormac J. Sreenan, UK
Zoubir Mammeri, France
Guy Pujolle, France
Otto Spaniol, Germany
Jozef Wozniak, Poland

WSAN

Pedro Cuenca, Spain
Luis Orozco-Barbosa, Spain
Ali Miri, Canada
Al Agha Khaldoun, France
Otto Duarte, Brazil
Augusto Casaca, Portugal
Ivan Stojmenovic, Canada
Pedro Marrón, Germany

Technical Program Committee

Marek Amanowicz	Military Academy of Technology, Poland
Dominique Barthel	Orange Labs, France
Eesa Bastaki	Dubai Silicon Oasis Authority, UAE
Eric van den Berg	Telcordia Technologies, Inc, USA
Philippe Bertin	Orange Labs, France
Robert Bestak	Czech Technical University, Czech Republic
Raffaele Bruno	IIT-CNR, Italy
Wojciech Burakowski	Warsaw University of Technology, Poland
Carlos Cardeira	IST Lisboa, Portugal
Georg Carle	Universität Tubingen, Germany
Augusto Casaca	INESC, Portugal
Hakima Chaouchi	Télécom Sud Paris, France
Edward Chlebus	Illinois Institute of Technology, USA
Luis Correia	Technical University of Lisbon, Portugal
Pedro Cuenca	Universidad de Castilla-La Mancha, Spain
Amitabha Das	Nanyang Technological University, Singapore
Sajal Das	Nanyang Technological University, Singapore
Franco Davoli	Università degli Studi di Genova, Italy
Mieso Denko	University of Guelph, Canada
Michel Diaz	LAAS-CNRS, France
Eryk Dutkiewicz	Macquarie University, Australia
Luigi Fratta	Politecnico di Milano, Italy
Mario Freire	University of Beira Interior, Portugal
Rajit Gadh	UCLA, USA
Enrique Vázquez Gallo	Universidad Politécnica de Madrid, Spain
Adam Grzech	Wroclaw University of Technology, Poland
Zygmunt J. Haas	Cornell University, USA
Daryoush Habibi	Edith Cowan University, Australia
Sonia Heemstra de Groot	Ericsson EuroLab, The Netherlands
Geert Heijenk	University of Twente, The Netherlands
Zbigniew Hulicki	AGH University of Science and Technology Cracow, Poland
Youssef Iraqi	Dhofar University, Oman

Shengming Jiang	Norwegian University of Science and Technology, Norway
Yuming Jiang	Norwegian University of Science and Technology, Norway
Bożena Kamińska	School of Engineering Science, Simon Fraser University, Canada
Lukas Kencl	CTU-Ericsson-Vodafone R&D Centre, Prague, Czech Republic
Martin Klepal	Cork Institute of Technology, Ireland
Peter Komisarczuk	Victoria University, New Zealand
Ousmane Kone	Université Paul Sabatier – IRIT, France
Yevgeni Koucheryavy	Tampere University of Technology, Finland
Srdjan Krco	Ericsson IE Research Centre, Ireland
Francine Krief	LABRI, France
A. E. Krzesinski	University of Stellenbosch, South Africa
Sławomir Kukliński	Warsaw University of Technology, Poland
Marek Kwiatkowski	DSTO, Australia
Houda Labiod	ENST Paris, France
Shou-Chih Lo	National Dong Hwa University, Taiwan
Pascal Lorenz	University of Haute Alsace, France
Wiesław Ludwin	AGH University of Science and Technology Cracow, Poland
Maryline Laurent-Maknavicius	GET/INT Evry, France
Zoubir Mammeri	Paul Sabatier University, Toulouse, France
Vicenzo Mancuso	University of Palermo, Italy
Pietro Manzoni	Universidad Politécnica de Valencia, Spain
Dora Maros	Budapest Polytechnic, Hungary
Ami Marowka	Shenkar College of Engineering and Design, Israel
Rudolf Mathar	RWTH, Germany
Antonin Mazalek	University of Defence in Brno, Czech Republic
Steve McLaughlin	University of Edinburgh, UK
Ali Miri	University of Ottawa, Canada
Jelena Misic	University of Manitoba, Canada
Michal Mrozowski	Gdansk University of Technology, Poland
Mieczysław Muraszkiewicz	Warsaw University of Technology, Poland
Marek Natkaniec	AGH University of Science and Technology Cracow, Poland
Ioanis Nikolaidis	University of Alberta, Canada
Sotiris Nikoletseas	Research Academic Computer Technology Institute, Patras, Greece
Thomas Noel	University of Strasbourg, France
Andrew Odlyzko	University of Minnesota, USA
Luis Orozco-Barbosa	Universidad de Castilla La Mancha, Spain
Andrzej Pach	AGH University of Science and Technology Cracow, Poland
Elena Pagani	University of Milan, Italy
Gerard Parr	University of Ulster, UK

Krzysztof Pawlikowski	University of Canterbury, New Zealand
Dirk Pesch	Cork Institute of Technology, Ireland
Ana Pont-Sanjuan	Universitat Politècnica de València, Spain
Ramón Puigjaner	Universidad de las Islas Baleares, Spain
Guy Pujolle	University of Paris 6, France
Utz Roedig	Lancaster University, UK
Michael Rumsewicz	University of Adelaide, Australia
Dominik Rutkowski	Gdansk University of Technology, Poland
Djamel Sadok	Federal University of Pernambuco, Brazil
Debashis Saha	Indian Institute of Management Calcutta, India
Tadao Saito	Toyota Info Technology Center, Japan
Venkatesh Sarangan	Oklahoma State University, USA
Thomas Schmidt	Hamburg University of Applied Sciences, Germany
Patrick Senac	ENSICA, France
Bartomeu Serra	Unversitat de les Illes Balears, Spain
Michelle Sibilla	IRIT Paul Sabatier University, France
David Simplot-Ryl	INRIA, France
Wojciech Sobczak	Gdansk University of Technology, Poland
Otto Spaniol	University of Technology of Aachen, Germany
Dirk Staehle	University of Wuerzburg, Germany
Andrzej Stepnowski	Gdansk University of Technology, Poland
Ivan Stojmenovic	University of Ottawa, Canada
Yutaka Takahashi	Kyoto University, Japan
Andreas Timm-Giel	University of Bremen, Germany
Petia Todorova	Fraunhofer Institut FOKUS, Germany
Phouc Tran-Gia	University of Würzburg, Germany
Victor A. Villagra	Technical University of Madrid, Spain
Luis Javier Garcia Villalba	Universidad Complutense de Madrid, Spain
Luis Villasenor-González	CICESE Research Center, Mexico
Zuzana Vranova	University of Defence in Brno, Czech Republic
Krzysztof Wesołowski	Poznan University of Technology, Poland
Joerg Widmer	DoCoMo Communications Laboratories Europe GmbH, Germany
Bernd Wolfinger	Hamburg University, Germany
Konrad Wrona	NATO C3 Agency, The Hague, The Netherlands

Reviewers

Marek Amanowicz	Robert Bestak
Constantinos Marios Angelopoulos	Raffaele Bruno
Luis Barbosa	Wojciech Burakowski
Dominique Barthel	Augusto Casaca
Eric van den Berg	Edward Chlebus
Philippe Bertin	Luis Correia

Pedro Cuenca
Franco Davoli
Mieso Denko
Michel Diaz
Gengfa Fang
Julien Fasson
Mario Freire
Jorge Gomez-Montalvo
Adam Grzech
Geert Heijenk
Alberto Hernandez
Zbigniew Hulicki
Youssef Iraqi
Yuming Jiang
Ryszard Katulski
Lukas Kencl
Martin Klepal
Jerzy Konorski
Andreas Koutsopoulos
Srdjan Krco
Francine Krief
Marek Kwiatkowski
Maryline Laurent-Maknavicius
Shou-Chih Lo
Pascal Lorenz
Wieslaw Ludwin
Zoubir Mammeri
Vincenzo Mancuso
Pietro Manzoni
Dora Maros

Rudolf Mathar
Ali Miri
Jozef Modelski
Marek Natkaniec
Ioanis Nikolaidis
Thomas Noel
Dimitra Patroumpa
Dirk Pesch
Josef Pieprzyk
Ana Pont
Guy Pujolle
Utz Roedig
Michael Rumsewicz
Dominik Rutkowski
Djamel Sadok
Debashis Saha
Tadao Saito
Venkatesh Sarangan
Thomas Schmidt
Patrick Senac
Ivan Stojmenovic
Petia Todorova
Le Chung Tran
Victor Villagra
Javier Garcia Villalba
Luis Villasenor
Joerg Widmer
Widyawan Widyawan
Bernd Wolfinger
Jozef Wozniak

Table of Contents

Network Design

Sensor Networks

Trust Management and Competitive Networking

Location Algorithms I

Evolution of 3G, 3G/4G and Future Generation Systems

Location Algorithms II

Handover Mechanisms

Unified Charging and Billing Solution.
Unified – Next Generation of Charging Systems in Mobile Networks

Daniel Donhefner

Nokia Siemens Networks
Business Support Systems Research and Development

1 Motivation and Overview

The mobile market evolves from commodity voice and simple messaging services to value-added data and multimedia services. This not only implies to move from pure telecom to IT/IP- environment, but to exploit their markets with innovative and differentiated offerings to keep the churn rate low and attract new customers. Communication Service Providers (CSP) must focus increasingly on meeting individual needs and higher expectations of their subscribers. They expect service packages that can be tailored to meet the specific demands of their personal situation, preferences and lifestyle. This requires a flexible customer-centric approach instead of the legacy historical grown and diversed system architecture and organizations of CSPs.

The CSPs are facing the challenge to set up a complete operational environment in a very short time frame. However this is also the opportunity to reach a truly future proof and modern solution and gain competitiveness if the optimal architectural decisions are made now. Nokia Siemens Networks offers a full turnkey solution based on a modern architecture and world market leading products.

The success factors for CSPs are:

- *prepaid-postpaid convergence* to offer attractive end use services and streamline processes
- *solid CRM strategy* to improve customer insight and customer loyalty by customer centricity
- state of the art IT architecture via *middleware* approach
- focus on *productized services* (rather than project specific solutions) to decrease time to market and reduce OPEX
- stepwise approach to start with basic functions and the option to extend the products in the future evolution

With this Nokia Siemens Networks addresses all these issues and thus considerably extends the scope of competing solutions on the market that only introduce convergence.

A convergent charging and rating platform bundled with value-added applications is an elemental part to fulfill all market requirements. CSP`s marketing teams need flexible tools to enable their efforts for creating new business and defend the existing customer base. Many CSPs worldwide are currently developing their own convergent strategy to be clearly ahead of their competitors that are limited with their traditional

J. Wozniak et al. (Eds.): WMNC 2009, IFIP AICT 308, pp. 1–7, 2009.

legacy environment. The method of paying for services is not any longer a reason to separate offers and customers.

Nokia Siemens Networks *Unified Charging and Billing* solution allows CSPs to overcome the traditional restrictions when shaping innovative and differentiated services that target different customer types, payment methods or network types.

Success factor: Operational efficiency
Overcoming business silos

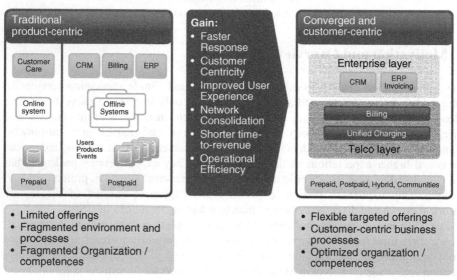

Fig. 1. Success factor: Operational efficiency – overcoming business silos

Traditionally CSPs had to update their legacy billing systems to cope with every new service. This often leads to a multiplicity of incompatible systems and databases, enlarging the complexity of the IT environment at the CSP. The reward for reducing the number of systems and databases and the simplification of the service introduction process represents the significant reduction in operational costs.

Consequently, simply segmenting customers into prepaid and postpaid is no longer valid. Offering attractive services and stimulating service usage is currently coupled with mastering a multitude of workflows and legacy interdependencies. It is not economically justifiable to have two completely separate systems for users who simply have different payment methods.

Shifting to a customer-centric approach brings significant competitive advantages to CSPs in terms of improved business flexibility and faster time-to-market of new services. It optimizes operational efficiency and reduces costs, due to reduced number of systems and interfaces, and using synergies of formerly separated pre- and postpaid systems and organizational structures.

Due to consolidated online information on subscriber's service usage and context, it enables the CSP to create attractive cross-bundling and cross-promotion offers across the whole subscriber basis. It enables CSPs to offer the same functionality, services, customer care and cost monitoring to all their customers, regardless, if they pay in advance before using the services, or if they will be invoiced on a regular basis after usage. With the possibility of having a complete customer insight across all service usage, offers can be made even more personal and targeted. Offering all combinations of voice, data and content services with prepaid and postpaid pricing plans creates loyalty, attracts new subscribers and can be used to promote higher revenue generating services like data.

Unified 2-layer architecture
Charging and Billing & Customer Care and Financials

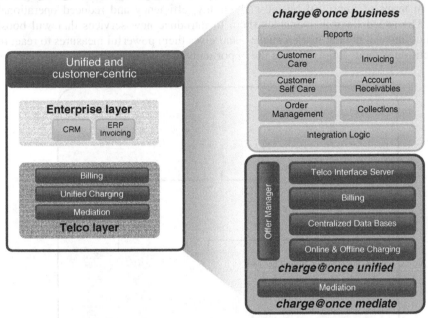

Fig. 2. Unified 2-layer architecture – Charging and Billing & Customer Care and Financials

Instant, online charging information is a vital tool for stimulating service usage. One of the major factors inhibiting the uptake of new value-added services is the concern of customers that they may be committing themselves to open-ended expenditure. In an end user research carried out by Nokia Siemens Networks end of 2006, more than 80% of respondents said, that knowing the cost of a service in advance would make them much more likely to use it. Furthermore, online charging allows CSPs to make balance-related offers and improves customer interaction. Prepaid customers in the fastest growing markets appreciate the opportunity of always having an up-to-date prepaid balance visible in their phone display. This is an example of the customer-centric features supported by our solutions.

Online charging and credit control implemented to postpaid customers also reduces the financial risks for CSPs. If a subscriber overruns his credit limit by using a voice service, the loss to the CSP is small. But new content-based services with higher tariffs significantly increase the financial risk. The exposure is even higher when the service or application is provided in partnership with a content provider who is entitled to a percentage of the amount paid by the user. If the customer fails to pay his bill, then the CSP has not only lost his own margin, but also has to pay the content provider. Online charging reduces the CSP's exposure to such losses.

With unified charging and billing the CSP can now benefit from having online and offline charging in the same solution. The possibility to use the optimum charging method, independent of the payment method means full flexibity for the offered services. At the same time it is possible to maximise the return on investment by choosing the cheaper offline charging method when the online control is not needed.

Making the shift to a customer-centric approach brings significant advantages to CSPs in terms of improved business flexibility, efficiency and reduced operational costs. The added flexibility enables them to introduce new services that will boost revenues and win greater market share – and gives them powerful measures to react to competition without simply having the opportunity to reduce prices.

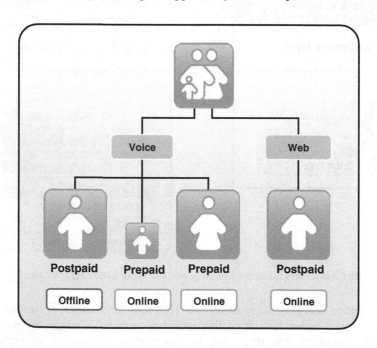

Fig. 3. Customer-centric approach

Nokia Siemens Networks believes to have the best solution for the CSPs for the following reasons:

- *Rely on Market Leading Solution Providers*. The solution is built on proven and world market leading products. Nokia Siemens Networks provides convergent charging and billing charge@once unified. In addition charge@once business consists of rebranded and pre-integrated Oracle products for CRM (Siebel), Invoicing (eBS) and Middleware (SOA Suite).

- *Profit from Nokia Siemens Networks End-to-End Commitment and Strong Regional Presence.* Nokia Siemens Networks has a strong regional footprint in all regions of the world, including an own system integration branch CSI.
- *Enhance Competiveness by Distinguished Productized Approach and Flexible Customization.* The productized approach together with high flexibility through on-site adaptation (adapt@once) allows CSP to profit from frequent product enhancements and to quickly introduce new services.
- *Go With the Pre-Integrated Two-Layer Architecture, the Future Architectural Standard for BSS.* The two layer approach is replacing the traditional billing systems by splitting its functionalities between the Telco specific charging and enterprise generic CRM and ERP including Financials. This allows to deploy best in class products and to gain from economy of scale.
- *Put CRM in the Proper Place.* Unlike some competitor offerings, which are centered around billing, Nokia Siemens Networks builds on a CRM as the master of the main customer data.
- *Gain with a Stepwise Evolution to Reach Your Business Targets with Minimal Risk.* Nokia Siemens Networks offers a smooth stepwise evolution. This minimizes risks, addresses the business targets of the CSP in a prioritized order and allows to grow in a flexible way.

2 Solution Overview

The *Unified Charging and Billing* solution offers rating, charging and billing calculation of content, events, sessions in SS7 and IP networks independent of the payment method. It is a modular, highly scalable and flexible solution, offering open interfaces for a smooth integration into the CSP's operational environment. Based on the long-term experience of Nokia Siemens Networks in the real-time prepaid market, it provides superior connectivity, scalability, capacity and performance as well as the carrier-grade availability needed for business-critical online charging solutions.

This solution is an evolutionary step towards a two-layer architecture work split between a transaction-latency optimized "Telco Layer" and an "Enterprise Layer" which is telecom agnostic and dedicated to fast enterprise process integration. With this horizontally layered architecture every function applies to the entire customer base and to the entire product portfolio - overcoming the historical grown diversity of systems and processes between pre- and postpaid and resulting vertical borders for fast, flexible and consistent product shaping and customer segmentation.

As a CSP, you win flexibility, data consistency and reduce your OPEX, credit risks and revenue leakage due to the consolidation of system environment and business processes.

This approach simplifies and accelerates operational workflows and puts CSPs into the position to focus on fast roll-out of any offering to any member of the customer base i.e. to offer all services, product bundles, and quality of services to all customer and access types.

Add. CRM Modules
Add. ERP Modules
charge@once business
Customer Care
Financials
Reports
Integration Logic
Provisioning
Telco Interface Server
Billing
Online Charging
Offline Charging
Session Control
charge@once unified
Topup Management
CAP
INAP
Diameter
CDR
Mediation
SS7 Network
IP Network
CRM = Customer Relationship Management
ERP = Enterprise Resource Planning
*examples

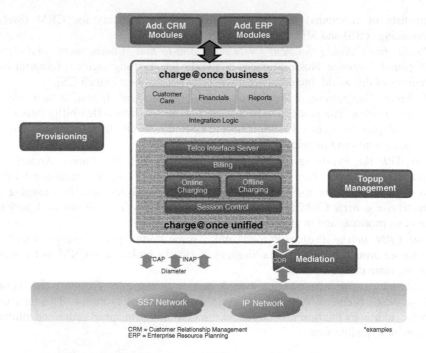

Fig. 4. Unified Charging and Billing – modular architecture

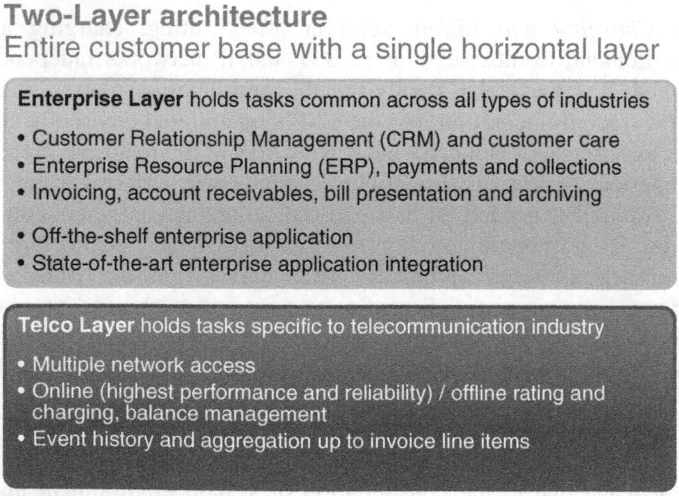

Two-Layer architecture
Entire customer base with a single horizontal layer
Enterprise Layer holds tasks common across all types of industries
• Customer Relationship Management (CRM) and customer care
• Enterprise Resource Planning (ERP), payments and collections
• Invoicing, account receivables, bill presentation and archiving
• Off-the-shelf enterprise application
• State-of-the-art enterprise application integration
Telco Layer holds tasks specific to telecommunication industry
• Multiple network access
• Online (highest performance and reliability) / offline rating and charging, balance management
• Event history and aggregation up to invoice line items

Fig. 5. Two-Layer architecture – entire customer base with a single horizontal layer

The Nokia Siemens Networks *Unified Charging and Billing Solution* supports all of these scenarios and enables CSPs to create attractive bundles of services and charge for these efficiently, regardless of whether they were designed for a prepaid or postpaid subscriber. This will allow CSPs to differentiate themselves in the market place, thus driving up the ARPU and increase loyalty of subscribers.

3 Benefits at a Glance

The optimized architecture for all applications, all networks and all users enables fast adaptation to market needs, and significantly reduces the operational cost of managing multiple systems and workflows for Prepaid and Postpaid customers. The *Unified Charging and Billing* solution allows to eliminate legacy billing systems and covers the different availability requirements for online and offline processing with dedicated functional components in a cost-optimized way.

For a CSP, the evolution to a *Unified Charging and Billing* solution offers following benefits:

* CSPs can rapidly deliver differentiated and targeted solutions to meet market and individual customer demands and thus increase their service revenues
* Introduction of services regardless of payment method of the subscriber
* The ability to deliver attractive pricing schemes, increased service selection to all customers, and an improved personalized care and self-care will attract new customers and maximize the retention of existing customers.
* Online charging, cost control and price notification services that help end-users to keep control of their spending will actually make them confident to use value-added services, and above all it reduces the entry barriers for new offerings
* Financial risks for the CSP can also be minimized as all services can be charged in advance or supervised in real time and customers can easily be moved to credit supervision and vice versa.
* Online or near-online rating and charging for all new and innovative applications as well as for traditional services in mobile, fixed, SS7 and IP networks prevents from revenue leakages.
* A consolidated system generates significant reductions in operational costs through reduced complexity, improved data management and reduced correction efforts.
* The solution is designed around a complete set of CSP business process descriptions using an end-to-end view for customer interaction, product definition and billing/charging.
* Integration is simplified due to use of same logical data definition across the solution, the open APIs and the powerful Telco Interface Server ensure the seamless integration with Enterprise Layer components and external systems as well – be it a financial system, CRM, Inventory management system or any other commercial off-the shelf system.
* Efficient new hardware reduces OPEX due to less floor-space, increased calculation performance and less energy consumption

Presence Management and Merging Presence Information for NGN Services

Sebastian Schumann[1], Eugen Mikoczy[1], Pavol Podhradsky[1],
Feliciano Muruchi[2], and Michael Maruschke[2]

[1] Slovak Technical University, Ilkovicova 3, 812 19 Bratislava, Slovak Republic
{schumann,mikoczy,podhrad}@ktl.elf.stuba.sk
[2] Deutsche Telekom AG, Hochschule fuer Telekommunikation Leipzig,
Gustav-Freytag-Str. 43-45, 04277 Leipzig, Germany
{feliciano.muruchi,maruschke}@hft-leipzig.de

Abstract. This paper describes an approach for interworking scenarios between Session Initiation Protocol (SIP) based and non SIP based frameworks (e.g. web services) in case of the presence management service. The characteristics of the concept of a centralized presence management will be introduced.

Based on the Parlay X web service framework a presence management service solution has been developed, deployed and tested. The re-usability of the developed presence management service solution within an IP Television (IPTV) framework and the future potential of the presence service in various IPTV applications are analyzed.

1 Introduction

The presence service is one of the key services that next-generation architectures like the IP Multimedia Subsystem (IMS) can provide already.

Based on a simple SIP infrastructure (non-IMS), this paper will show the usefulness of interconnecting frameworks around the presence service. The authors deployed the SIP based presence management service interconnected with web services successfully. An existing testbed has been extended with a web service framework as first step. After the web service framework has been deployed in a stable state, standards have been discussed and solutions proposed. Moreover, prototypical applications for both presence client scenarios (presentity and watcher) have been defined, deployed and tested.

The paper will also discuss the re-usability of a once deployed presence service from the IPTV framework. The concepts are similar and proof again that service enablers and framework interconnections have a great future. Standardization is currently on-going within this area, the authors contributed here as well.

2 Presence Architecture within the SIP Framework

In a SIP framework, the proxy server passes all presence related messages (SIP SUBSCRIBE and SIP PUBLISH) to the presence server. It handles the publications and stores the presence information of the publishing user [1,2,3].

J. Wozniak et al. (Eds.): WMNC 2009, IFIP AICT 308, pp. 8–19, 2009.

In contrast to the presence server, which is usually reachable for the users only through the SIP core, the Extensible Configuration Access Protocol (XCAP) server advertises its interface directly towards the User Equipment (UE). The XCAP protocol allows clients to read, write and modify XML data on the server.

In combination with a presence server, the XCAP server manages data like buddy lists, presence policy and presence information and makes it accessible through the XCAP interface to the presence clients.

2.1 The Role of Presence

Aggregation of presence information results in many presence states from many sources altogether collected by a single source in the network. To organize the presence states and also to create useful presence information for each service, a presence state model that prioritized the information will be proposed. It is up to the presence server to create a final state from all this information. Some major characteristics of this status model are listed below.

The state model should group all states that cannot be or have not been mapped into service specific groups. Attributes (see Table 2) that can be mapped and their respective communication means should be aggregated and prioritized according the users preferences.

The following list is an extract of possible service presence states:

- Voice presence (wired UE): online, present at phone, busy, on the phone
- Voice presence (wireless UE): online, profile (silent/on meeting - normal - work - at home - on road), out of reach, do not disturb, on the phone, current location
- Calendar presence: activity, time frame, location
- GPS/RFID enabled devices: location

Common attributes have been outlined already. An automatized privatization should take part acc. the following presentities:

- User set state preferred over automatic state. Example: mobile state, e.g. on the road, overwritten by active call, e.g. on the phone
- Real location information preferred over static assumption. Example: GPS information modifies calendar presence state
- Communications ranking.

All attribute ranking are proposals that the user should be able to modify, if required.

2.2 Merging Presence Sources for Real-World Scenarios

Nowadays, operators do not only operate telco domains (e.g. plain telephony using VoIP or some NGN), but offer various communication services through different domains (e.g. web 2.0 applications, web services, IPTV). The challenge appears now when different domains should be interconnected and share presence information. Concretly, different frameworks will be taken into account to explain the merger of handling the presence information in detail: The SIP domain, which will be the presence management domain, and the web service and IPTV domains respectively, which will access the presence information.

One presence management service within different domains requires two conditions to be full-filled:

– The methods that SIP uses to publish/subscribe to presence information must be mapped to methods from the other domain. This ensures that the presence information can be accessed from the other domain.
– The actual presence information must be mapped towards the other domain. This ensures that the other domain understands the information it exchanges.

If the requirements from above are implemented, the domains can communicate with each other and also understand the content. The non-SIP domain can request presence information, process it and also publish states.

3   Parlay X Web Service Framework

The Parlay X framework has been defined by the standardization bodies ETSI, 3GPP and Parlay Group. Its specifications describe, how application developers can use different telecommunication services through an open standardized interface to generate new, innovative applications. The used technology is the web service architecture, which is based on SOAP[1] and Web Service Definition Language (WSDL).

3.1   Presence Service within the Parlay X Framework

The main functionalities of the presence service within the Parlay X framework are specified through logical interfaces that are standardized in [4]. This specification defines namespaces, sequence diagrams, data definitions, fault definitions, a detailed description of these logical interfaces and its corresponding WSDL descriptions. The specified interfaces are:

– The PresenceConsumer interface
– The PresenceNotification interface
– The PresenceSupplier interface.

The PresenceConsumer interface and the PresenceNotification interface define the watcher side of the presence architecture. The latter interface is a call back interface, which allows the presence web service to send notification requests to the web service client. Hence, the watcher does not only act as client, but also as server in the web service context by offering this interface. This approach is not common or practical for a web service architecture. It is however the only

---

[1] Originally defined as Simple Object Access Protocol but acronym was dropped with Version 1.2.

possibility to receive information in a subscription mode. This is essential for the presence service.

The PresenceSupplier interface is used for the presentity side in order to publish presence information and to manage the authorization of its presence information.

3.2   Interconnecting Presence Web Services with the SIP Framework

Each interface supports different methods in order to meet its function. These methods have to be mapped to the corresponding methods, which are used in SIP for the presence service. [5] proposes presence mapping scenarios between the Parlay X web service framework and the SIP/IMS framework.

The analysis of [5] revealed some issues regarding the exact sequences of protocol messages. These sequence diagrams do not appear standard conform regarding [1,2,3] particularly the use of the SIP message ACK and the lack of the final SIP 200 OK responses. Furthermore, this dialog is not practical in terms of using the startPresenceNotification method.

Fig. 1 shows, how the mapping for the watcher side of the Parlay X specification could work successfully with an existing SIP presence core. Therefore, the PresenceConsumer interface and the Presence Notification interface are considered. Finally, the proposed diagram is oriented on [5] and is modified for practical reasons.

For the sake of completeness, all response messages (SIP and SOAP) were added. The main differences towards [5], sec. 5.1 are the following:

1. SIP ACK message is not implemented.
2. Implicit SIP NOTIFY after 200 class response for SUBSCRIBE is necessary.
3. SOAP startPresenceNotificationRequest requires a new SIP subscription.

The authors explain those changes with the following reasons:

1. The SIP ACK method is not valid within the SIP Event framework. It is only part of the SIP three-way handshake for session set-up. The ACK (nowledgement) described in [5], sec. 6.1.1 that changes state (S1) to (S2) is the 200 class reply for SUBSCRIBE.
2. The SIP NOTIFY is omitted in [5], sec. 5.1, however mentioned in sec. 6.1.1. As it is implictly required after the 200 class message ([3], sec. 3.1.4.1), it has been added to the figure.
3. The startPresenceNotificationRequest issues a new subscription on the SIP side. This major change towards [5] is caused by the fact that the first SUBSCRIBE (whose NOTIFY response could be used acc. [5], sec. 6.1.3) could be a one-time polling. StartPresenceNotification expects however a SIP 'subscriber-watcher' behavior, i.e. expiry not equal zero and hence ongoing notification. This is not explicitly excluded for the subscribePresence event. Limiting subscribePresence for expiry not equal zero does not make sense, as the notifySubsciption is not used to actually elaborate the presence information and only indicates a successful subscription.

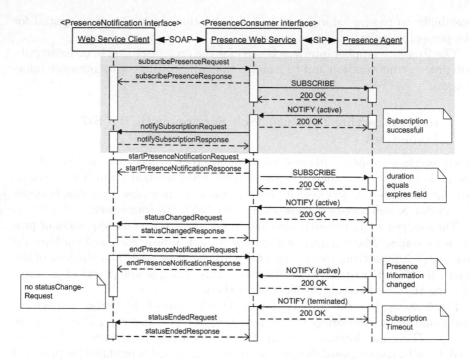

Fig. 1. Signaling flow overview of watcher side (as web service client)

For clarity, the getUserPresence fetching from [5], sec. 5.1 has been omitted in Fig. 1. The initial presence subscription (SOAP differentiates between subscription and actual presence information exchange) has been highlighted in light grey in Fig. 1. The following procedure shows the exchange of the actual presence information.

[5] does also not yet consider the mapping of attributes, which are used within the methods of both sides. Table 1 shows the proposed approach for the attribute mapping as a possible solution.

This proposal was implemented and tested successfully and will be described in section 3.3.

In addition to the previous steps, the presence status information needs to be mapped from the Presence Information Data Format (PIDF) and its extension Rich Presence Extension to PIDF (RPID), which is used by the SIP domain, to the defined structure within the Parlay X specification. Table 2 shows the mapping of basic presence information.

This mapping shows that all of the activity values, which are specified in the Parlay X specification, can be mapped clearly to PIDF/RPID. As not all activity elements of the RPID standard can be found within the Parlay X specification, some information will still be lost in case two entities of different domains (SIP and web service) would interact with each other.

Table 1. Parlay X attributes/SIP parameter mapping table for watcher side

SOAP methods	Parlay X attributes	SIP parameters
subscribePresence	presentity	TO header field
	attributes	PIDF[a] document
	reference (endpoint, interfaceName, corre-lator)	TO header field, FROM, CONTACT, watcher identifier
notifySubscription	presentity	FROM header field
	decisions	PIDF document
startPresenceNotification	presentity	TO header field
	attributes	PIDF document
	reference (endpoint, interfaceName, corre-lator)	TO header field, FROM, CONTACT, watcher identifier
	duration	Expires header field
endPresenceNotification	reference (endpoint, interfaceName, corre-lator)	TO header field, FROM, CONTACT, watcher identifier
subscriptionEnded	presentity	FROM header field
	reason	optional REASON header field
statusEnded	correlator	Take correlator from reference from startPresenceNotification
statusChanged	correlator	Take correlator from reference from startPresenceNotification
	presentity	TO header field
	changedAttributes	PIDF document

[a] PIDF: Presence Information Data Format

Table 2. Presence status information mapping table for watcher and presentity side

PIDF + RPID	Parlay X
status=open	AVAILABLE
RPID = busy	BUSY
RPID = on-the-phone	ON_THE_PHONE
RPID = away OR status=closed	AWAY
RPID = meeting	MEETING
RPID = lunch	MEAL
RPID = holiday	HOLIDAY
RPID = steering	STEERING
RPID = in-transit	INTRANSIT
RPID = travel	TRAVEL
RPID = sleepingv	SLEEPING
RPID = permanent-absence	PERMANENTABSENCE

3.3   Testbed Solution

By offering a presence web service, which is described in chapter 3.2, any possible system or service can be extended directly with a web service client or through a proxy in order to communicate its presence information towards the presence service. Within the laboratories of STU, an example communication scenario has been deployed (Fig. 2).

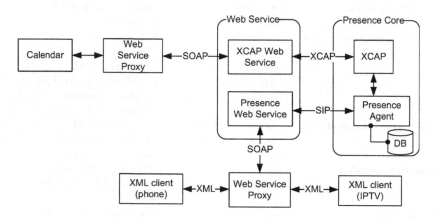

Fig. 2. Test platform presence web service

The tested scenario shows the usability of the discussed interconnection between the web service framework and the SIP framework. Part of this scenario are two non-SIP based entities. One entity is an IP phone with an XML interface (for provisioning or service integration) and acts as a watcher over a proxy towards the presence web service. The other entity is represented by an application, which retrieves calendar data from a calendar service and acts as presentity by publishing it. Since the pure publication of presence information can be done through pidf-manipulation over XCAP as well, the web service access offered by the XCAP server is used. Only the PresenceConsumer interface was implemented within the testbed presence web service.

The example sequence diagram in Fig. 3 shows the signaling flow for the watcher side, which receives presence information in a subscription mode.

1. The IP phone triggers the subscription mode with an HTTP GET message, which executes a script on the web service proxy.
2. The web service proxy creates a startPresenceNotificationRequest SOAP message, which is send to the presence web service.
3. The presence web service subscribes the presence information.
4. The notification is received by the presence web service.
5. In case the notification is active and contains PIDF data, this information is sent to the web service proxy using the statusChangedRequest SOAP message.

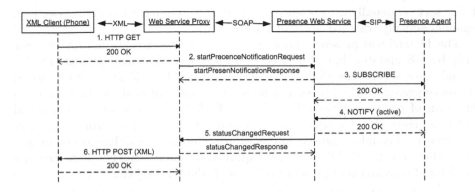

Fig. 3. Sequence diagram of subscription mode

6. As a result the presence information is revealed on the phone over a XML document.

Note: The 200 OK messages and the SOAP response messages acknowledge the previous requests (like HTTP GET) on the application layer.

4 NGN Based IPTV Frameworks and Use Cases for Presence

The following section describes IPTV services and related aspects for the usage of presence for NGN based IPTV, which have been defined within ETSI TISPAN. Both NGN based IPTV concepts were specified by TISPAN: NGN dedicated IPTV and IMS based IPTV [6] (where the IMS presence framework from 3GPP and OMA could be reused) will be explained, focussing on the role of presence in particular.

4.1 Services within the IPTV Frameworks

Basic IPTV services and their implementation within the NGN framework can communicate with the presence framework to update the presence states:

– Broadcast TV (BC with or without trick modes) – delivery of linearly broad-casted TV channels. Trick Modes enable control playback and pause, forward, rewind content. Presence helps service control to follow user actions.
– Pay Per View (PPV) – users pay e.g. only for particular show or time period, not whole TV channel or package. Presence could be used to trigger PPV usage.
– Content on Demand (CoD) – user requests content consumption on demand.
– Personal Video Recording (PVR) – user can record content in network (network or n-PVR) or locally in STB (client or c-PVR). Presence could be used for decision, if content is recorded locally or in network (if User Equipment (UE) is offline, only n-PVR is used).

Basic services usually use presence to observe the actual state of the service (e.g. channel actually watched, state of VoD preview).

The hierarchical presence model can be used to aggregate information not only by UE updates, but also directly from IPTV functional elements (its internal state, service state or service action data). The utilization of web based presence aggregation concepts and relevant correlations of data allow the combination of presence with IPTV services. Perfect examples are combinational services, where IPTV services are combined with presence to improve the user experience (through enhanced interaction, profiling, personalization or targeting). More advanced IPTV services require also a more complex presence state model and correlation of presence states with the service behavior:

- Interactive TV (iTV) – service providing interactivity between provider/ broadcaster and end user or between several users. Presence helps to identify available users for such an interactivity.
- User Generated Content (UGC) – content produced by the end user with the intention to share it with other users. Distribution of UGC makes sense only to the available (online) party, that is willing to receive content (user can set presence to busy to avoid any push services like UGC or PushCOD).
- Content Recommendation (CR) – service advisory for favorite shows based on user's preferences and behavior (here presence is crucial to define right recommendation because if system fails in identification of correct presence state for exact user, it will dismiss correct recommendation which will be in contradiction with user expectation and real preferences).
- Personalized channel (PCh) – user specific list of programs that are scheduled as play list for personalized preview. When the user pauses a show (e.g. moves out of the room with STB), it can be picked up on other device based on presence.
- Targeted advertising (Ad) – advertising mechanism which is targeted to specified group of users based on their user profiles. Together with user profiling and presence, the operator (or third party advertiser) can identify user interests and target type/form/place of advertising.
- Profiling and personalization (P&P) – feature which enables personalized IPTV services based on user preferences and user profile. The provider can also use the information about the user behavior and content consumption.
- IPTV and NGN Service Interaction (e.g. presence based game, incoming call notification, sharing the remote control).

4.2   Presence Concept in NGN Integrated IPTV

In the TISPAN NGN integrated IPTV subsystem (formerly NGN dedicated IPTV), any IPTV services may be combined with the presence service capability. Presence is provided via a presence server, which is presented as NGN Application Server Function (NGN ASF) or the NGN ASF could be used as gateway to an external presence server (previously described web service presence for example). More information is available in [7], where the authors contributed details to the particular IPTV procedures towards TISPAN already.

The following specific IPTV attributes may be fetched:

- Service currently accessed/activated.
- Content currently accessed/watched.
- Presence filtering for buddy list members and buddy management.

4.3   Presence Concept in NGN IMS Based IPTV

TISPAN specified presence [8] via an endorsement from the main presence standard from 3GPP (which in fact uses [9]).

The publication of IPTV specific information in presence documents depends on user-configurable data stored in the UE.

If the presence service is used, the UE shall implement the role of a Presence User Agent (PUA) as specified in TS 283 030. The UE may send a SIP PUBLISH request to update the presence status based on reception of a final SIP 200 OK concerning any service session initiation procedure (after any IPTV service, e.g. VoD, n-PVR, UGC, etc.). Additionally, the UE may also send a PUBLISH request during a BC session after having performed a channel-change (i.e. sending request for a particular BC service) or just after a timer expires send the presence information about really watched channel (e.g. when zapping, immediate change to next channel is not send). Fig. 4 shows the message exchange in detail.

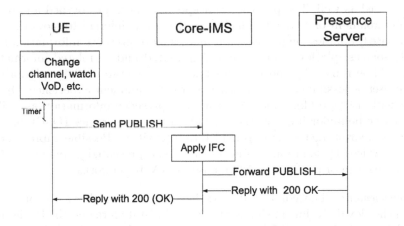

Fig. 4. Example of presence information update after service status change (from UE)

Another more efficient option is the direct interaction of the Service Control Function (SCF) with the presence server to update the presence status for user[2]. This solution could minimize the need for presence signaling traffic and transmit SIP messages across the IMS for each service initiation a second time[3].

---

[2] The SCF serves in fact service initiation requests.
[3] Additional to INVITE messages with service initiation.

5   Future Prospects

The described SIP presence framework and the Parlay X testbed and prototypes were deployed with NGNlab [10]. Between main services (IMS core, SIP AS), we count multimedia communication, collaborative tools, video streaming services, enhanced messaging and also plenty of different web services.

One part of the further work will concentrate on realizing the interaction with more services. Besides extending the platform with measurement tools, the following integration scenarios are planned:

- NGN web based applications (e.g. Presence Web 2.0 integration), web and e-learning portal integration;
- NGN multimedia streaming services (e.g. Web streaming, IPTV applications) will try to communicate with the presence service in described manners.

6   Conclusion

The centralized presence management concept presented in paper can reduce overhead within the provider network and anticipates redundancy. Users or services can use the presence management, and benefit from the information that is already contributed towards the system. Due to different meanings, i.e. mappings, it might not be able to interpret all attributes, but mapped states can be interpreted as well. The proposed concepts have been prototyped and tested in real environment of NGNlab. The integration cross different frameworks (SIP or Web based) enables, aggregates and merges more presence information from multiple sources, platforms or devices. Aggregated and correlated information then should be reused by end user, his buddies or operator service platform to improve user accessibility, communication or entertainment experience. Operator can with user profiling and collecting user presence information as well as service usage behavior improve the user Quality of Experience (QoE) via personalized or more target services provided to subscribers. Possible improvements of IPTV services by mentioned concept have been presented in paper together with existing role of presence in NGN based IPTV frameworks.

Acknowledgement. The conducted work has been the result of a cooperation between the Slovak Technical University (STU) in Bratislava and the Hochschule fuer Telekommunikation (HfTL) in Leipzig. The testbed in the STU has been extended with the web service framework and all described interaction scenarios have been deployed there. STU and HfTL are working together on common topics around Next Generation Networks since 2006 and participate in ongoing projects involving their core network interconnection. This paper also presents some of the results and acquired experience from various research project such as NGNlab project [10], European Celtic-EURECA project Netlab [11], Leonardo da Vinci projects InCert [12] and Train2Cert [13], AV project: Converged technologies for next generation networks (NGN), Slovak National basic research projects VEGA No. 1/3094/06 and VEGA 1/4084/07.

References

1. Rosenberg, J. et al.: SIP: Session Initiation Protocol, RFC 3261 (2002)
2. Rosenberg, J.: A Presence Event Package for the Session Initiation Protocol (SIP), RFC 3856 (2004)
3. Roach, A.B.: Session Initiation Protocol (SIP)-Specific Event Notification, RFC 3265 (2002)
4. ETSI ES 202 391-14 v1.2.1 (2006-12) Open Service Access (OSA); Parlay X Web Services; Part 14: Presence (Parlay X 2)
5. ETSI TR 102 397-14-1 v1.1.1 (2005-08) Open Service Access (OSA); Mapping of Parlay X2 Web Services to Parlay/OSA APIs; Part 14: Presence Mapping; Subpart 2: Mapping to SIP/IMS Networks
6. ETSI TS 183 063 V2.3.0 (2008-11), TISPAN; IMS based IPTV, Stage 3 (2008)
7. ETSI TS 183 064 V2.1.1 (2008-10),TISPAN; Dedicated IPTV subsystem, Stage 3 (2008)
8. ETSI TS 182 008 V2.1.1 (2008-10), TISPAN; Presence Service; Architecture and functional description (Endorsement of 3GPP TS 23.141 and OMA-AD-Presence_SIMPLE-V1_0)
9. OMA Presence SIMPLE, Architecture Document OMA-AD-Presence_SIMPLE-V1_0-20060725-A, Open Mobile Alliance (July 2006)
10. NGNlab - NGN laboratory at STU in Bratislava, http://www.ngnlab.eu
11. NetLab - Use Cases for Interconnected Testbeds and Living Labs, http://www.celtic-initiative.org/Projects/NETLAB/
12. InCert Next Generation Network Protocols Professionals certification in InCert, International Certificates of Excellence in Selected Areas of ICT, http://incert.eu
13. Train2Cert, Vocational Training for Certification in ICT, http://train2cert.eu

Interoperability of IP Multimedia Subsystems: The NetLab Approach

Andrés Marín-López[1], Daniel Díaz-Sánchez[1], Florina Almenárez-Mendoza[1],
Francisco Rodríguez García[2], Joaquín López Rizaldos[2], Alfonso Cubero Vega[2],
Itziar Ormaetxea[3], Juan Luis Lopez[4], Sergio Fernández[4], Fernando Ortigosa[4],
Ivan Kotuliak[5], Eugen Mikoczy[6], Timo Lahnalampi[7], Timo Koski[8],
and Sussana Avessta[8]

[1] Universidad Carlos III de Madrid, Spain
{amarin,dds,florina}@it.uc3m.es
[2] Telefónica Investigación y Desarrollo, Spain
{frg,jolo,acubero}@tid.es
[3] Software Quality Systems, Innovalia Group
iormaetxea@sqs.es
[4] Ericsson Network Services, Spain
{juanluis.lopez,sergio.fernandez,fernando.ortigosa}@ericsson.com
[5] Slovak University of Technology in Bratislava, Slovakia
ivan.kotuliak@stuba.sk
[6] Slovak Telekom, Slovakia
eugen.mikoczy@t-com.sk
[7] Digital Media Service Innovations, Finland
timo.lahnalampi@dimes.fi
[8] University of Turku, Finland
{timo.koski,annys}@utu.fi

Abstract. Standardization bodies have spent lots of efforts and have
extensively defined IP Multimedia Subsystem interfaces. While Telcos
have started its deployment, IMS-based applications boosting user needs
are still to come. Such applications must not suffer from interoperability
issues caused by different vendors and different administrative domains.
NetLab project aims at exploring such interoperability issues by interconnecting together testbeds at three different countries. Use cases will
be defined to find out about interoperability, and also to search for a user
appealing application in collaboration with the LivingLabs community.
This paper introduces the project vision on the architectural, security,
QoS, and interoperability issues, together with new services illustrating
the interconnection of testbeds.

Keywords: IMS interoperability, authentication and QoS in IMS, Living
Labs.

1 Introduction

Next Generation Networks (NGN) are considered as a major step in the telecommunication world evolution. The novelty consists in the transition of the core

J. Wozniak et al. (Eds.): WMNC 2009, IFIP AICT 308, pp. 20–31, 2009.

network from the PSTN (Public Switched Telephone Network) to the all-IP with QoS management. Advantages for the operators go from simplifying the core network using less complex equipment to the cost saving (both CAPEX and OPEX). Telecom operators have rapidly understood these advantages and started to replace PSTN by NGN.

The deployed equipment has been conformed to the soft-switch architecture from the beginning. The soft-switch represents a monolithic central block, which handles whole signalization and controls gateways. Such approach is rather inflexible. During the time, 3GPP has started standardization of the NGN implementation in the mobile networks as IP Multimedia Subsystem (IMS) [1]. It appeared in release 5 and now it has become a widely accepted standard and architecture for the NGN implementation (also for fixed networks). The main enhancement of the IMS compared to the soft-switch is the distribution of the functions into different nodes and the definition of interface points. Thanks to this, Telcos can use different functions from respective vendors and the crucial point is that applications have become completely open to the 3rd party developers.

The IMS is already well standardized. However, the applications and their behavior in the various interconnect scenarios are still under heavy understudy. The aim of the NetLab project is to develop a platform of interconnected test beds involving three different countries together with different Living Labs. Living Lab is an experimentation environment in which technology is given shape in real life contexts and in which (end-) users are considered 'co-producers'. Living Lab involve human-centric development of new ICT-based services and products. Panlab project has developed a federation framework of interconnects testbeds belonging to regional clusters, hence partitioning the management of the resources to different levels. The main design principle in Panlab federation is the dedication of resources under the common federation governance. NetLab platform should be designed to tackle interoperability, scalability, complexity and mobility aspects holding security and QoS requirements. Such an approach allows refining Panlab/Living Lab concepts with respect to different issues (organization, business models, legal aspects, security, IP protection) and to develop 2 IMS and DVB-H related use cases involving Panlab and Living Labs scenarios. The NetLab project is complementary to both Panlab and LivingLab, where Panlab testbeds have a technological focus, and Living Labs a user centric focus.

The organization of the article is as follows: section 2 covers the architectural issues including security and QoS aspects, and the overview of the platform interconnection possibilities, section 3 describes Netlab proposal for new services and use cases, and section 4 presents the related works. Finally, we end with some conclusions.

2   Architectural Issues

The development of IMS services is currently ongoing, but the services of different vendors are still lacking on the interoperability. NetLab's interconnected testbed architecture will support IMS service and application developers by providing different testing environments under one agreement and system. The architecture shall support service and technology developers of both the owner parties of testbeds and external parties in need of testing environment. There are two main issues in approaching the architecture goal.

First, the interoperable parts of different testbeds need to be defined in order to be able to run end-to-end services across. This includes a study of service scenarios and use cases. Based on them, a profile of interoperable services and applications is defined. This activity will lead to a service catalog of possible applications and their availability in different testbeds, and definition of methods how to access them.

Second, both organizational and technical access methods to testbeds need to be defined. This includes interfaces for both interconnection between users of testbed organizations and external users. On the organizational level, usage rights are defined in an interdomain agreement between parties. The interdomain agreement is supported on the technical level by definition of methods for authorization, authentication and accounting (AAA). Based on the method selection, common protocols for both AAA and service access will be defined, from the network layer and up. Other technical issue in definition of the architecture is the requirement of maintaining a sufficient QoS (Quality of Service) level.

2.1 Security Aspects

TISPAN (Telecommunications and Internet converged Services and Protocols for Advanced Networking) and 3GPP (3rd Generation Partnership Project), defined different Transport Control Layers for IMS; this has not contributed to a full fixed-mobile convergence.

In the fixed domain TISPAN has adhered to traditional authentication solutions, where dial-up protocols (mainly Point-to-Point Protocol, PPP) are used as authentication methods in the physical line where user is connected and identified. 3GPP also maintains, in mobile networks, Global System for Mobile communications (GSM) mechanisms, improving slightly security issues. These different approaches lead to having two disparate authentication models based on different protocols, different identities, different infrastructure and different user provision processes.

NetLab project main results will be in the area of interoperability and interconnection of testbeds. Besides, we will explore the compatibility issues of different terminals and IMS clients with different IMS cores. Some of our first results have lead us to explore in the security issues and requirements of the system. Once IMS deployments and IMS services become as widespread as 3G services today, we will be prepared to give answer to security issues we will explore in NetLab such as: achieving fast user authorization with IMS application services, specially for nomadic and roaming users, protecting users from rogue visiting networks, security configurations guaranteeing users and Telcos the quality of authentication, and integration of legacy conditional access systems protecting multimedia contents in IMS enabled IPTv services.

NetLab project proposes a novel AAA convergent architecture. Using the same authentication algorithms is needed in order to achieve a common platform. Since authentication algorithms for fixed networks have not evolved in the last decades and drag out some drawbacks (for example, user identity is attached to the physical line, not allowing nomadism and adding complexity to provisioning systems), mobile networks authentication mechanisms are proposed. Two main methods are used nowadays in mobile networks: Subscriber Identity Module (SIM [3]) and Authentication and Key Agreement (AKA [4]), both of them based on Universal Integrated Circuit Card (UICC), using . Nevertheless, due to the different physical access mediums used by fixed and mobile networks, we will experiment with a different authentication transport protocol for fixed networks: 802.1X [2]. Fig. 1 shows the AAA architecture proposed.

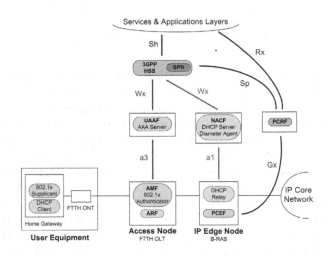

Fig. 1. Convergent AAA Architecture

2.2   QoS Aspects

Most ISPs pursue providing both user QoS requirements and QoS admission control. To provide user applications QoS requirements implies granting end-to-end resources in order to provide a packet delivery that meets the application requirements for every traffic flow, such as bandwidth, delay, packet loss, jitter, etc. Providing QoS admission control implies taking into account when the network is not going to be able to grant a given user application traffic flow QoS requirements and permits the operator executing the corresponding action according to a defined policy (for example, do not permit to establish the given traffic flow or inform to the end-user that the network is unable to assign the required QoS resources and the application quality could be degraded).

Standardization bodies, such as TISPAN in fixed networks and 3GPP in mobile ones, have done an attempt to achieve these targets. Since effective end-to-end QoS mechanisms (e.g. RSVP) do not provide operator-managed admission

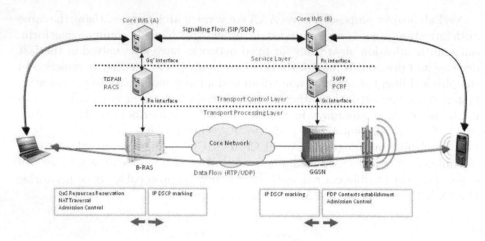

Fig. 2. QoS High-Level Description

control techniques, both TISPAN and 3GPP have considered that the access network is the bottleneck (assuming that the core network is over dimensioned) and have focused all his efforts on implementing both QoS resources reservation and QoS admission control in the access network. In the last years many efforts have been devoted in order to design a convergent QoS Architecture taking as a starting point both TISPAN and 3GPP proposals. Nevertheless, this goal has not been achieved yet, and the last attempt carried by 3GPP (PCRF definition) is still in progress. NetLab project will extend the current QoS platforms in order to achieve a convergent architecture, getting a common transport layer for both mobile and fixed networks enabling a network-agnostic service layer (see Fig. 2).

Regarding that new broadband access technologies, such as FTTH and LTE, will supply a dramatically increase in the user available bandwidth, standardization bodies' premises are not longer valid. The bottleneck could be located in different locations, such as inside the user home network or even in the core network.

Traffic flows IP DSCP marking is proposed in NetLab project, so that real-time traffic flows (used in conversational services) will be prioritized over some other background flows (such as web navigation) in the whole network (home user network, access network and core network).

Bottlenecks could also appear in layer-2 devices, such as Wi-Fi access points or PLC devices, which do not implement layer-3 features and are not able to prioritize high-priority traffic since will not detect DSCP flags. To grant user QoS requirements in these devices, layer-2 flags will be also marked, classifying traffic flows according to standards such as WMM and Ethernet 802.1p. NetLab project defines 4 QoS traffic classes (Voice, Video, Best-Effort and Background) and every traffic flow will be categorized in one of these classes, providing QoS resource reservation and admission control for IMS services in the access network and prioritizing high-priority traffic flows in the whole network in order to provide end-to-end QoS in case the bottleneck is located in other network segment.

2.3   Platform Interconnection

Interconnection between testbeds provides end user access to all available services which are provided also by partners testbeds over home network (own operator).

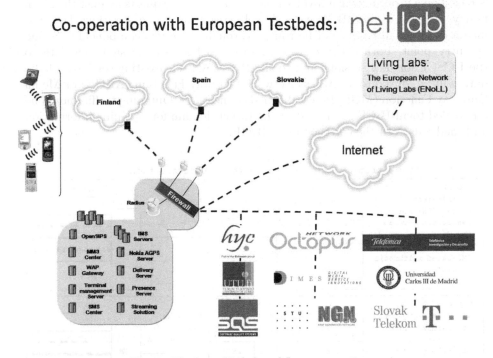

Fig. 3. Platform High-Level Interconnection

There are several possible way of IMS or NGN interconnection in the standards (e.g. those from ETSI TISPAN, 3GPP). Additional to this there are some specifications for IMS interconnect which should be adapted and considered. 3rd Generation Partnership Project 3GPP: Interconnection IPv4 vs. IPv6 (dual models) also deals with NAT and IP address incompatibility [6]; interconnection to pure SIP domain[7]; ETSI Telecommunications and Internet converged Services and Protocols for Advanced Networking (TISPAN): new work items for NGN interconnection (NGN architecture R2 [9], IMS based IPTV[8]), and GSM Association: IMS Roaming and interworking [10,11].

A interconnection scenario fully based on standards requires some additional components: located at the transport layer, an Interconnection Border Gateway Function (IBGF) is needed; at the signaling layer, an Interconnection Border Control Function (IBCF) with co-located Interworking Function (IWF) are used. The main functions in the transport layer are traffic control, Topology Hiding (THIG) and Bearer Control (BC). The IBCF uses THIG for signaling information and controls the IBGF through the Service based Policy Decision Function

(SPDF). This SPDF performs a coordination function acting as Policy Decision Point for Service-based Policy control (see also [5]). It makes policy decisions depending on service policy rules defined by the network operator. Furthermore, the underlying network technology will be hidden from applications.

The authorization of application functions requesting transport control service is based on a process, which involves the check of these requests against the given policy. The IBCF and IBGF control each session including signaling data and media data, as all traffic will be routed through these two components. Acting as entry point, both entities are configured to forward the signaling data to the I-CSCF for new sessions and to the P/S-CSCF for existing sessions. Media data will be routed to an Access Network Gateway Function (ANGF) or Media Gateway Function (MGF) for established connections. Outgoing traffic should be forwarded to an IBCF connected to the target domain for signaling information and media data to the corresponding IBGF.

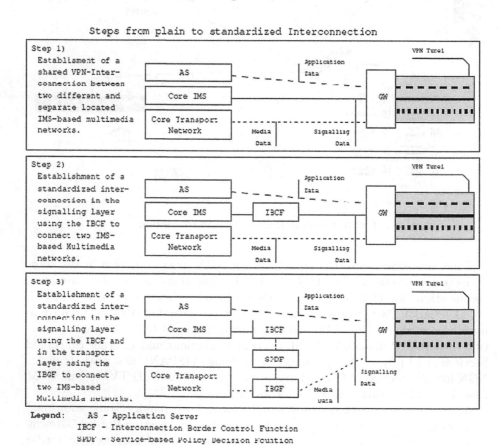

Fig. 4. Possible interconnection of IMS testbeds, GEANT connections not shown

As first steps the interconnection of partners testbeds will depend on available components, but most probably an interconnection without all the entities as described in standards can be established first and later extended. This is seen as first and simplest approach to show, so that servers of interconnected testbeds can communicate in spite of the difference of environments and that the standard compliant interconnection has a solid base. All required steps for a grateful changing of the actual running configuration to a standardized interconnection in the future are shown in Fig. 4. Take into account that the VPN interconnection will be later substituted by MPLS tunnels over the GEANT network.

The proposed set-up allows sharing of Application Servers (AS), which are located in different domains. Using the presence service, two different interfaces are shared and forwarded between the testbeds. ISC/Ma is the first interface. It will be used to exchange information between the Serving (S)-CSCF and the Presence-AS. The latter interconnection is established on the Ut interface, used by a User Agent (UA).

3   Services and Use Cases

Currently operators are able to run IMS and DVB-H services on their own platforms, but when time comes to interconnect infrastructures, problems have to be solved to deal with service interoperation, orchestration, roaming and end-to-end QoS (Quality Of Service). The objective of NETLAB is to have services working on top of the multi-platform and multi-operator interconnected environment; the result is to enable real life tests and to provide field materials to conduce theoretical work on service modeling and testing.

To define the services, focus has been put on both the application servers side and the terminal clients side. The application servers implement the service logic, and in a multi-platform and multi-operator environment, interoperation, orchestration or roaming problems have to be solved. The terminal clients negotiate end-to-end QoS, and serious interoperability problems appear when interconnecting two access networks with different technologies (most notably, mobile and fixed), because the QoS signalling mechanisms are not totally compatible.

To deliver adapted services to the real-life experiments and to address interesting problems for theoretical work, suitable use cases have been defined. The focus was put on IMS services that have to run on top of a multi-platform and multi-operator environment. Basic IMS services such as "VoIP", "Presence" or "Instant Messaging" are required, but other services such as "Push to Talk" or "IP TV" were addressed.

Once services have been deployed on interconnected platforms, they have to be tested and benchmarked at the operator level. The objective is to validate that a given service interoperates properly, that it works in roaming context and that end-to-end security and QoS are achieved. Before services are delivered for real-life testing, the use case services must be adapted to the targeted platform environments, the terminal clients should be adapted to fix QoS interoperability problems; and the services are deployed on the IMS platforms; the services are recursively tested, at the operator level, to fix interoperability problems.

NetLab project proposes different services to improve the interoperability among different testbeds (under different administrative domains, using different implementations, different vendors, equipment, etc.). Some of the proposed services on the interconnected testbeds are the following:

Integrated rich communication with IPTV

The rich communication services provide basic set of services expected by operators from IMS based NGN in our scenarios will be evaluated potential use cases where show advantages for integration of voice/video communication combined with messaging, presence, buddy list cross domains and beyond usual IMS roaming scenarios. Main difference will be achieved by accessibility of services also outside home network where applications can be provided and combines by multiple platforms application servers and enable use e.g. presence cross domains. We would like to show also possible integration of rich communication with content services and IPTV.

IPTV service remote control

This service will allow remote controlling IPTV services (e.g. parental control). The kids will be able to get the permission to watch some content even if the parents are not at home. The aim is to show how the traditional IPTV service can be integrated with other networks and services (e.g. SMS) through the IMS core network to improve user experience and conformity in usage. At the same time, this service also represents a scenario where the interconnection of different domains is required, especially when user want control his home services from external domain.

Advanced PVR

The main aim of this scenario is to allow the interconnection of two different accesses networks, in this case we will interconnect the mobile phone network together with the IPTV network to allow a new innovative and interesting service which consists on being able to schedule and access a recordings remotely.

QoS over heterogeneous networks

The target of this scenario is to demonstrate how end-to-end QoS resources could be granted in an IMS-based service where traffic flows across heterogeneous networks. In this scenario, IMS service layer invokes the corresponding QoS resources reservation procedures in different interconnected domains in order to grant end-to-end QoS.

Unified authentication

The main goal to achieve in this scenario is to show how a unified authentication and authorization mechanism can be used for both mobile and fixed accesses. In this way, a unique identity will identify the end user and all services will be provided independently whether the user is accessing through a fixed or mobile network and from home or visited network.

Extended DLNA

DLNA standards have been very successful and nowadays most consumer electronic devices are incorporating the functionalities proposed by DLNA, especially those regarding to content and media sharing. Nevertheless, DLNA standards are reduced to local area networks and features to share contents through wide area networks are not included. The main goal of this innovative service is to extend DLNA scope to reach contents share in remote devices. IMS capabilities will be used to publish DLNA media sharing features and to authorize and control remote access to local contents. IMS signaling stage will be extended to request the control layer to apply the required network configuration which allows DLNA media sharing through WAN.

4 Related Works

ENoLL (www.openlvinglabs.eu) was launched in November 2006 in Helsinki with 19 members at the first phase, the crew grew next year into 50 during Portuguese presidency and finally French presidency in November 2008 embraced altogether 129 Living Labs across the whole world as a result of the third wave of applications and their evaluation.

Living Lab is the concretisation of human-centric involvement for development of new ICT-based services and products. This is done by bringing different stakeholders together in a co-creative way. This is normally a very local set-up; an area or a region is establishing a living lab to foster its development and to improve the life of its citizens.

Since Living Lab is about people together for better tomorrow, scaling up the concept and jointly benefiting from the power of co-creation is fundamental for the philosophy to be put in action. Hence the Network of Living Labs. Despite the idea of living labs born in US, the network is a European concept, which now in the third wave grows global and beyond European borders. This again is only coherent with the ideology of joining relevant communities fostering co-creation.

Despite the need from the market, to the date Living Labs have not been connected to run comparable tests across multiple Living Labs. NetLab will pioneer in this task and utilize the ENoLLnetwork.

Interconnecting testbeds for a richer research and development environment is currently a global initiative, but especially advanced in Europe. National clusters have emerged and Framework Programme has targeted resources for creating large-scale research infrastructure and federating testbeds. Fig. 5 compares different related initiatives.

OneLab2 builds on PlanetLab (US) federation principle, runs independent management entity PlanetLab Europe, and gradually extends and broadens the offerings of the global networking research infrastructure by federating more nodes, introducing novel tools, but also new heterogeneous resources via vertical interconnection and new incentive models for sharing and running the testing facilities.

PII, also known as Panlab II, implements the federation framework developed previously and interconnects testbeds belonging to regional clusters, hence partitioning the management of the resources to different levels. The main design principle in PII federation is the dedication of resources under the common federation governance. PII framework addresses heterogeneous resources offered to

Project landscape

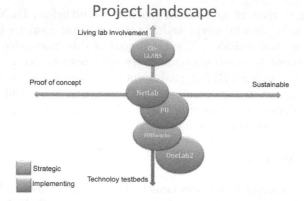

Fig. 5. Related initiatives

customers meeting also more business specific needs and criteria, such as SLAs. The proof of concept federation is implemented on IMS testbeds at three sites in Europe. The concept of User-Driven Innovation is included as PII offerings and the whole implementation process is analyzed techno-socio-economically. **NetLab** project approach is more pragmatic by connecting and piloting IMS test beds interconnectivity and testing new concepts related to Living Lab and Panlab concepts.

CO-LLABS and **FIREworks** are coordinating projects enabling more effective dissemination and resource allocation within testbed and living lab community.

5   Conclusions

NetLab is a cross domain platform of interconnected testbeds from three EU countries (Finland, Spain, Slovakia), together with interconnection to different European tested environments (projects like Panlab, Living Labs or overall framework FIREworks). NetLab is designed to tackle converged services over IMS based testbeds with focus on interoperability, scalability, complexity and mobility aspects as well as security and QoS requirements, coupled with validation in large scale testing environments. Use cases have been defined within the project to find out the interoperability issues from a more practical perspective.

Acknowledgments. This work has been partially supported by CELTIC Net-Lab Project ID CP5-018 under Spanish Science and Innovation Ministry Grant No:TSI-020400-2008-108.

References

1. 3GPP recomendation 23.228: IP Multimedia Subsystem (IMS), stage 2 (2009)
2. IEEE Standard for Local and metropolitan area networks; Port-based Network Access Control, IEEE Standard 802.1X (December 2004)

3. Haverinen, H. (Nokia), Salowey, J. (Cisco Systems): RFC 4186; Extensible Authentication Protocol Method for Global System for Mobile Communications (GSM) Subscriber Identity Modules (EAP-SIM) (January 2006)
4. Arkko, J. (Ericsson), Haverinen, H. (Nokia): RFC 4187; Extensible Authentication Protocol Method for 3rd Generation Authentication and Key Agreement (EAP-AKA) (January 2006)
5. Aboba, B. (Microsoft), Blunk, L. (Merit Network, Inc.), Vollbrecht, J. (Vollbrecht Consulting LLC), Carlson, J. (Sun), Levkowetz, H. (ipUnplugged): RFC 3748; Extensible Authentication Protocol (EAP) (June 2004)
6. 3GPP TSG Services and System Aspects: Interworking aspects and migration scenarios for IPv4-based IP Multimedia Subsystem (IMS) implementations (Release 8) TR 23.981 V8.0.0 (December 2008)
7. 3GPP TSG Core Network and Terminals: Interworking between the IM CN subsystem and IP networks (Release 8) TS 29.162 v8.1.0 (December 2008)
8. ETSI, TISPAN IPTV Architecture; IPTV functions supported by the IMS subsytem RTS 182 027 v2.2.0 (November 2008)
9. ETSI, TISPAN NGN Functional Architecture. Section 7 "NGN Interconnection". ETSI ES 282 001 v2.0.0 (March 2008)
10. GSM Association: IMS Roaming & Interworking Guidelines v3.6 (November 2006)
11. GSM Association: A guide to GSMA's SIP Trials focusing on practical interworking of IMS over the evolved GRX (February 2007)
12. Lopez, J., Rodriguez, F., Fandino, A., Palacios, J.M., Garcia, A., Gonzalez, F.: Smart Card based Authentication Mechanism for FTTH Access Networks. In: The 16th IEEE Workshop on Local and Metropolitan Area Networks, LANMAN 2008, Cluj-Napoca (September 2008)
13. Lopez, J., Fandino, A., Garcia, A., Palacios, J.M., Rodriguez, F., Gonzalez, F.: Proposal for a Convergent Authentication and Authorisation Infrastructure in NGN. In: 47th FITCE Congress, London (September 2008)

A General IMS Registration Protocol for Wireless Networks Interworking

Daniel Díaz-Sánchez[1], Davide Proserpio[1], Andrés Marín-López[1],
Florina Almenárez-Mendoza[1], and Peter Weik[2]

[1] Universidad Carlos III de Madrid, Avda de la Universidad 30, E-28911, Leganés, Spain
{dds,dproserp,amarin,florina}@it.uc3m.es
[2] Fraunhofer Institut FOKUS, Kaiserin-Augusta Allee 31, 10589 Berlin, Germany
peter.weik@fokus.fraunhofer.de

Abstract. One of the most critical tasks when accessing services through the
IP Multimedia Subsystem is the registration process. The process involves two
registrations, the first with the access network, the second with IMS. This leads
to an overhead authentication that introduces a big delay. This article proposes
an improvement for IMS registration protocol able to relate IMS registration to
an access network registration by cryptographically binding both of them. This
approach provides a general solution, saves time during registration and avoids
several attacks.

1 Introduction

IP Multimedia Subsystem (IMS) is an approach for specifying the evolution of circuit switched to packet switched networks with an special focus on fixed-mobile convergence. IMS services can be accessed independently from any access network as
fixed networks, GPRS, UMTS, LTE, WIMAX and WiFi. IMS might cooperate with
access network providers, for instance, in authentication duties. Future network scenarios present a single core network (IMS) that handles requests from clients connecting
through access networks administrated by other companies. In such scenarios a user
might spontaneously perform a vertical handover to increase bandwidth triggered by an
application while others might be connected to different services through different access network technologies. For instance, a user watching a movie through its last-mile
high speed optical fiber can be reading emails at the same time in a PDA attached to a
GPRS station; both services might be provided by the same IMS core network. These
scenarios are appealing since subscribers can remain attached to a GPRS or UMTS network almost everywhere while they can connect to any available 802.11b/g/n or Wimax
hotspot on demand. The process of connecting to IMS services starts with the Mobile
Equipment (ME) acquiring connectivity through an access network. This process involves the execution of an authentication mechanism that typically engages ME and its
home network in a challenge response message exchange. However, the IMS home network might not be involved in the authentication process, for instance, a friend can give
us the L2 authentication key of his wireless router. Once the ME has Internet connection it must register with IMS by exchanging again challenge-response messages. As the

J. Wozniak et al. (Eds.): WMNC 2009, IFIP AICT 308, pp. 32–43, 2009.

reader might infer, this double authentication process leads to a very time-consuming overhead, specially if the ME connects from a visited IMS network. For example, 3GPP requires for WLAN clients to execute an L2 authentication mechanism to authenticate with access network and then to authenticate to IMS (SIP-Digest-AKA). Interworking have been widely studied and many solutions have been proposed in order to reduce registration time giving as a result solutions ranging from those specifying incremental changes for a specific access network technology, that can be considered closed solutions, to those providing general tunneled protocols able to carry any authentication mechanism, that might suffer from Man-In-The-Middle attacks (MITM) if are not well defined. To overcome the problem, this article presents a general registration protocol for IMS that accelerates the process, while prevents MITM attacks by cryptographically relating access network registration to IMS registration.

2   Authentication for Accessing IMS Services

This section summarizes current IMS interworking with access networks, analyzes its benefits and drawbacks and recapitulates requirements for a general authentication scenario for interworking with non cellular access networks.

2.1   Authentication in the IMS

IP Multimedia Subsystem provides a control plane using Call Session Control Function (CSCF) servers by means of the Session Initiation Protocol (SIP) [1]. Subscriber data is managed by the Home Subscriber Server (HSS) and the Authentication Center (AuC). IMS defines the following types of CSCF servers. A proxy-CSCF (P-CSCF), the first hop in a visited network, that redirects SIP messages from ME to ME's home network. It also establishes an IPSEC security association with the ME. The confidentiality and integrity keys, C_K and I_K respectively, are derived as a result of the authentication performed with the HSS and conveyed to the P-CSCF using signaling. The I-CSCF, is an interrogating-CSCF located at home network that locates a server able to manage subscriber originated SIP messages. Finally, the **S-CSCF**, a serving-CSCF, authenticates subscribers retrieving authentication vectors from the HSS.

IMS authentication is based on HTTP Digest Authentication [2], using AKAv1-MD5 [3] as algorithm, which requires exchanging four messages (2 Round Trip Times-RTT) between the subscriber, at the visited network, and the subscriber's home network. IMS authentication leans on an UICC (Universal Integrated Circuit), a smart card located at ME, that contains an ISIM application (virtual subscriber module) which shares a long term secret (K_I) with the HSS. The IMS registration protocol works as follows (see also Fig. 1):

In the step 1, the ME is registered with the access network and has discovered a P-CSCF. In step 2, the ME uses the UICC to obtain subscriber information as the registration URI (to locate home network), the public/private identity and the Contact Address to build a SIP REGISTER message. Moreover, the ME includes a Security-Client header indicating which IPSEC algorithms supports. In step 3, the ME sends the aforementioned REGISTER message to the P-CSCF which inserts a P-Visited-Network

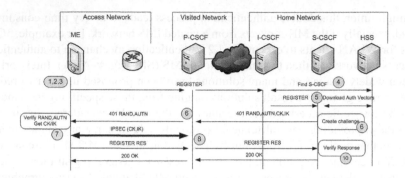

Fig. 1. Message exchange for a successful IMS registration

identifier in the REGISTER message. The P-CSCF reads and removes the Security-Client header and redirects the REGISTER message to the discovered I-CSCF (home network). The I-CSCF locates an appropriate S-CSCF to handle ME's messages by sending, in step 4, a Diameter User-Authentication Request (UAR) to the HSS. The REGISTER message reaches the S-CSCF at step 5. The S-CSCF downloads authentication vectors for the subscriber from the HSS. Those vectors are quintuplets containing parameters for authentication and key derivation using AKA [3] as: a random challenge (RAND), an authentication token (AUTN), the expected ME response (XRES), an integrity key (I_K) and a confidentiality key (C_K). AUTN is derived by the HSS using the long-term secret it shares with the ME (K_I) and a sequence number (SQN). The step 5 finish with the S-CSCF building a 401 Unauthorized message adding a WWW-Authenticate header containing AUTN and RAND. The S-CSCF includes also C_K and I_K in the message to be consumed by P-CSCF. Finally, the S-CSCF sends this message back to the ME. In step 6, the 401 Unauthorized message reaches the P-CSCF at visited network. The P-CSCF extracts and removes C_K and I_K from the message (the ME can derive C_K and I_K from AUTN and RAND using the UICC) and adds a Security-Server header selecting one IPSEC algorithm from those proposed by the client in step 2.

In step 7, the ME receives the 401 Unauthorized message and uses the UICC to calculate a response to the challenge (RES), C_K and I_K from AUTN and RAND. Then the ME establishes a security association with P-CSCF using C_K/I_K and composes a new REGISTER message containing RES and a Security-Verify header. Then it forwards the message to the P-CSCF over the brand new IPSEC security association. The P-CSCF, upon the reception of the message over a protected channel $(C_K$ and $I_K)$, implicitly authenticates the ME (step 8). Then, it redirects the message to the I-CSCF. The message is forwarded to the S-CSCF in step 9. Finally, in step 10, the S-CSCF receives the REGISTER message and checks if RES matches XRES to legitimate the subscriber. If the user is successfully authenticated, the S-CSCF builds a 200 OK message and sends it back to the ME finishing the IMS registration process.

2.2   Authentication in Access Networks

Cellular access networks, as UMTS, provide good coverage almost everywhere but data rates are far from being appropriate for several applications. As a result, interworking with other access technologies providing higher bandwidths is appealing: enables on demand or opportunistic vertical handovers when consuming IMS services. Authentication in 3G networks uses a secret stored in a smart card (UICC) to perform authentication and key derivation (AKA). The HSS also vouches for identities, being responsible of challenging a supplicant, verifying supplicant responses and deriving keys to protect radio channel. The virtual subscriber module handling the long term secret used for 3G authentication (K_U) is called USIM and is collocated at the UICC together with ISIM (K_I). Thus, authentication mechanisms for Packet Switched domain (3G) and IMS are independent so a double authentication is performed.

The authentication process in other access networks varies with the technology. For instance, IEEE 802.11 relies on L2 access control and security mechanisms specified by IEEE 802.11-1999 Part 11 that specifies L2 encryption, IEEE 802.1X describes a Network Access Control that uses Extensible Authentication Protocol (EAP), and IEEE 802.11i which supersedes WEP and WPA. Besides, WiFi public services authentication is often performed against a web site leaving L2 unprotected. Others, as Long Term Evolution provides security features in MAC level. 3GPP enforces some requirements for 3G interworking with other access networks: first, the UMTS security architecture must not be compromised and second, authentication must be mutual, based on a challenge-response mechanism using the long term secret stored in UICC and must result in a key derivation. These requirements reduces the amount of authentication algorithms that can be used during the authentication with access networks, in practise, lead to the adoption of AKA.

There are several drawbacks in this interworking scenario. Regarding the time spent during authentication, AKA requires 2 RTTs to convey the challenge to the ME and to receive the response, thus, when combined with IMS authentication yields to 4 RTTs. Moreover, the adoption of AKA requires implementing EAP-AKA for every access network technology (even in L2) in contrast with the benefits, for instance, scalability, cost-effectiveness and early adoption of new technologies, of using protocols above network layer, as PANA [4] that carries EAP payloads.

2.3   General Tunneled Authentication Mechanisms

UMTS and IMS authentication protocols are independent from each other and their subscriber modules, USIM and ISIM respectively, handle different key material. However this key material might be reused for access network authentication preventing the UICC from implementing a subscriber module for any upcoming technology. This obvious simplification introduces a feasible MITM attack since the algorithm has no way to know the purpose of the authentication, for instance, a MITM can use victim's network authentication messages to impersonate the subscriber obtaining access to the same or other network expecting those credentials. This attack can be prevented only if the authentication mechanism results in a key derivation and the key is used to protect the channel between supplicant (ME) and authenticator (typically the Network Authentication Service - NAS).

There are many works proposing general tunneled authentication mechanisms that enable carrying EAP payloads over other protocols. The idea behind is to reuse legacy authentication protocols for other purposes creating a tunnel that authenticates the NAS before starting the inner authentication mechanism. Once the NAS is correctly authenticated, it will forward authentication messages (inner protocol) to a back-end authentication service (typically the HSS). In this way, an authentication protocol inside the tunnel can be reused in a secure fashion since it is executed over a tunnel between the client and the NAS. Moreover, this tunneled protocols help to alleviate the problem of having implemented authentication protocols in L2 by providing an alternative transport over a higher protocol. Nevertheless, the feasibility of aforementioned MITM attack is already present. The reason is that this kind of tunnel protocols require to distribute credentials to every NAS, since the NAS should be authenticated. The problem is explained in [5] and [6] and appears when an authentication protocol is designed as the combination of two protocols: an outer protocol and an inner protocol. The outer protocol, for instance TLS [7], is used to protect the exchange of messages of the inner protocol. The inner protocol is used to authenticate the user to the network and the outer to authenticate the network to the user. Among the protocols affected by this attack we can find PEAP, EAP-TTLS, PIC and PANA over TLS [8]. The problem appears under any of the following conditions. 1) The inner protocol can be used in other environments. It happens when the inner protocol has no way to know if it is used inside a tunnel or not. 2) The client fails to verify the server certificate in the outer protocol. This might be frequent when connecting to access networks since the ME lacks of connection to Internet to download Certificate Revocation Lists (CRLs) or any other information necessary to verify NAS certificate. Besides, despite this is an unacceptable error from client side, the network must provide mechanisms to overcome the fact that a single client error can compromise security (specially when non professional users are involved).

The attack works as follows: the MITM waits until a legitimate device (ME) starts an untunneled legacy remote authentication protocol. Then, the MITM starts a tunnel with an authentication agent (access network) and starts sending legitimate user's authentication messages over the tunnel until the legitimate client is successfully authenticated. Then, the MITM derives keys to protect the channel from the outer tunnel keys stealing service to the legitimate client. To overcome this attack there are two simple solutions. In the first, the inner protocol must provide not only authentication but also must result in a key derivation. Those keys should be used to protect a channel between the client and the server. Thus, there is an implicit authentication since only the client knows those keys. However, it can be solved if the outer tunnel keys are derived from the long-term secret used in the inner protocol or both inner protocol and outer tunnel are somehow related.

2.4 Security Analysis of the IMS Registration

A new registration protocol must provide, at least, the same degree of security as the previous protocol. For that reason we perform a basic security analysis of the standard IMS registration using BAN logic [9] to identify the believes and how the different entities authenticate each other (implicitly or explicitly). The initial conditions are: the ME has connection to Internet through an access network, L2 is protected and the ME shares a long term secret called K_I with the HSS.

$$ME \Leftrightarrow_k HSS$$

Step 3 : The P-CSCF includes a P-Visited-Network in REGISTER asserting its identity to subscriber's home network. The REGISTER message is transmitted over a security interface among providers (Za).

S-CSCF believes (P-CSCF said REGISTER)

Step 6 : The P-CSCF receives I_K and C_K from S-CSCF over secured inter-provider network.

P-CSCF beleives (S-CSCF said I_K,C_K)

Step 7 (A) : The ME is able to extract I_K, C_K, AUTN and RAND from WWW-Authenticate, so it is able to **authenticate the home network**.

Since ME $\Leftrightarrow_k$ HSS and
ME sees $\{I_K,C_K,AUTN,RAND\}$ from $\{I_K,C_K,AUTN,RAND\}_K$
then ME beleives (S-CSCF said $\{I_K,C_K,AUTN,RAND\}_K$).

Step 7 (B) : The ME believes that P-CSCF is trusted since its home network accepts it as valid. Step 7 (C) : The ME creates a security association with an endpoint X using I_K and C_K. Since C_K and I_K are provided by ME's home network to a trusted P-CSCF, the ME is sure the endpoint X is the P-CSCF.

Since ME believes (ME $\Leftrightarrow_{C_K,I_K}$ P-CSCF) and
ME sees $\{$IPSEC-Payload$\}_{C_K,I_K}$
then ME believes (P-CSCF said IPSEC-Payload).

Step 8: The P-CSCF receives the REGISTER message with a response to the challenge over the security association so ME is implicitly authenticated by P-CSCF.

Since ME believes (ME $\Leftrightarrow_{C_K,I_K}$ P-CSCF) and P-CSCF sees $\{REGISTER\}_{C_K,I_K}$
then P-CSCF believes (ME said REGISTER).

Step 10: The S-CSCF checks the response message from the ME by comparing RES with XRES. If both are equal, S-CSCF **authenticates the ME**. The security analysis is the same as for Step 7 (A).

3   Registration for Secure Interworking with Access Networks

The objectives of the proposed registration framework are to reduce the registration time, to provide a general authentication framework for any upcoming technology, to prevent attacks, to fulfill 3GPP requirements, and to maintain backwards compatibility. For that reason, the requirements for the registration protocol are: 1) The access network registration must be performed over any layer, thus the authentication must be based on an EAP method suitable to be used over L2 or an upper level (PANA) EAP carrier. 2) The ME must be able to both authenticate the NAS and cryptographically prove access network registration to IMS. Moreover, L2 must be protected. For that reason, the EAP method must securely derive keys (EAP-TLS) and a cryptographic proof of the message exchange (TLS Exporter[10]). 3) The IMS registration should be cryptographically related to access network registration. 4) The security association between ME and P-CSCF should be protected with a key derived from the long term secret K_I and the access network registration proof. The scenarios are shown in Fig. 2. In the first

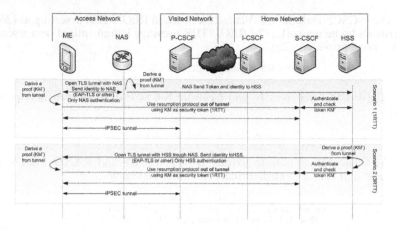

Fig. 2. Scenarios for resumption protocol with cryptographic relation to access network tunnel

scenario, the ME opens a tunnel with NAS using EAP-TLS over L2 or PANA. The ME provides an identifier to the NAS so the NAS can resolve ME's home network address. The NAS extracts key material from the tunnel using TLS exporter. It derives two proofs (P_N and P_I) from the TLS key material using a Pseudo Random Function (PRF) over the result of concatenating the master key with two different texts. The ME extracts key material and derives the same proofs. The ME then starts the IMS authentication process. It includes P_N as a token in a SIP REGISTER message and protects message integrity with a signature (that signature can be checked by the HSS). Simultaneously, the NAS sends P_N and P_I to the HSS using Diameter. The signature is used to authenticate the ME and P_N is used by the HSS to relate access network to IMS network. Finally, the HSS provides IPSEC keys for confidentiality and integrity. Those keys are derived as follows $C_K' = PRF(C_K|P_I)$ and $I_K' = PRF(I_K|P_I)$ relating both authentication processes to prevent MITM attacks. In this way, the HSS can explicitly authenticate the ME (signature), the ME can implicitly authenticate the NAS and the NAS can implicitly authenticate the ME. This process saves 3 RTT without compromising security and avoiding MITM attacks. In the second scenario, the EAP-TLS tunnel is opened directly with the HSS saving only 1 RTT.

We rely on an asymmetric ephemeral key that must be registered by the ME before being used. This key should be able to generate signatures. Providers might enforce minimum length and a validity period policies. The key, called R_k (resumption key), should be stored in the HSS (with a key index) together with the token provided by the access network. Moreover, the token might be used to resume a previous IMS session in other contexts.

3.1   Protocol Definition and Security Analysis

In this section we describe the proposed registration protocol in detail. The description assumes that the user has already derived a public/private key pair and registered the public key (R_k) under his profile at HSS. The ME opens a EAP-TLS tunnel with the

NAS (1st scenario) using PANA or a L2 protocol. Once both the HSS and the ME have available the cryptographic proofs P_N and P_I (step 1), the ME registers with IMS as follows (see Fig.3).

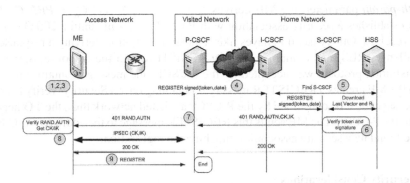

Fig. 3. Message exchange for a successful IMS registration resumption

In step 2, the ME builds a REGISTER message, as in standard IMS registration (described in Sect. 2.1) except for the inclusion of an *Authorization* header. Then, in step 3, the ME includes a **nonce** field as part of the *Authorization* header. This field contains a token obtained from a tunnel with the NAS (P_N), that will be used to relate both registrations. The ME also adds an *auth-param* to the *Authorization* header with the text *"resume@idx"*, indicating the index under R_k is registered. Finally, the ME generates a S/MIME body including: information from headers as explained in Authenticated Identity Body Format [11], the *Authorization* and Security-Client headers and a signature generated with R_k.

The ME sends the REGISTER message outside the tunnel to the P-CSCF which inserts a P-Visited-Network identifier. The message is forwarded to the I-CSCF at ME's home network (step 4). The I-CSCF finds a S-CSCF (step 5). The REGISTER message reaches the S-CSCF in step 6. Then, the S-CSCF extracts **key index** and **nonce** parameters from REGISTER headers, contacts the HSS to download the tokens (P_n and P_I), R_k and new authentication vectors. The S-CSCF builds a 401 Unauthorized message containing a WWW-Authenticate header with a nonce (AUTN—RAND) obtained from a **new authentication vector**. The S-CSCF checks if the nonce contains the same P_N received from the NAS and the signature. If both the signature and the token (P_N) are valid, the S-CSCF **authenticates the subsriber**. To inform the ME about the successful authentication, the S-CSCF includes an *auth-param* parameter in the WWW-Authenticate header of the 401 message with the text *"resume@idx"*. Moreover, the S-CSCF derives two keys for integrity and confidentiality, $I'_K = PRF(I_K|P_I)$ and $C'_K = PRF(C_K|P_I)$ and include them in the message for the P-CSCF. The S-CSCF sends it to the ME through the I-CSCF and P-CSCF. Then it builds a 200 OK message that will be sent immediately after the 401 Unauthorized message. Otherwise, if the S-CSCF either does not support this registration protocol or the ME can not be authenticated, it sends a 401 message to the ME without modifying the WWW-Authenticate header (as standard registration).

In step 7, the P-CSCF extracts and removes C_K' and I_K' from the 401 Unauthorized message and adds a Security-Server header selecting an algorithm for IPSEC. In step 8, the ME extracts AUTN, RAND, I_K and C_K from the WWW-Authenticate header **authenticating explicitly the home network**. If the WWW-Authenticate header contains the *auth-param* parameter, the ME derives $I_K' = PRF(I_K|P_I)$ and $C_K' = PRF(C_K|P_I)$. Then it establishes a security association with P-CSCF and waits until a 200 OK message is received. Otherwise, it behaves like standard registration deriving a response to the challenge (RES) and composing a new REGISTER with the response. Step 9: if the registration protocol was accepted by the S-CSCF (the message contains an auth-param) the ME creates a REGISTER message containing only a Security-Verify header. This message must be consumed by the P-CSCF at visited network thus, the **TO** header points to the P-CSCF. The ME sends this REGISTER message to the address of P-CSCF over the brand new security association finishing the registration.

3.2   Security Considerations

In this section we analyze the proposed protocol showing that is as secure as standard protocol. Initial conditions: the ME and the HSS shares a long term secret called K_I, an ephemeral resumption key called R_k and two security tokens (P_I and P_N):

$$ME \Leftrightarrow_{K_I, R_k, P_I, P_N} HSS$$
$$HSS \text{ believes (ME has jurisdiction over } R_k)$$

Step 4 : The P-CSCF includes a P-Visited-Network in REGISTER asserting its identity to subscriber's home network. The REGISTER message is transmitted over a security interface among providers. (equivalent to standard registration, step-3).

Step 6: The S-CSCF receives the REGISTER message from the ME, it downloads the R_k, P_I, P_N and a new authentication vector. It first checks the date and P_N contained in the message against the information downloaded from the HSS. Then it checks the signature against R_k. If the signature is valid and the date and P_N are valid, the S-CSCF authenticates the subscriber.

$$HSS \text{ believes (ME has jurisdiction over } R_k) \text{ and}$$
$$S\text{-}CSCF \text{ believes (ME believes nonce}^1)$$
$$\textbf{then } S\text{-}CSCF \text{ believes nonce.}$$

$$S\text{-}CSCF \text{ believes (ME said nonce) and}$$
$$S\text{-}CSCF \text{ believes nonce is fresh}$$
$$\textbf{then } S\text{-}CSCF \text{ believes (ME believes nonce).}$$

Step 7 : The P-CSCF receives I_K' and C_K' from S-CSCF over secured inter-provider network. (equivalent to standard registration, step-6). Step 8 (A) : The ME is able to extract I_K', C_K' from AUTN and RAND, so it is able to **authenticate the home network**. (equivalent to standard registration, step-7A).

Step 8 (B) : The ME believes that P-CSCF is trusted since its home network accepts it as valid.Step 8 (C) : The ME creates a security association with an endpoint X using I_K' and C_K'. Since C_K' and I_K' are provided by ME's home network to a trusted P-CSCF, the ME is sure the endpoint X is the P-CSCF. (equivalent to standard registration, step-7C).

---

[1] This nonce is the base64 representation of the security token.

C'_K and I'_K depends on P_I; P_I was derived from the tunnel with the NAS, thus there is no MITM.

Step 9: The P-CSCF receives the REGISTER message with a Server-Verify header over the security association so ME is implicitly authenticated by P-CSCF (equivalent to standard registration, step-8). Table 1 summarizes both protocols authentication processes showing which entities are explicitly or implicitly authenticated.

Table 1. Authentication among entities during registration for both protocols

Authentication	Standard	Proposed
Home network authenticates visited network (federation)	step 3	step 4
Visited network authenticates home network (federation)	step 6	step 7
Home network authenticates ME (explicit)	step 10	step 6
ME authenticates home network (explicit)	step 7A	step 8A
ME authenticates visited network (implicit)	step 7b-7C	step 8B-8C
Visited network authenticates ME (implicit)	step 8	step 9

4   Related Work

[12] proposes a solution for secure authentication in a heterogeneous wireless access scenario. This solution requires moving part of the P-CSCF functionality (including security association) to the access network. This WLAN P-CSCF redirects ME's REGISTER messages to the visited network inserting a header that indicates the type of authentication demanded by the ME (WLAN and IMS or WLAN only). This header is not protected thus it can compromise security. Moreover, the home network must allow its key material to be populated to access networks thus requires strong trust relations between IMS operators and access networks. Our proposal does not modify IMS architecture. Moreover we provide an IPSEC association protected with keys derived from two authentication processes avoiding masquerades even from unknown access networks.

In [13] the authors propose a one pass authentication procedure to obtain access to IMS services over GPRS access networks. The author proposes a modification to SGSN that adds the IMSI (associated with a PDP context) to any register message so the S-CSCF can check if the IMPI matches the corresponding IMSI authenticating the user. This solution can not be considered general. Besides, the authentication in IMS is performed without cryptography thus any user might impersonate other just by manipulating the IMSI. Other solutions as [14], propose to move the authentication to layer two using 802.1x with EAP-AKA, thus removing authentication at service level. This kind of solutions are not independent from the specific access technology so requires defining an specific procedure for any incoming technology. Our solution is a general approach to the interworking problem. It can be used with any upcoming access technology since it can be used either over L2 or PANA. [15] proposes a make-before-break handover scheme under IMS that defines a set of new SIP headers to negotiate a security association with the new visited access network speeding up the registration process. The modifications we propose are compatible with this solution also. In [16], the registration is accelerated by reducing the amount of messages exchanged between the I-CSCF and the HSS to find an appropriate S-CSCF. Those improvements are also compatible with our solution.

5 Conclusions

This article describes a registration improvement for IMS that allows using a security token to relate network access registration with IMS registration (or to resume an older IMS registration). The proposed protocol can be used by any upcoming access network technology since the access network registration can be done over L2 or PANA. Moreover, it avoids several attacks since UICC credentials are not exposed during access network registration but the entire registration process (access network and IMS) depends on the successful registration with IMS. We achieve this goal by cryptographically relating IMS registration to access network registration in such a way that only the owner of the UICC will be able to derive the final IPSEC keys if knows P_l. We analyzed the security of our proposed registration protocol and compared to the standard IMS registration security showing that both of them provide the same degree of security. Moreover, we save up to 3 round trip times during the registration.

References

1. Rosenberg, J., Schulzrinne, H., Camarillo, G., Johnston, A., Peterson, J., Sparks, R., Handley, M., Schooler, E.: Sip: Session initiation protocol. Technical Report RFC3261, IETF (2002)
2. Franks, J., Hallam-Baker, P., Hostetler, J., Lawrence, S., Leach, P., Luotonen, A., Stewart, L.: Http authentication: Basic and digest access authentication. Technical Report RFC2617, IETF (1999), http://www.ietf.org/rfc/rfc2617.txt
3. Technical report, 3GPP, Third Generation Partnership Project, Technical Specification Group Services and Systems Aspects, 3G Security, Security Architecture, Technical Specification 3G TS 33.102, V3.7.0 (2000)
4. Parthasarathy, M., et al.: Protocol for Carrying Authentication and Network Access (PANA) (RFC 5191 - 2008), PANA Threat Analysis and Security Requirements (RFC 4016 - 2005), PANA Requirements (RFC 4058 - 2005)
5. Puthenkulam, J., et al.: The Compound Authentication Binding Problem. Technical report, IETF (2003)
6. Asokan, N., Niemi, V., Nyberg, K.: Man-in-the-middle in tunneled authentication protocols. Technical report. In: 11th Security Protocols Workshop (2002)
7. Dierks, T., Allen, C.: The transport layer security (TLS) version 1.0. Technical Report RFC2246, IETF (1999), http://www.ietf.org/rfc/rfc2246.txt
8. Ohba, Y., Baba, S.: Pana over tls. Technical report, IETF (2002)
9. Burrows, M., Abadi, M., Needham, R.: A logic of authentication. ACM Transactions on Computer Systems 8, 18–36 (1990)
10. Rescorla, E.: Keying material exporters for transport layer security (tls). Technical Report draft-ietf-tls-extractor-05.txt, IETF (2009),
    http://tools.ietf.org/html/draft-ietf-tls-extractor-05
11. Peterson, J.: Session initiation protocol (sip) authenticated identity body (aib) format. Technical Report RFC3893, IETF (2004), http://www.ietf.org/rfc/rfc3893.txt

12. Veltri, L., Salsano, S., Martiniello, G.: Wireless lan-3g integration: Unified mechanisms for secure authentication based on sip. In: IEEE International Conference on Communications, 2006. ICC 2006, vol. 5, pp. 2219–2224 (2006)
13. Lin, Y.B., Chang, M.F., Hsu, M.T., Wu, L.Y.: One-pass gprs and ims authentication procedure for umts. IEEE Journal on Selected Areas in Communications 23, 1233–1239 (2005)
14. Celentano, D., Fresa, A., Longer, M., Robustelli, A.: Improved authentication for ims registration in 3g/wlan interworking. In: IEEE 18th International Symposium on Personal, Indoor and Mobile Radio Communications, 2007. PIMRC 2007, pp. 1–5 (2007)
15. Lee, H., Moon, B., Aghvami, A.: Enhanced sip for reducing ims delay under wifi-to-umts handover scenario. In: Next Generation Mobile Applications, Services and Technologies. NGMAST 2008, pp. 640–645 (2008)
16. Farahbakhsh, R., Varposhti, M., Movahhedinia, N.: Transmission delay reduction in ims by re-registration procedure modification. In: Next Generation Mobile Applications, Services and Technologies. NGMAST 2008, pp. 142–146 (2008)

Analysis of Session Establishment Signaling Delay in IP Multimedia Subsystem

Melvi Ulvan and Robert Bestak

Czech Technical University in Prague
Technicka 2 166 27, Prague 6, Czech Republic
{ulvanm1,bestar1}@fel.cvut.cz

Abstract. This paper investigates and analyzes SIP delay in the session establishment signaling procedure in the IMS system. We investigate the delay for end-to-end link scenarios such as WiMAX-to-WiMAX, UMTS-to-UMTS, UMTS-to-WiMAX and vice versa. The analyses consider three types of delays: transmission delay, processing delay and queuing delay. The obtained results show that the main delay of session establishment signaling process is due to the processing delay. In addition, the lower channel rate in the UMTS network as well as IMS service rate has significant impact to the session establishment signaling delay.

Keywords: Wireless networks, IMS, SIP messages, Session delay.

1 Introduction

The IP Multimedia Subsystem (IMS) is the next generation IP based infrastructure enabling convergence of data, speech, video and mobile network technology. It is the foreseen solution that will provide new multimedia communication services by combining voice and data in an access independent IP based architecture. The IMS architecture is defined in 3GPP, 3GPP2 and IETF standards.

The Session Initiation Protocol (SIP, [1]) is used as a signaling protocol in the IMS environment. The SIP protocol provides functionalities such as terminal location, session establishment, session management and participant invocation, including creating, modifying, and terminating sessions with one or more participants. Sessions can contain any combination of services such as voice, data, audio, video, etc, and they can be modified at any time by adding new parties or changing the nature of session. In this paper, we shall consider the session establishment, particularly the signaling delay during the session establishment process.

The SIP-based IMS signaling delay for the IMS session establishment procedure is analyzed in [2]. The author analyzes the end-to-end delay if the source is a UMTS terminal and the destination is a WiMAX terminal and vice versa. The signaling delay is analyzed separately for transmission delay, processing delay and queuing delay. However, the paper just investigates the total delay,i.e. there is no information which delay part mostly contributes to the total delay. The optimization of SIP session setup delay for voice over IP (VoIP) service in 3G networks is studied in [3]. The authors

J. Wozniak et al. (Eds.): WMNC 2009, IFIP AICT 308, pp. 44–55, 2009.

evaluate the SIP session setup performances considering various underlying protocols (such as Transport Control Protocol, User Datagram Protocol, Radio Link Control) as function of frame error rate (FER). An adaptive retransmission timer is proposed to be implemented in order to optimize the delay. Analysis of SIP-based mobility management in 4G network is carried out in [4]. Though, the authors do not focus on the session establishment procedure, some delay issues are discussed, particularly the delay on radio link control (RLC) and non-RLC. Some of the considered values in [4] are used in our analyses.

In this paper, we review and analyze the delay of session establishment signaling process. Structure of session establishment signaling process that is based on the standard is presented. We analyze the delay, not only during the IMS processes where the S-CSCF, P-CSCF, I-CSCF and HSS take part, but we also take into account the transmission delay and delay when the SIP messages being queued in the network. The transmission delay is analyzed as end-to-end delay where the source terminal (ST) and the destination terminal (DT) are either UMTS or WiMAX terminals. In term of transmission delay, the RLC (Radio Link Control) delay at the UMTS network and the non RLC delay at the WiMAX network are considered. The queuing delay is analyzed for the M/M/1 model. We also examine which delay among the transmission, processing and queuing delays mostly contribute to the total delay.

The rest of the paper is organized as follows. In Section 2, we describe the IMS signaling messages for session establishment, as well as their features at different entities and different channel rates, and also the description of transmission, processing and queuing delays. The delay analyses and results are given in Section 3. Finally, Section 4 concludes our paper.

2 IMS Session Establishment Signaling

2.1 Session Establishment Procedure

The IMS architecture contains multiple SIP proxies called Call Session Control Functions (CSCFs) with following roles: i) P-CSCF (Proxy-CSCF) which is the first contact point in the IMS architecture and it interacts with GGSN (Gateway GPRS Support Node), ii) I-CSCF (Interrogating-CSCF) which acts as a SIP Registrar and is responsible for routing sessions to appropriate S-CSCF (Serving-CSCF), and finally iii) S-CSCF that performs session control and service trigger [4].

Figure 1 shows the session establishment flows in case a source terminal wants to establish a session with the destination terminal. The source terminal generates a SIP INVITE request and sends it to the P-CSCF. The P-CSCF processes the request; for example, it decompresses the request and verifies the user's identity before forwarding the request to the S-CSCF. The S-CSCF processes the request, executes service control which may include interactions with Application Servers (ASs) and eventually determines the entry point of home operator of user B based on user B's identity in the SIP INVITE request.

The I-CSCF receives the request and contacts the HSS (Home Subscriber Station)to find out the S-CSCF that is serving user B. The request is passed to the S-CSCF. The S-CSCF is in charge of processing the terminating session which may

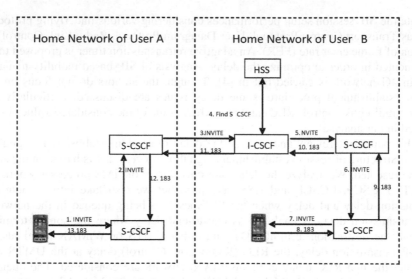

Fig. 1. Session establishment flows [5]

include interactions with ASs and eventually transmits the request to the P-CSCF. After further processing (e.g., compression and privacy checking), the P-CSCF transmits the SIP INVITE request to destination terminal (DT). The DT generates a response "183 Session Progress" that is sent back to the source terminal following the same route (i.e., DT → P-CSCF → S-CSCF → I-CSCF → S-CSCF → P-CSCF → ST). After a few more round trips, both, ST and DT, complete the session establishment phase and they are ready to start the application (e.g., voice call).

The whole call flow diagram of IMS session establishment procedure is shown in Figure 2. More details about the IMS session establishment procedure can be found for example in [5].

2.2 Session Establishment Messages

In general, there are three forms of IMS messaging; i) immediate messaging, ii) session-based messaging and iii) deferred delivery messaging [5]. Each form of IMS messaging has its own characteristics. The immediate messaging and session-based messaging operate in the IMS architecture directly. Moreover, the deferred delivery messaging form runs in the Packet-Switched (PS) domain that is an independent network infrastructure separated from the IMS. In this paper, we only consider the second case, session-based messaging, particularly the SIP messages that are involved in the session establishment procedure.

The SIP is an application protocol that is designed to establish communication *sessions* in request/response message model [1]. The SIP request message defines the operation requested by client whereas the SIP response message provides information from the server to the client indicating the status of that request. According to [1], there are six types of request messages: REGISTER, INVITE, ACK, CANCEL, BYE, and OPTION. As can be seen in figure 2, INVITE and ACK messages can only be involved in the session establishment procedure. Other request messages such as PRACK and UPDATE are defined in standard [6] and [7].

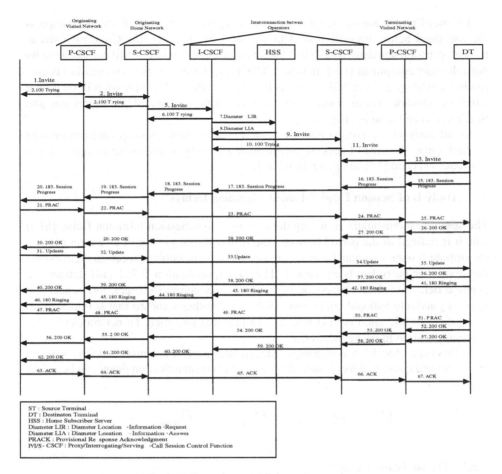

Fig. 2. IMS session establishment procedure

The INVITE and UPDATE messages have similar functionalities. The first one, INVITE, indicates user or service to be invited to participate in a session. The structure of message includes a description of the session to which the destination terminal is being invited. The ACK message confirms that the destination terminal has received a final response to an INVITE request. ACK is only used with INVITE request. Whereas the UPDATE message allows a client to update parameters of session (such as the set media streams or used codec) but has no impact on the state of dialog. In that sense, it is like the re-INVITE message, but unlike re-INVITE, the transmission of UPDATE message can be preceded by the INVITE message. This makes it very useful when updating session parameters within the early stage of dialogs [8].

The PRACK request plays the same role as the ACK message, but it is for provisional responses. However, there is an important difference, the PRACK message is a normal SIP message such as BYE. .Its own reliability is ensured hop-by-hop through each stateful proxy. Also like BYE message, but unlike the ACK message, PRACK message has its own response [6].

To speed up the session establishment, the application protocol should compress the messages before transmission. The signaling compression (SigComp) is used as the compression standard. The 3GPP has mandated the support of SIP compression by both the user equipment (UE) and the P-CSCF [9]. However, the compression is currently mandatory and the 3GPP intends to eliminate the SIP compression in the future wireless networks (local, metropolitan). More details about SIP compression and SigCom can be found in [10][11] or [12].

In our analysis, we assume that the session establishment messages are compressed by SigComp. The size of session establishment messages, according to figure 2 and references [1] and [13], are given in table 1.

2.3 Analysis of Session Establishment Signaling Delays

The session establishment signaling delay is known as Session Initiation Delay (SID) which is defined as the period between the instant the originator of a session triggers the initiate session command and the instant the session initiator receives the message that the other party has been alerted. The ITU specification E.721 [14] defines the average delay for three connection types: local connection (3.0 sec), toll connection (5.0 sec) and international connection (8.0 sec). Another standard that is important to taken into consideration is ITU Rec. G.114 [15] that specified the network delay for voice application in packet networks.

In this paper, the delay is decomposed into three parts: transmission delay, processing delay and queuing delay. Thus, the end-to-end communication delay can be calculated as [2]:

$$D_{total} = D_{transmission} + D_{processing} + D_{queue} \tag{1}$$

2.3.1 Transmission Delay

In this paper, we consider four wireless end-to-end scenarios: UMTS-UMTS, WiMAX-WiMAX, UMTS-WiMAX and WiMAX-UMTS. The transmission delay is affected by the underlying protocols used by SIP (e.g. UDP, TCP, or RLC) that influence the session establishment time. Another affect may arise from the error recovery strategy (e.g. ARQ, FEC, HARQ, etc.).

The following channel data rates (B/W) are considered in our simulation scenarious: 19.2 kbps and 128 kbps for UMTS network, and 4 Mbps and 24 Mbps for WiMAX.

The number of frame in a packet (k) is required to be calculated for every specified channel rates. In UMTS, the RLC frame duration (τ) is assumed to be 20 ms. In case of WiMAX, the frame duration is set to be 2.5 ms. Additionally, the frame duration in WiMAX is independent on the channel bit rate. The number of bytes in a frame is equal to $B/W \times \tau$. The value of k for particular signaling messages (shown in table 1) can be calculated as:

$$k = number \ of \ byte \ / \ message \ size \tag{2}$$

Hence for instance, let's consider the INVITE message and the channel rate 19.2 kbps. Than, the number of byte per frame equals $19.2 \times 10^3 \times 20 \times 10^{-3} \times 1/8 = 48$ bytes and value of k, given by expression 2, equals $810/48 = 17$.

By using the same method for other SIP messages and the given channel rates, we can determine the value of k corresponding to different types of messages involved in session establishment (Table 1).

Table 1. k value of SIP messages for specified channel rates

Session establishment message	Compressed size (byte)	Channel rate			
		19.2 kbps	128 kbps	4 Mbps	24 Mbps
INVITE	810	17	3	1	1
100 TRYING	260	6	1	1	1
183 SESSION PROGRESS	260	6	1	1	1
PRACK	260	6	1	1	1
200 OK	100	3	1	1	1
UPDATE	260	6	1	1	1
180 RINGING	260	6	1	1	1
ACK	60	2	1	1	1

2.3.1.1 Transmission Delay in UMTS

To analyze the delay to transmit SIP messages over the UMTS network, we exploit the delay model for frame and packet transmission over a wireless link which is proposed in [16][17]. The analysis of SIP transmission delay when transmitting a packet over the UMTS is given as:

$$D_{UMTS} = D + (k-1)\tau + \frac{k[P_f - (1-p)]}{P_f^2} \times \left| \sum_{j=1}^{n} \sum_{i=1}^{j} P(C_{ij}) \left[2\,jD + \left(\frac{j(j+1)}{2} + i \right) \tau \right] \right| \quad (3)$$

The open-air operation of UMTS radio access network is vulnerable to noise influenced that generate packet loss. In equation 3 above, the effective packet loss is noted by P_f and can be calculated as follow:

$$P_f = 1 - p + \sum_{j=1}^{n} \sum_{i=1}^{j} P(C_{ij}) = 1 - p[p(2-p)]^{n(n+1)/2} \quad (4)$$

Where n is the maximum number of RLC retransmission trials and C_{ij} (representing the first frame received correctly at destination) is the ith retransmission frame at the jth retransmission trial.

2.3.1.2 Transmission Delay in WiMAX

In case of SIP re-transmission in WiMAX network, the SIP retransmission is considered to be provided by upper layer protocols (e.g., TCP) until the successful transmission is completed. The upper layer protocol packet loss rate (q) in this case is given as $q = 1 - (1-p)^2$, where p is the probability a frame and k is the number of frames per packet. Let's the number of retransmission denoted as N_m, then the average delay of transmitting a packet over the WiMAX network (D_{WiMAX}) is calculated as follow :[8][17]

$$D_{WiMAX} = (k-1)\tau + \frac{D}{(1-q^{N_m})(1-2q)} + \frac{1-q}{1-q^{N_m}} \times D\left[\frac{q^{N_m}}{1-q} - \frac{2^{N_m+1} \times q^{N_m}}{1-2q}\right] \qquad (5)$$

The description of parameters involved in equation (3), (4) and (5) and their typical and assumed value are expressed at table 2.

Table 2. Parameters, description and values

Symbol	Parameter description	Value
ρ	Utilization	0.7 for HSS; 0.4 for other entities
T, τ	Frame Duration	20 ms (UMTS); 2.5 ms (WiMAX)
μ	Processing rate for each SIP message	250 packet/s
p	Probability of a frame being in error	0.02 (constant)
D	Propagation delay	100 ms for UMTS; 0.27 (4 Mbps)
		0.049 (24 Mbps)
k	Number of frames	5 (constant)
n	Maximum number of RLC retransmission trials	3
L	IP address length in bits	32
S	Machine word size in bits	32
N_m	Number of User	5000

2.3.1.3 Total Transmission Delay

According to figure 2, it can bee seen that the session establishment processes in the IMS involve 12 message exchanges between the source terminal and P-CSCF of the visited IMS network. In addition, there are another 12 message exchanges which are involved between P-CSCF of the terminating IMS network and destination terminal.

In our first scenario, the source and destination terminals are UMTS terminals. Thus, the IMS session establishment transmission delay is given as:

$$D_{trans-UMTS} = 24\,messages \times D_{UMTS} \qquad (6)$$

By using the same approach, we can determine the IMS session establishment transmission delay in case of WiMAX terminals as:

$$D_{trans-WiMAX} = 24\,messages \times D_{WiMAX} \qquad (7)$$

The results are depicted at figure 3. (a). It shows that the UMTS network has higher delay in transmission compare to WiMAX. The detail description can be found at section 3.

The third (resp. forth) scenario considers the source terminal to be UMTS terminal (resp. WiMAX) and the destination terminal represents WiMAX terminal (resp. UMTS). Thus, the session establishment transmission delay is given as:

$$D_{trans-UW/WU} = 12\,messages \times D_{UMTS} + 12\,messages \times D_{WiMAX} \qquad (8)$$

The obtained results are shown in figure 3 (b). It can be seen that the UMTS's channel rates affected the transmission delays. The lower channel rate contribute to the most significant delay.

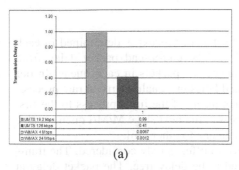
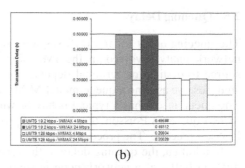

(a)   (b)

Fig. 3. (a) Session establishment transmission delay UMTS-UMTS and WiMAX-WiMAX ;
(b) Session establishment transmission delay UMTS-WiMAX

2.4 Processing Delay

The processing delay is determined in all entities in the IMS signaling path, i.e. P-CSCF, I-CSCF, S-CSCF in both home and visited networks, plus the Home Subscriber Server (HSS) where the subscriber's profile is stored. The processing delay included the address lookup table delay. It is a delay when a query is sent to HSS for a particular IP address, then HSS has to lookup its table for the given IP address. The HSS table contains the list of N subscribers managed by the IMS network. Our work assumes $N = 5000$ subscribers.

The HSS lookup is an important part of the processing delay ($D_{proc\text{-}HSS}$), it can be approximated as [2][8]:

$$D_{proc-HSS} = D_{proc-ED} + 100\left(Log_{k'+1} N + \frac{L}{S}\right)$$ (9)

Where L is the length of IP address in bits (L=32 bits for IPv4 and L=128 bits for IPv6), S is the size of server's processor architecture (e.g., 32 or 64 bits), k' is a system-dependent constant, and $D_{proc\text{-}ED}$ represents the fixed processing delay due to the encapsulation and de-capsulation of packets. In our analyses, $D_{proc\text{-}ED}$ = 4 ms [2]. Since the encapsulation and decapsulation of packets are the only process that takes place in other entities, therefore we assumed the processing delays in rest of entities are equal to $D_{proc\text{-}ED}$.

The multiplication factor of 100 [3] in the equation 9 is used since the lookup time is increased by around 100 ns for each memory access. By using the given value of parameters, the multiplication part gave the result in nanoseconds. Since, the increase of multiplication part is too small, the value of $D_{proc\text{-}HSS}$ is assumed to equivalent to $D_{proc\text{-}ED}$.

The processing delay for the IMS session establishment process (D_{proc}) is given as:

$$D_{proc} = 7D_{proc-ST} + 24D_{proc-PCSCF} + 24D_{proc-SCSCF} + 6D_{proc-ICSCF} + D_{HSS} + 5D_{proc-DT}$$ (10)

where $D_{proc\text{-}ST}$, $D_{proc\text{-}PCSCF}$, $D_{proc\text{-}SCSCF}$, $D_{proc\text{-}ICSCF}$, $D_{proc\text{-}DT}$ denote the packet processing delay at source terminal, P-CSCF, S-CSCF, I-CSCF and destination terminal. The coefficients in equation (10) are determined based on the number and type of messages that each network entity has to process (see Figure 2). With respect to our assumptions, the value of D_{proc} is approximately 0.67 ms.

2.5 Queuing Delay

The queuing delay for the IMS session establishment process is determined in each network entity involved in the IMS signaling. The end-to-end packet delay from source to destination terminal depends on number of packets in each queue. In our analyses, we assume a queue model M/M/I and Poisson signaling arrival rate process. The queue model has a typical behavior which means if the input process to the first M/M/I queues is Poisson, then the input process to the next stage M/M/I queue is also Poisson process. The processes are independent one to other [16].

In addition, the queuing delay at the receiver buffer is only considered, The transmission buffer at a network node is supposed to be delay free. The packet delay at source terminal queue is approximated as:

$$D[\omega_{ST}] = \frac{\rho_{ST}}{\mu_{ST}(1-\rho_{ST})} \tag{11}$$

Where $\rho_{ST} = \frac{\lambda_{e-ST}}{\mu_{ST}}$ represents the utilization at the source terminal, μ_{ST} denotes the service rate at the source terminal, λ_{e-ST} represents the effective arrival rate at the ST:

$$\lambda_{e-ST} = \sum_{i \in N_{ST}} \lambda_i \tag{12}$$

Where N_{ST} denotes the number of active sessions including the considered IMS session. By determination of the utilization at a network node, the effective arrival rate λ_e at that node can be obtained. In the same way, the λ_e can be calculated at the other network entities. Thus, the queuing delay D_{queue} can be approximated as:

$$D_{queue} = 7D[\omega_{ST}] + 24D[\omega_{PCSCF}] + 24D[\omega_{SCSCF}] + 6D[\omega_{ICSCF}] + D[\omega_{HSS}] + 5D[\omega_{DT}] \tag{13}$$

where $D[\omega_{ST}]$, $D[\omega_{PCSCF}]$, $D[\omega_{SCSCF}]$, $D[\omega_{ICSCF}]$, $D[\omega_{HSS}]$ and $D[\omega_{DT}]$ denotes the expected value of a unit packet queuing delay at Source Terminal, P-CSCF, S-CSCF, I-CSCF, HSS and Destination Terminal. Based on equation (13) the queuing delay in our analyses equals to 0.255 ms.

2.6 Total Delay for Session Establishment Procedure

Based on the previously deduced equations, the total, end-to-end, delay for the IMS session establishment procedure can be now calculated. In case, the source and destination terminal are UMTS terminals, resp. WiMAX ones, the total delay is given by equation (14), resp. (15):

$$D_{total-UMTS} = D_{trans-UMTS} + D_{proc} + D_{queue} \tag{14}$$

$$D_{total-WiMAX} = D_{trans-WiMAX} + D_{proc} + D_{queue} \tag{15}$$

If the source terminal is a UMTS terminal and the destination terminal is a WiMAX terminal, and vice versa, the total end-to-end delay is given by:

$$D_{total-UMTS/WiMAX} = D_{trans-UMTS/WiMAX} + D_{proc} + D_{queue} \tag{16}$$

The results obtained from equations (14) and (15) are shown in Figure 4.(a). In addition, the results of equation (16) can be seen in Figure 4.(b). Based on both figures, it can be observed that UMTS and WiMAX networks have different profile to deal with IMS session establishment delay. The higher delay may be a factor of transmission delay in UMTS network as seen in Figure 6, but perhaps it is not the most delay contributor.

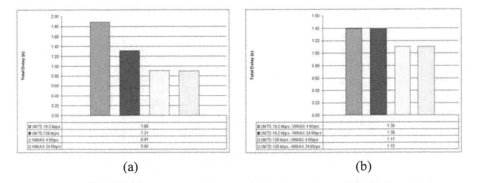

(a)   (b)

Fig. 4. (a) The total delay of IMS session establishment for UMTS-UMTS and WiMAX-WiMAX, (b) The total delay of IMS session establishment for UMTS-WiMAX

3 Numerical Results

Based on results obtained in the previous section, this section presents description of the delay analysis within the IMS session establishment procedure. Chosen values of parameters for the analysis are listed in Table 2.

Figure 3. (a) shows the transmission delay when the source and destination terminals are UMTS terminals, resp. WiMAX terminals. It can be observed that the session establishment delay is significantly affected by the channel data rate, since the delay considerably decreases as the channel rate increases. The WiMAX network scenario with higher channel rates outperforms the UMTS network scenario. Additionally, the session establishment delay is negligibly affected by modifying the WiMAX channel rates. The value of k **remains** the same for higher channel rate.

The network links have a slight impact on both delays since the processing and queuing delays are influenced by the processes that take place in each network entities such as P-CSCF, S-CSCF, I-CSCF and HSS, source and destination terminals. The processing and queuing delays are more affected by IMS service/processing rate, signalling arrival rate and by number of subscribers. Figure 5 shows the impact of different service rates on the session establishment delay. The delay decreases as the service rate increases, different channel rates has slight impact on the delay.

Typical values of transmission delay depend on the end-to-end delay scenario. If the source and destination terminals are WiMAX terminals, the typical transmission delay is very small and can be neglected (Figure 3. (a)). On the other hand, in case of UMTS, the typical transmission delay is approximately 1ms (at 19.2 kbps) and 0.4 ms (at 128 kbps). In case of UMTS-WiMAX scenario, as seen in Figure 3.(b), the typical transmission delay value is about 0.5 ms (UMTS-19.2 kbps) or 0.2 ms (UMTS-128 kbps).

The typical processing and queuing delay values are approximately 0.67 ms and 0.225 ms, respectively. Therefore, based on Figure 4 (a) and (b), the total value of IMS session establishment delay in all end-to-end scenarios can be observed as follow:

o   UMTS - UMTS: 1.31 ms (at 128 kbps) and 1.88 ms (at 19.2 kbps).
o   WiMAX - WiMAX : around 0.9 ms (for the given channel rates).
o   UMTS to WiMAX: approximately 1.4 ms and 1.1 ms at 19.2 kbps and 128 kbps.

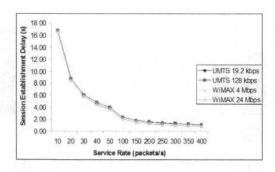

Fig. 5. Effect of changing service rate (μ) on IMS session establishment delay

4   Conclusions

This paper analyses signalling delay of session establishment procedure in the IMS environment. There are considered 4 scenarios UMTS to UMTS, WiMAX to WiMAX, UMTS to WiMAX (and vice versa) and different types of delays are taken into account. The results indicate that the processing delay contributes the major delay of the session establishment procedure. In addition, lower channel rates of UMTS network have a major impact on the delay. The session establishment delay decreases as the service rate increases. Obtained results are in the range that is specified by standards.

Since the IMS should support various technology platforms, therefore the optimization of session establishment procedure is required to be examined to allow the lower channel rates work within the grade of service. In addition, the investigation of delay in IMS applications, e.g. VoIP application and messaging, has benefit to increase the system performance.

Acknowledgement. The research work was further supported by a grant of Czech Ministry of Education, Youth and Sports No. MSM6840770014.

References

1. Rosenberg, J., Schulzrinne, H., Camarillo, G., Johnston, A., Peterson, J., Sparks, R., Handley, M., Schooler, E.: SIP: Session Initiation Protocol. RFC 3261 (June 2002)
2. Munir, A.: Analysis of SIP Based IMS Session Establishment Signaling for WiMAX-3G Networks. ICNS. In: Fourth International Conference on Networking and Services (ICNS 2008), pp. 282–287 (2008)

3. Fathi, H., Chakraborty, S.S., Prasad, R.: Optimization of SIP Session Setup Delay for VoIP in 3G Wireless Networks. IEEE Trans. Mob. Comput. 5(9), 1121–1132 (2006)
4. 3GPP. Signaling flows for the IP multimedia call control based on Session Initiation Protocol (SIP) and Session Description Protocol (SDP); Stage 3 (Release 5). TS 24.228 (v5.15.0) (September 2006)
5. Katz, M.D., Fitzek, F.H.P.: WiMAX Evolution Emerging Technologies and Applications. © 2009 John Wiley & Sons, Ltd. (2009) ISBN: 978-0-470-69680-4
6. Rosenberg, J., Schulzrinne, H.: Reliability of Provisional Responses in the Session Initiation Protocol (SIP). RFC 3262 (June 2002)
7. Rosenberg, J.: The Session Initiation Protocol (SIP) UPDATE Method. RFC 3311 (September 2002)
8. Banerjee, N., Wu, W., Basu, K., Das, S.K.: Analysis of SIP-based mobility management in 4G wireless networks. Computer Communications 27, 697–707 (2004)
9. 3rd Generation Partnership Project; Technical Specification Group Services and System Aspects; Architectural requirements (Release 5).TS 23.221
10. Price, R., Bormann, C., Christoffersson, J., Hannu, H., Liu, Z., Rosenberg, J.: Signaling Compression (SigComp). RFC 3320 (January 2003)
11. Camarillo, G.: Compressing the Session Initiation Protocol (SIP). RFC 3486 (February 2003)
12. 3rd Generation Partnership Project; Technical Specification Group Core Network; IP Multimedia Call Control Protocol based on Session Initiation Protocol (SIP) and Session Description Protocol (SDP); Stage 3 (Release 5), TS 24.229 (January 2005)
13. Melnyk, M., Ajukan, A.: On Signaling Efficiency for Call Setup in all-IP Wireless Networks. In: Proc. of IEEE International Conference on Communication (ICC 2006), Istanbul, Turkey (June 2006)
14. ITU-T E.721. Series E : Overall Network Operation , Telephone Service, Service Operation and Human Factor
15. ITU-T G.114. Series G: Transmission Systems and Media, Digital Systems and Networks
16. Chee-Hock, N., Boo-He, S.: Queuing Modeling Fundamentals with Applications in Communication Networks. 2nd © John Wiley & Sons, Ltd., ISBN : 978-0-470-51957-8
17. Das, S.K., Lee, E., Basu, K., Sen, S.K.: Performance Optimization of VoIP Calls over Wireless links Using H.323 Protocol. IEEE Transactions on Computers 52(6) (June 2003)

Adaptive Reservation MAC Protocol for Voice Traffic in Wireless Ad-Hoc Networks

Ghalem Boudour, Cédric Teyssié, and Zoubir Mammeri

IRIT, Paul Sabatier University, Toulouse, France
{boudour,cedric tyssie,mammeri}@irit.fr

Abstract. Due to the On/Off nature of voice traffic, resource reservation for voice traffic in ad-hoc network is a very challenging task. On one hand, allocating resource to voice sources for all the duration of the connection results in bandwidth wasting because of the Off periods. On the other hand, releasing resources during the Off periods and reallocating them during active periods causes large and variable access delay and increases jitter. In this paper we propose an adaptive reservation protocol for voice/data support. The proposed scheme allocates slots to voice sources each time they wake up, and gives them high priority to send their reservation requests over data sources. The bandwidth unused by idle voice sources is rendered temporarily available for reservation. Simulation results show that the proposed scheme improves the performances of voice traffic in terms of dropping rate.

1 Introduction

Voice over Mobile Ad-Hoc Networks (MANETs) has attracted a great attention in recent years. However, the delay constraints of voice applications makes the MAC protocols design for MANETs a very challenging task. It has been shown that the IEEE 802.11 standard is not able to meet delay guarantees of such multimedia applications. Recently, some protocols based on bandwidth reservation start attracting the interest of research community. In reservation MAC protocols each source reserves a number of slots which fulfill its bandwidth requirements, and uses the reserved slots in subsequent super-frames with no contention. The key idea in time-slot reservation is the exchange of control packets between nodes. Each node is required to maintain a coherent view of reservations made by its neighbors through monitoring reservation control packets transmitted by its neighbors. Despite these protocols alleviate efficiently the effects of packet collision during the reservation phase thanks to the use of control mini-slots and collision resolution schemes, a solution that takes into account the dynamic characteristics of VBR traffic -like voice- remains problematic.

As commonly known, voice sources are not active during all the connection. They are equipped with a voice activity detector (VAD), and follow an alternating pattern of talkspurts and silence periods (On/Off). On one hand, if we allocate resource to voice sources for all the duration of the connection we can ensure minimal access delay and satisfy their delay requirements in high traffic load conditions. However, the scarce bandwidth is wasted during the Off periods. On the other hand, making

J. Wozniak et al. (Eds.): WMNC 2009, IFIP AICT 308, pp. 56–68, 2009.

voice sources release their reservations during the idle periods and reallocating them during active periods results in large access delay and increases delay and jitter. Another issue with this scheme is that a voice source may not find available slots when it switches again to the active period, especially at high traffic load.

The main idea of our reservation scheme, which we call Adaptive Reservation Protocol for Voice (ARPV), is to release temporarily resources reserved by voice sources when they go to the sleep mode, and giving them the opportunity to restore their reservations when they wake up. Neighbor nodes which have data traffic are allowed to reserve slots released temporarily. ARPV is designed to be embedded on mobile nodes with time synchronization capabilities such as GPS.

The rest of this paper is organized as follows. In section 2, we give an overview of reservation MAC protocols proposed in the literature. In section 3, we present the ARPV protocol. Section 4 presents some simulation results. We end the paper by conclusions and future work.

2 Background and Related Work

The IEEE 802.11 [1] standard is considered as the de-facto MAC protocol for wireless networks. Its simplicity and reduced cost have contributed in its wide deployment. However, it suffers serious throughput degradation and unfairness due to the hidden and the exposed terminals and the binary exponential backoff, the thinks which make it unable to fit the requirements of multimedia applications over multihop networks. As alternative to this scheme, reservation-based schemes were proposed to provide delay guarantees for multimedia applications.

In FPRP [4], the super-frame is composed of a reservation frame (RF) followed by several information frames (IF). A node, which wants to reserve a slot in the IF, is required to follow a five-phase reservation process during the RF. These five phases are used by each node to compete to reserve a slot, and to inform neighbors about the result of the competition (reservation success of failure).

In CATA [12], the super-frame is composed of slots and each slot is composed of four control mini-slots and one Data mini-slot. Control mini-slots are used to reserve the slot, and prevent neighbors from reserving this slot when it is reserved.

In R-CSMA [5], each super-frame is composed of a contention period (CP) and a set of TDMA slots. A node, which wants to establish a reservation, follows a three way handshake during the CP in order to negotiate reservation of slots with the receiver. Neighbor nodes which hear the reservation packets record the reservation thus preventing any collision during reserved slots. In [8], we have proposed the ER-CSMA protocol, which is an extension of R-CSMA to resolve the reservations clash due to mobility. The reservation clash happens when two nodes which are far away from each other and which have reserved the same slot move. If one of them enters in the transmission range of the other, collisions happen in reserved slots and one or both of them loses its reservation.

RTMAC [3] is a reservation scheme that doesn't need global synchronization between nodes. Each node has its own super-frame which consists of reservation-slots (resv-slots). To carry its real-time packets, a node reserves a block of consecutive resv-slots, which is called connection-slot and uses the same connection-slot to transmit in successive super-frames. Reservation of a connection-slot is achieved following a three-way handshake like in R-CSMA.

DARE [2] extends the concepts used in RTMAC and R-CSMA for point to point reservations to establish reservations on each hop along a path.

3   Adaptive Reservation Protocol for Voice Traffic (ARPV)

The super-frame of ARPV is composed of a SYNC slot, followed by a Reservation Sub-Frame (RSF), followed by a Data sub-frame composed of S Data slots. The SYNC slot is used for synchronization. All nodes are synchronized on the super-frame basis. Data slots are used to carry voice packets and data packets, and their length is set to the transmission time of one voice packet. The ACK mini-slot is used to acknowledge successful reception of voice and data packets through the transmission of ACK frame. The RSF is composed of R Collision Resolution Slots (CRS) used for the reservation requests and reservation release requests. Each *CRS* is composed of five control mini-slots used to reserve slots, and inform neighbor nodes about reservations.

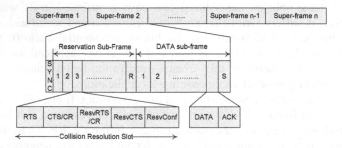

Fig. 1. Super-frame structure in ARPV

Each node maintains a Slot State Table (SST) which is updated each time a slot is reserved locally or by neighbor nodes. Unlike the other reservation schemes where data sources use contention to send each data packet on available slots, data sources in ARPV can reserve Data-slots at low voice traffic load, and when voice sources go to silence mode. For a node i, a slot may be in one of the following states:

- *Available for transmission*: No neighbor had reserved the slot for reception and node i can reserve it for transmission.- *Available for reception*: No neighbor had reserved the slot for transmission and node i can reserve it for reception.
- *Transmission reserved for data:* a neighbor of node i has reserved the slot to transmit data packets. Node i can grab the slot to reserve it for voice reception.
- *Reception reserved for data:* a neighbor of node i has reserved the slot to receive data packets. Node i can grab the slot to reserve it for transmission for a voice source.
- *Transmission/Reception reserved for voice:* a neighbor has reserved the slot for voice packet transmission (reception respectively).

- *Temporarily transmission released:* the slot is reserved by a voice source, but is temporarily released because the voice source is in the idle phase. The voice source can restore its reservation when it switches again to the activity period.
- *Temporarily reception released:* the slot is reserved by a voice receiver, but is temporarily released because the voice source in idle phase. A data source can reserve this slot for transmission, but it may lose it when the voice connection wakes up.

3.1 Basic Reservation Scheme

A node (voice or data source), which needs to establish a reservation, chooses a CRS following the scheme described in section 3.2, sends an RTS packet in the 1st mini-slot of this CRS and waits for a Collision Report packet from one of its neighbors or CTS from the intended receiver. The aim behind these two first control mini-slots is to ensure that there is no two-hop neighbor contending for reservation in the current CRS, and that the reservation request will be received by all one-hop neighbors. When the receiver receives correctly the RTS packet it replies with CTS in the 2nd mini-slot. If a collision of the RTS occurs at any neighbor of the sender, the neighbor sends a Collision Report packet in the 2nd mini-slot to indicate that the reservation request could not be recorded. In this case, the sender cancels its reservation request and restarts the collision resolution process. Otherwise, if no collision occurred during the 1st mini-slot, the sender receives correctly the CTS. In this case, the sender sends a ResvRTS packet in the 3rd control mini-slot. Beside source and destination addresses, the ResvRTS contains the list of available slots at the sender, and the number of requested slots, and the class of service (voice or data). The list of available slots specifies the list of slots of the data sub-frame available for transmission from the sender viewpoint. The requested bandwidth is derived from the number of slots.

Any neighbor of the receiver which senses collision during the 2nd mini-slot sends a Collision Report (CR) packet during the 3rd mini-slot. The aim behind this second CR is to indicate to the receiver that more than one node is establishing reservation at the same time and that the reservation could not recorded. When the receiver receives the CR or senses collision during the 3rd mini-slot, it remains silent during the 4th mini-slot and the reservation process is stopped. In the event of not receiving ResvCTS in the 4th mini-slot, the sender concludes that the reservation has failed and retries the reservation process in another CRS. If no collision occurred during the 2nd mini-slot, the receiver will correctly receive the ResvRTS. It checks its SST and replies with a ResvCTS if there are common free slots between available slots list specified in the ResvRTS and its local reception available slots. ResvCTS specifies the set of slots which will be reserved with the sender. Each node that receives the ResvCTS updates its SST and prevents reserving the specified slots for transmission. When the sender receives ResvCTS, it replies with ResvConfirm packet during the 5th mini-slot indicating the set of slots which have been reserved for transmission. Sender neighbors that hear ResvConfirm update their SST and will not accept reservation requests for the reserved slots.

3.2 Reservation Requests Transmission

However, the access scheme during the RSF has an important effect on the perform-ance of the reservation protocol. In this section we propose two access schemes for contention resolution.

Static priority access scheme. In this scheme, voice sources have higher priority to send their requests. They send their reservation requests on CRSs with permission probability p^v, while data sources transmit their requests with probability p^d, such that $p^d < p^v$. The advantage of this scheme is its simplicity. However, it doesn't adapt to the dynamic traffic loads changes.

Dynamic priority access scheme. In this scheme, the permission probability of both voice sources and data sources is adapted to traffic load and collision rate in the net-work. We adapt the Binary Feedback collision resolution algorithm of Mikhailov [9] which was proposed to stabilize the Slotted-ALOHA scheme in wireless cellular net-works. The algorithm defines a recursive function $S(t)$ and updates it slot by slot ac-cording to the channel state. $S(t)$ is given by:

$$S(t+1) = \begin{cases} \max\{1, S(t) - e + 1\}, & \text{if slot } t \text{ has } E \text{ feedback} \\ S(t) + 1, & \text{if slot } t \text{ has } NE \text{ feedback} \end{cases} \quad (1)$$

Where e is the base of the natural logarithm, $t \geq 1$, $S(1) = 1$, and E and NE are feed-backs sent by the base station at the end of each slot to indicate that the slot was empty (E) or nonempty (NE). A slot is in the E state if nodes did not transmit packets during this slot, and is NE otherwise. A node with a packet for transmission transmits its packet on slot t with a permission probability $p(t) = 1/s(t)$.

We propose to adapt this algorithm to be used for contention resolution during the *RSF* of our protocol. Every node monitors the channel during the control mini-slots of each CRS in order to detect transmissions of its neighbors. The permission probability is increased if idle channel is detected in the current CRS, i.e. idle channel detected in the 1st and the 2nd control mini-slots. A node reduces its permission probability if reservation failure is detected in the current CRS. The reservation failure is detected when a collision is sensed in the 1st or 2nd control mini-slots, or when a Collision Report is received in the 2nd mini-slot. The node maintains the same permission prob-ability it used in the previous CRS if a successful reservation is established by a one or two-hops neighbor. Successful reservation in a CRS is detected when a ResvCTS or ResvConfirm is received indicating that one or two-hop neighbor is successfully establishing a reservation. The new S function is expressed as follows:

$$S(t+1) = \begin{cases} \max\{1, S(t) - bonus\}, & \text{if idle channel detected in the } 1^{st} \\ & \text{and } 2^{nd} \text{ mini} - slot \\ S(t) + penality, & \text{if reservation failure detected} \\ S(t), & \text{if successful reservation established} \end{cases} \quad (2)$$

Where: *Penality = 1, bonus = e* for voice sources and *Penality = e, bonus = e-1* for data sources.

A node, which has a reservation request to transmit in the CRS t, calculates $S(t)$ based on the estimate of $S(t-1)$ and transmits its reservation request following the scheme described in section 3.1 with a probability $p(t)=1/S(t)$.

3.3 Adaptive Bandwidth Reservation for Voice and Data Sources

Voice packets waiting for reservation are queued and dropped if their access delay exceeds the voice delay bound D_{max}. Such a bound is derived from delay constraint of speech communication. Since data sources have less stringent delay requirements, packets generated by data sources without reservation are never dropped. New voice calls are allowed to reserve slots available for transmission, and grab reception reserved slots for data sources. The ResvRTS sent by a new voice source includes slots in *"available for transmission"* and *"reception reserved for data"* states. The receiver can accept a reservation request on slots in the *"available for reception"* and *"transmission reserved for data"* states.

When a voice source switches to the silence period, it releases temporarily its reserved slot so that neighbor data sources can reserve it for reception. A slot reserved by a voice source is considered to be temporarily transmission-released when left empty. When neighbor nodes detect clear channel during this slot, they mark it as temporarily transmission-released and are allowed to reserve it for data reception. However, new voice calls are not allowed to reserve slots in the *"temporarily reception released"* state. This ensures that new voice calls are admitted into the system only if there is an available slot, and that already admitted voice sources can restore their resources when they wake-up from the silence state. A slot reserved for voice reception is considered temporarily reception-released when clear channel is detected during the ACK mini-slot of the slot. Unlike voice sources where slots are temporarily released during inactivity periods, a slot reserved by a data source is considered to be definitively released (available) if not used for transmission for one time. Data sources are required to reserve slots each time they have packets.

A voice source in the *Off* state checks permanently its packet queue. When packets are present in the queue, the node concludes that it switches to the activity period, and temporarily released slots must be restored. The node chooses a CRS, and follows the five reservation phases in order to restore its reservation with the intended receiver. Two reservations restoration strategies can be adopted.

Strategy 1. The sender and the receiver restore their previously reserved slots and don't take care of slot availability at their neighbors. All neighbor data sources, which have eventually reserved these slots, lose their reservations, and are required to reserve other slots. We call this scheme the aggressive reservation recovery. This scheme may lead to an increase of the delay of data packets.

Strategy 2. The waking-up voice source attempts to reserve other available slots than the previously reserved slots if these ones have been reserved by some neighbor data source. If slot availability at the sender and the receiver doesn't permit to cancel the old reservation, the sender and the receiver restore their old reserved slots. This can be achieved through piggybacking the state of each slot on the ResvRTS. The receiver in its ResvCTS grants reservation on either the old reserved slots, or on other available slots. This strategy has the advantage that it avoids data sources to establish new reservation each time neighboring voice sources switch to the activity period. Consequently, the delay of such sources is optimized.

However, if a voice source which has temporarily released its slot moves far away, nodes which were in its neighborhood will not be able to update the state of the slot. The slot will remain indefinitely in the temporarily released state, and cannot be reserved by other voice sources. A solution to this issue consists in associating a timeout to each temporarily released slot as explained in section 3.5.

3.4 Reservation Release Scheme

We consider four cases to release reserved slots.

i) A voice traffic source has finished its transmission.
At the end of a voice session, the sender executes the collision resolution algorithm during the reservation sub-frame and sends a ResvRelease packet in the first control mini-slot of a CRS. The ResvRelease specifies the set of slots to be released. Any neighbor of the sender that hears a collision during the first mini-slot of the CRS sends a Collision Report (CR) in the 2nd mini-slot. If the sender hears a CR or collision in the 2nd mini-slot it chooses another CRS and continues sending the ResvRelease until no CR is received in order to ensure that all neighbors have received the ResvRelease. When the receiver receives the ResvRelease, it sends a ResvRelease in the same way as the sender in order to inform its neighbors about the reservation release. The ResvRelease sent by the receiver specifies the same set of slots that the sender has requested to release.

ii) A data source has lost its reservation because a voice source has restored its reservation or a new voice source grabs the reservation.
When a data source receives a ResvCTS which indicates a slot that it has reserved, the data source sends a ResvRelease packet to inform the receiver and release the reservation at neighbor nodes. Similarly, if the receiver of a data source receives a ResvConfirm for a slot which has been locally reserved, it releases the slot with the sender and with its neighbors.

iii) Connectivity is lost because sender or receiver movement.
As previously noticed, a slot reserved for voice is considered to be temporarily released when the voice source switches to the silence period. However, the slot may be also detected idle when the sender moves far away. The slot may stay indefinitely in the temporarily released state as it will never be restored. Consequently, the slot cannot be recycled. In order to avoid this situation, a connection timeout is associated with each temporarily released slot. If the node, which has reserved the slot, does not restore it before the timeout expires, the reserving node is considered moved far away, and the slot is made available for reception. Both neighboring voice and data sources will be able to reserve this slot for reception.

ARPV provides also a mechanism to allow reserving nodes to detect connectivity loss due to mobility of the receiver. When reserving node transmits a packet in its reserved slot, it is expected to receive either a positive or a negative acknowledgement during the corresponding ACK mini-slot. Loss of connectivity with the receiver is detected through monitoring the channel during this ACK mini-slot. If the sender does not receive any response to its transmission from the intended receiver, it concludes that the receiver is no more in the neighborhood. It releases its reserved slot, and stops sending voice packets during its reserved slot. The slot is made available for reception at the sender neighbors either when the connection timeout associated with the slot expires, or by making the sender sending an explicit ResvRelease packet to release the slot at its neighbors.

iv) Reservation is lost at the receiver side because collisions during reserved slots.
If a node detects collision during its reception-reserved slot, it concludes that the reservation is lost. It applies the reservation recovery process which consists in two steps. The first step consists in releasing the lost slot with the sender through sending a ResvRelease packet. The second step consists in the reservation repair. When the sender receives the ResvRelease it cancels its reservation and stops sending packets during reserved slot. Afterward, the sender restarts the reservation negotiation process with the receiver in order to reserve a new slot.

4 Simulation

We evaluate the performance of ARPV protocol and compare its behavior with the two contention resolution schemes using the NS-2 simulator. Performances of ARPV are also compared with the IEEE 802.11e EDCF scheme. For this purpose, we use the EDCF simulation model of Wiethölter [7]. Our protocol can be easily coupled with a QoS routing protocol to establish multi-hop end-to-end connections. However, in order to isolate the effectiveness of ARPV from that of the QoS routing protocol, we consider only single-hop connections in our current simulations.

Table 1. Simulation and protocol parameters

Parameter	Value
Channel bit rate (Mbps)	2
Transmission range (m)	150
MAC header (bytes)	38
PHY layer overhead (PLCP header+preamble) (bits)	4+48
Data slot payload size (bytes)	160
Data Slot length (ms)	1.18
Guard time between slots (μs)	2
Super-frame length (ms)	20
Maximum voice packets tolerable delay D_{max} (ms)	200
Connection timeout (s)	10
(p^v, p^d) (for static priority scheme)	(0.35, 0.25)
Number of data slots per super-frame	11
Number of CRS per super-frame	15
Simulation time (s)	1000s

Table 2. Contention parameters for EDCF and DCF

Traffic category	CWmin	CWmax	AIFSN
Data traffic	31	1023	7
Voice traffic	7	15	2
DCF (both voice and data)	31	1023	2

4.1 Simulation Model

Each node can be the source of a G.711 voice flow that generates packets of 160 bytes payload at 64 kbps. The talkspurt and silence period durations are exponentially distributed with a mean of 1 s and 1.35 s respectively. Voice sessions have duration of 200 s, and are randomly started. The *connection timeout* value is 10 s.

The super-frame length is set to the inter-arrival time of voice packets (i.e. 20 ms). Each Data slot consists of the transmission time of one voice packet. Each voice source is required to reserve a slot per super-frame. Following the ITU G.114 recommendation [6], a one-way delay of 300-800 ms is acceptable for some users, while others would not accept a delay above 200 ms. Voice packet loss in the range of 5-10 % are acceptable. In this paper, we assume a 200 ms voice delay bound. For data traffic, we consider FTP sessions that transfer 10 MB files started at random instants. Simulation and protocol parameters are summarized in table I. As the performances of the static priority scheme are sensitive to p^v *and* p^d values, we did several experiments in order to choose the best values. Following these experiments, a choice of p^v=0.35 *and* p^d=0.25 gave the best performances for both data and voice traffics. Two topologies are considered: a grid topology without mobility and random topology with mobility. The first one is used to evaluate the global performance of the protocol. The second one is used to evaluate the effects of node mobility.

4.2 Performance Evaluation in Grid Topology Scenarios

We consider a topology where 64 nodes are placed on a grid of 700m×700m size with distance between pair of adjacent nodes equals to 100 m. Nodes are considered stationary. We uniformly increase the load by increasing data and voice sessions in equal numbers. Figure 2 shows the voice traffic delay achieved by ARPV, DCF, and EDCF with the increase of traffic load. The figure shows that ARPV achieves lower delay than EDCF and DCF at high traffic load, and it is less affected by the increase of traffic load. The low impact of traffic load on voice traffic delay with ARPV is provided owing to the protection of voice traffic streams from the fluctuation of traffic load. A slot is reserved to each voice source and voice packets are transmitted at regular interval regardless of the traffic load. The high delay of EDCF and DCF is due to the increase of contention and collision rates at high traffic load. However, at low traffic

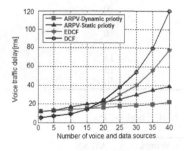

Fig. 2. Voice traffic delay

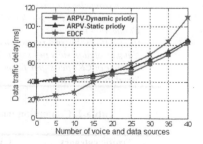

Fig. 3. Data traffic delay

load, EDCF and DCF outperform ARPV because the low level of contention implies a small number of collisions and small backoff. The same conclusions are drawn for data traffic delay (figure 3) with higher delay in comparison to voice traffic delay.

The dynamic priority scheme achieves lower voice traffic delay than the static priority one. This is due to the adaptive access scheme of the dynamic priority scheme during the *RSF*. As nodes adapt their permission probability according to the collision rate, the chance of successful access during the *RSF* is increased.

Figure 4 shows that all schemes achieve low dropping rate at low traffic load, while ARPV achieves better dropping rate than the EDCF at high load. At high traffic load, the DCF achieves the highest dropping rate (8%). The dropping rate with EDCF increases up to 5%. With the static priority it grows up to 2%, while the dynamic priority achieves the lowest dropping rate. Packet dropping with ARPV occurs mainly when voice sources switch from silence period to activity period at the beginning of talkspurts. Packets generated between the time a talkspurt is generated and a reservation is established are dropped if this time exceeds the maximum tolerable delay. Figure 5 shows that while the delay jitter with ARPV remains around 6 ms, it increases linearly with the EDCF and DCF. Unlike with ARPV where voice packets are transmitted at regular intervals, the voice packets delay with EDCF and DCF is unpredictable and depends on the values taken by the backoff timer.

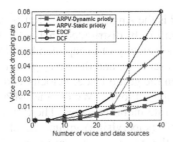 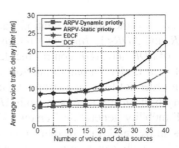

Fig. 4. Voice packet dropping with load increase   **Fig. 5.** Voice traffic jitter with load increase

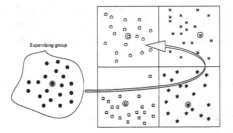

Fig. 6. Simulated area with 4 regions and 5 moving groups

4.3 Performance Evaluation in Mobile Environment

We consider a battlefield scenario where the army is assumed composed of groups of soldiers. Each group may have the task of scanning a particular region. To simulate soldier movements, we use the Reference Point Group Mobility (RPGM) model which is more appropriate to simulate such scenarios [10]. We consider 100 mobile nodes placed on a 1000m×1000m square area. Some nodes are pedestrians and others are vehicles. Simulated area is decomposed into four equal size regions as shown in figure 6. Nodes are partitioned into 5 groups of 20 nodes each, and each group has a leader. Four groups are assigned four regions. The 5th group is the supervising group that roams around the entire area.

In RPGM, the motion of a group leader characterizes the movement of group members. Hence, the group trajectory is determined by providing a path for the group leader in the form of a sequence of velocity vectors $V_{leader}(t)$. $|V_{leader}(t)|$ denotes the speed of group leader at time t, and $\theta_{leader}(t)$ denotes the angle made by $V_{leader}(t)$ with the X-axis. Initially, group leaders are randomly placed on the region to which they are affected. Then, nodes of a group are uniformly distributed in the neighborhood of their leader. Subsequently, each group member has a speed and direction that are derived by random deviation from those of the group leader. The velocity (speed and direction) of a group member at time instant t is calculated as follows:

$$|V_{member}(t)| = |V_{leader}(t)| + random() \times SDR \times max\_speed \qquad (3)$$

$$\theta_{member}(t) = \theta_{leader}(t) + random() \times ADR \times max\_angle \qquad (4)$$

Where SDR is the Speed Deviation Ratio, and ADR is the Angle Deviation Ratio, with $0 \leq SDR$ and $ADR < 1$. random() returns a value uniformly distributed in [-1,1]. The values of SDR and ADR control the deviation of the velocity of group members from that of their leader. In our simulation, max_angle is set to 2π and the max_speed is 10 m/s. In order not to have a strong deviation between the group leader velocity and the velocity of its group members, both ADR and SDR are set to 0.1. We use the IMPORTANT mobility generation framework [11] to generate mobility traces. The first four groups move randomly in there regions. For the 5th (supervising) group, the checkpoints list is chosen to give a circular motion to the group on the entire area. The number of checkpoints for each group leader is 20. The output trace file generated by IMPORTANT is used by the NS-2 in simulation.

We fixed the traffic load to 40 voice connections, 8 connections in each group, and increased the mobility from 0 m/s to 10 m/s with an increment of 1 m/s. The impact of mobility on the performance of ARPV is illustrated in figures 7 and 8.

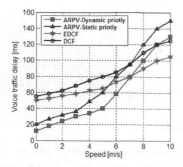
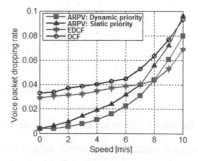

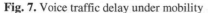

Fig. 7. Voice traffic delay under mobility   **Fig. 8.** Voice packet dropping under mobility

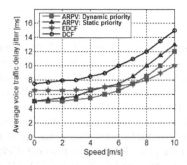

Fig. 9. Voice traffic delay jitter under mobility

ARPV behaves well until a mobility of 7 m/s where the EDCF outperforms ARPV. We observe also that the delay and dropping rate with ARPV increase faster than with EDCF. At 10 m/s, the delay with the dynamic priority scheme increases up to 130 ms, and up to 150 ms with the static priority scheme. It increases up to 100 ms with EDCF and to 120 ms with DCF. The fast delay increase with ARPV is due to the increase of queuing delay of voice packets when nodes lose their reservations due to mobility. At high mobility, nodes lose their reservations more frequently. When a node loses its reservation, its voice packets are queued until a new reservation is established. The increase of dropping rate of ARPV at high mobility is mainly due to two reasons: at high mobility, packets collide during reserved slots more frequently, and many voice packets waiting for reservation are dropped when their queuing delay exceeds the maximal tolerable delay D_{max}. However, for a speed lower than 7 m/s, ARPV achieves lower dropping rate and delay than EDCF because at low speed the network topology is more stable and reservation disruption occurs less frequently. As shown ob Figure 9, lower jitter is achieved by ARPV until a speed of 6m/s where higher delay jitter is observed due to the variation in queuing delay and frequent interruptions of the communications at high mobility.

5 Conclusion

In this paper, we propose a reservation MAC protocol for voice and data support in MANETs. In the proposed protocol, reservation priority is given to voice sources. The latter release temporarily their slots when they switch to the silence period so that other traffic sources can reserve them. In order to ensure that voice sources get enough available resources when they wake up from the silence period once they are admitted into the system, voice sources restore their temporarily released slots each time they wake up. It is feasible to use ARPV for transmission of voice traffic with minimal delay and dropping rate along with data traffic over mobile networks where the time synchronization is available. However, the results show that ARPV performs better in low-mobility and static scenarios. Thus, it seems to be suitable approach for voice traffic support in wireless mesh networks.

References

1. IEEE 802.11WG. Draft Supplement to Standard for Telecommunications and Information Exchange between Systems-LAN/MAN Specific Requirements-part11: MAC Enhancements for Quality of Service (QoS). IEEE 802.11e standard Draft/D13.0 (2005)
2. Carson, E., et al.: A Distributed End-to-End Reservation Protocol for IEEE 802.11-Based Wireless Mesh Networks. IEEE JSAC 24(11) (2006)
3. Manoj, B.S., Siva Ram Murthy, C.: Multimedia Traffic Support for Asynchronous Ad-hoc Wireless Networks. In: BROADNETS 2004, California, pp. 569–578 (2004)
4. Zhu, C., Corson, M.S.: A Five-phase Reservation Protocol (FPRP) for Mobile Ad hoc Networks. In: IEEE INFOCOM 1998, San Francisco, pp. 322–331 (1998)
5. Joe, I.: Qos-Aware MAC with Reservation for Mobile Ad-Hoc Networks. In: IEEE Vehicular Technology Conference, Los Angeles, pp. 1108–1112 (2004)
6. ITU Rec. G.114, One-way Transmission Time, ITU (February 1996)
7. Wiethölter, S., Hoene, C.: Design and Verification of an IEEE 802.11e EDCF Simulation Model in ns-2.26. TKN-03-19 Technical Report, Technische Universität Berlin (2003)
8. Boudour, G., Teyssie, C., Mammeri, Z.: Reservation Clash Handling to Optimize Bandwidth Utilization in MANETs. In: International Conference on Communication Theory, Reliability, and Quality of Service (CTRQ 2008), Bucharest, pp. 77–82 (2008)
9. Mikhailov, V.A.: Geometrical Analysis of the Stability of Markov Chains in R^n_+ and its Application to Throughput Evaluation of the Adaptive Random Multiple Access Algorithm. Problemy Peredachi Informatsii 24(1), 61–73 (1988)
10. Camp, T., Boleng, J., Davies, V.: A Survey of Mobility Models for Ad hoc Network Research. Wireless Communications & Mobile Computing 2(5) (2002)
11. Bai, F., Sadagopan, N., Helmy, A.: The IMPORTANT Framework for Analyzing the Impact of Mobility on Performance of Routing for Ad Hoc Networks. Ad-Hoc Networks Journal 1(4) (2003)
12. Tang, Z., Garcia-Luna-Aceves, J.J.: A Protocol for Topology-Dependent Transmission Scheduling in Wireless Networks. In: IEEE WCNC 1999, New Orleans, pp. 1333–1337 (1999)

On the Shortcoming of EDCA Queues in the Presence of Multiservice Traffic

Agnieszka Brachman and Lukasz Chrost

Silesian University of Technology, Institute of Informatics,
ul. Akademicka 16, 44-100 Gliwice
agnieszka.brachman@polsl.pl, lukasz.chrost@polsl.pl

Abstract. The paper aims at depicting certain limitations of EDCA queues related to the different traffic classes. Through simulation, we show main flaws of identical, medium size, Drop Tail queues used by default for each class, with reference to different traffic category's requirements. We propose some substitutes such as differentiated, basically smaller buffer sizes and Active Queue Management mechanisms that can be employed to improve the overall performance. We present a strategy for EDCA queues to maintain the desired quality of transmission for traffic belonging to different categories.

1 Introduction

Multimedia applications such as Voice over IP (VoIP), video streaming, Video on Demand (VoD) are gaining more and more popularity. At the same time data transmission performed by email, file transfer and web browsing applications is continually increasing. It brings out the necessity to provide different priorities to different applications, users or data flows or to guarantee a certain level of performance to a data flow; in short, to provide Quality of Service (QoS).

The wireless technology based on the IEEE 802.11 standard [1] remains the fundamental solution for wireless LANs. It is widely adopted in homes, offices and public areas causing ousting of wires, although it provides much slower and faultier transmission. High bit error rate along with slower transmission rates hamper the overall QoS and fairness provisioning. QoS must be provided at every stage of transmission and when considering 802.11 WLANs, the weak point lies at the link layer, in the contention-based medium access algorithm.

The IEEE 802.11e amendment introduced new enhancements to enable QoS support in wireless LAN applications through modifications to the Media Access Control (MAC) layer. The amendment has been incorporated into the IEEE 802.11 standard published in 2007. It defines new Hybrid Coordination Function (HCF), where there are two access methods within: HCF Controlled Channel Access (HCCA) and Enhanced Distributed Channel Access (EDCA). The two aforementioned methods represent two different kinds of QoS support: parametrized and prioritized.

HCCA provides contention-free access to the medium by polling the stations and granting so called transmission opportunity of fixed size. Polling is performed

J. Wozniak et al. (Eds.): WMNC 2009, IFIP AICT 308, pp. 69–79, 2009.

in a way that allows fulfilling negotiated QoS requirements. The scheduling process is fairly sophisticated. Despite the fact that the method seems a solution for the problem of providing QoS, it has not been implemented in any device so far. This is probably due to the complexity of the scheduling algorithm. HCCA is beyond the scope of the paper.

EDCA provides a prioritized QoS service by using an independent transmission queue for each traffic category called Access Category (AC). Traffic belonging to a higher priority class has a higher probability of accessing the medium, thus achieving a higher throughput when competing with lower priority traffic. It is achieved by differentiating inter-frame space (AIFS), minimum and maximum contention window size (CW_{min} and CW_{max}) and transmission duration time (TXOP). With EDCA, each station can have up to 4 queues, mapped to different traffic classes. Service differentiation can be achieved by assigning different set of values for every queue contending for access. Traffic with different AC contends independently for the medium access within a station. Performance evaluation of EDCA can be found in [2,3,4,5]. Simulation results along with the outcome in real testbeds confirm that EDCA provides prioritized QoS support.

However medium conditions can drastically change over time which can lead to throughput degradation. As the network becomes overloaded, contention window size increases which leads to long backoff counters and high queue occupation. This especially concerns lower priority queues. This may lead to poor network performance and significantly limit bandwidth utilization. In section 3 we show influence on delay, drop rate, goodput under medium and high congestion in WLAN. We depict how low priority queues are overloaded which leads to throughput oscillations and degradation.

In this paper, through simulation, we depict main flaws of identical, medium size, Drop Tail queues used for each class by default, with reference to different traffic category's requirements. In next section we describe substitutes such as differentiated, basically smaller buffer sizes and Active Queue Management mechanism that can be employed to improve the overall performance. Simulation results for proposed enhancements are presented in section 3.

2   Solutions for EDCA Queues

We investigated how queue management is solved at routers and tried applying the theory into EDCA queues. Routers avoid a serious throughput reduction in the case of a mild congestion, by employing queue management mechanisms, namely Drop Tail and various AQM (Active Queue Management) schemes. Queuing directly influences packet transmission delays, experienced drop rate and dropping scheme. The usage of Drop Tail or any AQM scheme raises many new issues one of which is the buffer sizing. Below we present a short description of evaluated mechanisms as well as discussion concerning buffer sizes.

Implementing a queue management algorithm to obtain a very low level of the buffer occupancy allows achieving high quality of offered services. However, it can also lead to the link underutilization. In this paper RED (Random Early

Detection) as an example of Active Queue Management (AQM) algorithm, is evaluated along with the Drop Tail strategy. A short description of each mechanism is presented below.

The idea of RED is to notify the TCP sender about incoming congestion by dropping packet before the buffer overflow occurs. RED calculates the current network load by counting the exponential moving average of the queue size - avg. If avg is smaller than min_{th} threshold all the packets are enqueued, if $avg > max_{th}$ all the packets are dropped. When avg is between the two thresholds, packets are dropped with linearly increasing probability. More detailed description of RED can be found in [6].

Although active queue management allows achieving the desired average queue length, the simple FIFO queue of fixed buffer size remains the most popular strategy applied at routers. The incoming packets are buffered until the queue is full. When there is no space left, packets are dropped.

The AQM mechanisms allow dynamically adjusting target queue length and packet drop probability and thus outperform Drop Tail mechanism in case of light and moderate congestions. The buffer size and the target queue size are two most important parameters considering queue management. While the use of very large buffers lowers the packet drop probability, it also significantly influences packet transmission delays. Truncation of the buffer size stabilizes the packet delays but it is believed to lead to bandwidth underutilization. However we think that small buffer sizes along with AQMs work for EDCA.

The idea of using RED for EDCA queues is not new. Yang et al. [7] presented a Priority Random Early Detection (PRED) algorithm, which integrated Random Early Detection (RED) with EDCA queues. PRED provides a queuing algorithm for the priority of packets within each node. Authors claim that PRED obtains higher throughput especially under heavy traffic load condition, allowing the packet with higher priority to acquire more throughput than Drop Tail. In the presented scenario queues size were set to 70 packets. Queueing delay was not analized and we believe that such large buffer sizes would lead to high delays and delay variations, especially for low priority traffic. The larger delay the more probable that packet will be considered lost and retransmitted. This maintains high throughput but significantly lower goodput, due to the unnecessarily retransmitted packets. Throughput is calculated as overall transmission and goodput as the effective transmission (without unnecessary retransmissions). Moreover, the usage of AQM algorithm for all traffic classes is in our opinion unjustified, since VoIP and video transmission can use UDP instead of TCP protocol and UDP does not react to early packet drops.

3   Performance Evaluation of EDCA Queues

The ns-2 simulator was used to study EDCA queues performance [8]. Since the EDCA model is not included in standard ns-2 release, we based on Sven Wiethlter and Christian Hoene's model, developed at Technical University of Berlin [9]. The model provides a dedicated, priority-driven queue management algorithm, implemented in set of classes. The main class used for the queue management operations (PriQ), stands for a configurable interface for 4 priority-specific Drop Tail queues. We have extended the model, providing a simple mechanism allowing utilization of any built-in AQM scheme for selected EDCA queue.

The paper focuses on simulation model where WLAN is used as a last-hop network. We proposed several scenarios to depict the main flaws and limitations of EDCA queues. Detailed simulation topology and scenarios are described below, along with simulation results.

3.1   Simulation Scenarios and Topology

The network topology is shown in fig. 1. The wired node is connected to the router with 10 Mbps links. The link propagation delay is set to 1 ms. The link from router R to Access Point AP has the propagation delay of 10 ms and the bitrate 10 Mbps. The wireless bandwidth is set to 11 Mbps which provides around 5-6 Mbps of effective bandwidth. The wireless nodes are uniformly distributed around AP. The wireless network is the bottleneck for each connection.

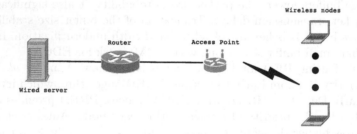

Fig. 1. Network topology

Firstly we investigated the behaviour of EDCA queues related to Access Categories where each queue is a simple FIFO (Drop Tail - DT) queue. The MAC parameters for EDCA, used in the simulations are summarized in table 1.

Table 1. Default EDCA values for AC queues

AC	Prio	$AIFS[AC]$	$CW_{min}[AC]$	$CW_{max}[AC]$	$TXOPLimit[AC]$
AC_BK	3	7	15	1023	0
AC_BE	2	3	15	1023	0
AC_VI	1	2	7	15	3.008
AC_VO	0	2	3	7	1.504

We focused on three basic scenarios, all with bidirectional traffic. Scenario I and II simulate heavy network congestion, scenario III represents mild network congestion.

Scenario I - Traffic is composed of FTP connections. Every wireless node transmits and receives data. Four uplink and downlink connections exist in the network, each belonging to different traffic category. TCP SACK is used at transport layer, packet size is 1500 bytes.

Scenario II - Traffic is composed of Constant Bit Rate (CBR) connections. Every wireless node transmits and receives data with constant bitrate set to 1 Mbps. Packet size is 1000 bytes. Four uplink and downlink connections exist in the network, each belonging to different traffic category.

Scenario III - Simulation traffic is composed of Variable Bit Rate (VBR), FTP and WWW connections. Two VBR, one FTP and one WWW connections transmit and receive on every wireless node resulting in eight connections for every node. Parameters for aforementioned connection set are summarized in table 2. VBR streams are generated using Pareto sources. WWW traffic is generated using Pareto-sized bursts of data, mean burst size is 50 KB and pareto shape parameter is 1.3. The burst occurrence is defined by poisson process with mean interval of 1.6 second.

Table 2. Parameters for each connection set in Scenario III

	Mean bitrate	Packet size	Protocol	Priority
VBR	64 kbps	210	UDP	0
VBR	300 kbps	1000	UDP	1
WWW	500 kbps	1500	TCP	2
FTP	-	1500	TCP	3

All connections are active during the entire simulation. Each simulation is run for 110 seconds. Statistics are collected during the last 100 seconds.

3.2 Simulation Results

We investigated one AQM algorithm - RED and compared the results with simple FIFO (DT) queue. The simulations were run twice for each scenario, setting a standard and a small buffer limit for DT queue as well as a short and a medium target queue size for RED. Comparing the gathered results, we proposed a hybrid solution for EDCA queues described in section 3.2.

Results for DT-EDCA Queues. Figures 2, 3 and 4 show the average queue length measured on the access point, in packets. The results are presented for all scenarios, when buffer size is set to 10 and 50 packets accordingly, for each EDCA queue. Queue limit of 50 packets is considered a standard limit for EDCA queues, therefore the limit of 10 packets should be regarded as a small buffer. 0 indicates the highest and 3 the lowest priority.

TCP connections tend to full link utilization which significantly influences the low priority traffic (fig. 2). The use of small buffers reduces the disparity among

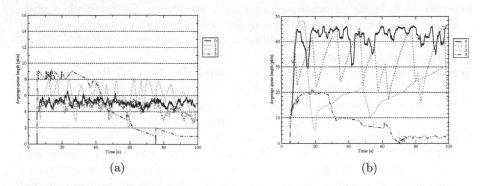

Fig. 2. Scenario I: Average queue length for DT-EDCA queues. The queue limit is set to: (a) 10 and (b) 50 packets.

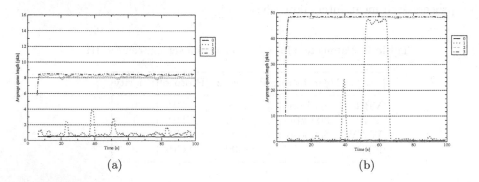

Fig. 3. Scenario II: Average queue length for DT-EDCA queues. The queue limit is set to: (a) 10 and (b) 50 packets.

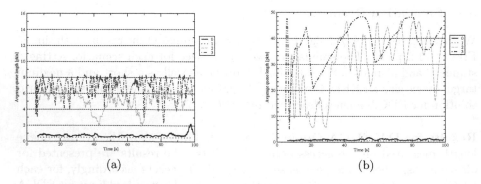

Fig. 4. Scenario III: Average queue length for DT-EDCA queues. The queue limit is set to: (a) 10 and (b) 50 packets.

flows of different priority. Also the higher priority the lower queue oscillations, which results in small delay variation.

UDP connections do not react to packet drops and do not reduce transmission rate, ergo lower priority connections always utilize the whole buffer (fig. 3). It can lead to excessive queueing delays and to the situation where although the packet arrives, it is no longer valid (i.e. in VoIP or video transmission). Packets are transmitted in both directions therefore network overload is high and download throughput is low. It results in some spikes in the average queue length for second highest access category and variation in queueing delay. It is worth mentioning that the large buffer results in higher spikes and consequently higher delay variation.

In scenario III a mixture of TCP and UDP connections is introduced. Average queue length is showed in figure 4, tables 3 and 4 summarize measured average transmission parameters for uplink and downlink connections belonging to different traffic categories.

UDP connections have the highest priority and they utilize approximately 20% of wireless bandwidth. The size of related queues does not influence delay neither drop rate. Both uplink and downlink flows achieve the maximum throughput. The buffer limit mainly affects transmission delay for the lowest priority traffic. For the 50 packets buffer size, average transmission delay of the lowest priority, downlink connections is over 3 seconds which is interpreted by most applications as non-availability. It is also ten times more than the transmission delay noticed by the uplink connections belonging to the same category. Similar situation occurs for the WWW traffic. Reducing the buffer limit to 10 packets decreases the transmission delay as well as the disparity in the uplink and downlink delay at the cost of a higher drop rate for the downlink FTP connections. This is because of the four connections contending for the lowest priority queue at the Access Point. The uplink connections have their own queue on every wireless node.

Results for RED-EDCA Queues. Figures 5, 6 and 7 show the average queue length on the Access Point, measured in packets for all scenarios accordingly, when the buffer size is set to 50 packets, for each EDCA queue. We set two different target queue lengths for the RED algorithm: 10 and 25. It it achieved by setting min_{th} and max_{th} values to: 5, 15 or 20, 30 accordingly.

When the small target queue length is used for the TCP connections, higher link utilization is achieved because of the higher average queue length in comparison to DT (figure 5). Higher target queue length lowers the queue length oscillations. Also the lowest priority traffic is allowed to maintain a higher transmission rate. The queue oscillations would be smaller if the number of connections increased.

AQM by no means influences UDP traffic sources, therefore the simulation results showed in figure 6 are very similar and comparable to the simulation results where Drop Tail is used (figure 3).

Measured, average transmission parameters for the uplink and downlink connections belonging to the different traffic categories when RED-EDCA queues are used are summarized in tables 5 and 6.

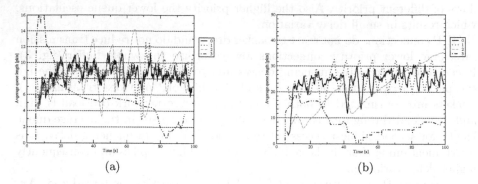

(a)   (b)

Fig. 5. Scenario I: Average queue length for RED-EDCA queues. The target queue length is set to: (a) 10 and (b) 25 packets.

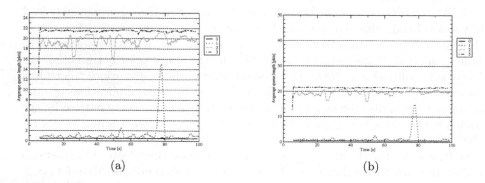

(a)   (b)

Fig. 6. Scenario II: Average queue length for RED-EDCA queues. The target queue length is set to: (a) 10 and (b) 25 packets.

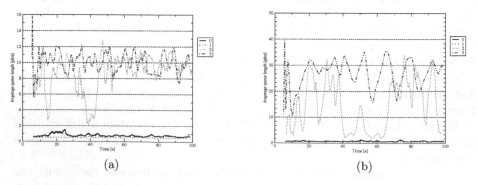

(a)   (b)

Fig. 7. Scenario III: Average queue length for RED-EDCA queues. The target queue length is set to: (a) 10 and (b) 25 packets.

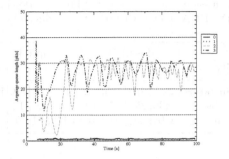

Fig. 8. Scenario III: Average queue lengths for hybrid EDCA queues

Table 3. Average transmission parameters for the uplink and downlink traffic categories in Scenario III and DT-EDCA queues with the queue limit equal to 10 packets

Priority:	downlink				uplink			
	3	2	1	0	3	2	1	0
avg thr [kbps]	206	759	318	107	995	1006	498	99
delay [ms]	460	101	15	11	177	52	14	11
drop rate [%]	12.20	6.86	0	0.12	6.54	0	0	0

Table 4. Average transmission parameters for the uplink and downlink traffic categories in Scenario III and DT-EDCA queues with the queue limit equal to 50 packets

Priority:	downlink				uplink			
	3	2	1	0	3	2	1	0
avg thr [kbps]	151	593	485	155	637	1180	530	141
delay [ms]	3125	602	17	12	276	68	15	11
drop rate [%]	4.70	3.30	0	0.03	6.73	5.37	0	0

Table 5. Average transmission parameters for uplink and downlink traffic categories in Scenario III and RED-EDCA queues with the target queue length set to 10 packets

Priority:	downlink				uplink			
	3	2	1	0	3	2	1	0
avg thr [kbps]	208	719	451	140	904	735	555	133
delay [ms]	638	163	16	11	192	61	14	11
drop rate [%]	8.74	5.13	0	0	6.27	4.86	0	0

Similarly to the results described in the previous section, the target queue length does not affect any transmission parameter of UDP connections. Correspondingly, the transmission delay of the lowest priority traffic drops with the

Table 6. Average transmission parameters for uplink and downlink traffic categories in Scenario III and RED-EDCA queues with the target queue length set to 25 packets

Priority:	downlink				uplink			
	3	2	1	0	3	2	1	0
avg thr [kbps]	227	670	485	122	972	727	710	116
delay [ms]	1436	255	15	10	183	52	14	11
drop rate [%]	4.16	0.81	0	0	6.42	4.41	0	0

Table 7. Average transmission parameters for uplink and downlink traffic categories in Scenario III and with hybrid EDCA queues

Priority:	downlink				uplink			
	3	2	1	0	3	2	1	0
avg thr [kbps]	254	935	514	121	1034	603	346	119
delay [ms]	1320	313	15	11	166	49	14	11
drop rate [%]	3.57	1.51	0	0.01	6.66	4.40	0	0

target queue length degradation, as well as the disproportion in the transmission delay of the downlink and uplink connections. RED also significantly improves the transmission delay along with the throughput for the larger target queue length. The drop rate is not affected (compare results for the lowest priority, downlink connections in tables 4 and 6). In comparison to DT, when to short target queue level is set, RED improves the drop rate and maintains slightly longer transmission delays and lower throughput for the lower priority traffic.

Hybrid EDCA Queues. We propose a hybrid solution for EDCA queues to achieve the best results for each traffic category. The high priority traffic is usually transmitted using UDP protocol, therefore no sophisticated AQM scheme is needed. Also the small buffer size results in low transmission delays which is very important for the real-time traffic. For this categories we propose to use the Drop Tail queues, limited in size to 10 packets. For WWW and FTP traffic RED queues with the target queue level of 25 packets are advised. The average queue length is presented in figure 8, the detailed simulation results for uplink and downlink connections in scenario III are summarized in table 7.

UDP connections of the highest priority are not affected. Transmission parameters are comparable to the results achieved in previous scenarios. For the lower priority traffic, the disparity in throughput of the downlink and uplink connections is decreased at the cost of the higher drop rate for the uplink connections. At the same time, the transmission delay is significantly longer for the downlink connections. However it is maintained in acceptable range.

4   Conclusions

We show, that the use of very small buffers (10 packets) for EDCA queues significantly reduces transmission delay especially for the low priority traffic. At the same time it can influence the throughput and the drop rate both in positive and negative manner. Through simulation we depict, that applying AQM in place of DT EDCA queues for the TCP traffic results in throughput improvement along with the drop rate reduction. Finally, we propose a hybrid solution for EDCA queues. The solution utilizes very short, DT queues for handling the UDP traffic, along with medium-sized RED queues for handling the TCP traffic. It proved to successfully deal with the described shortcomings of the standard DT EDCA model.

Acknowledgement. This work is supported by the Ministry of Science and Higher Education under grant KBN N N516 381134.

References

1. IEEE80211: IEEE Standard for Information technology-Telecommunications and information exchange between systems-local and metropolitan area networks-Specific requirements - Part 11: Wireless LAN Medium Access Control (MAC) and Physical Layer (PHY) Specifications. IEEE Std 802.11-2007 (Revision of IEEE Std 802.11-1999) C1–1184 (12 2007)
2. Choi, S., del Prado, J., Shankar, S., Mangold, S.: IEEE 802.11e contention-based channel access (EDCF) performance evaluation. In: IEEE International Conference on Communications, 2003. ICC 2003, May 11-15, vol. 2, pp. 1151–1156 (2003)
3. He, D., Shen, C.: Simulation study of IEEE 802.11e EDCF. In: The 57th IEEE Semiannual Vehicular Technology Conference, 2003. VTC 2003-Spring, April 22-25, vol. 1, pp. 685–689 (2003)
4. Ni, Q.: Performance analysis and enhancements for IEEE 802.11e wireless networks. IEEE Network 19, 21–27 (2005)
5. Xiao, Y.: IEEE 802.11e: QoS provisioning at the MAC layer. IEEE Wireless Communications [see also IEEE Personal Communications] 11, 72–79 (2004)
6. Floyd, S., Jacobson, V.: Random Early Detection Gateways for Congestion Avoidance. IEEE/ACM Transactions on Networking 1, 397–413 (1993)
7. Yang, C.C., Li, J.S., Ruei-Yi-Li, Lin, S.-Y., Wen, J.-H.: QoS Performance Improvement for WLAN Using Priority Random Early Detection. In: International Conference on Wireless Communications, Networking and Mobile Computing, 2007. WiCom 2007, September 2007, pp. 1996–1999 (2007)
8. ns2: The Network Simulator ns-2: Documentation (2009), http://www.isi.edu/nsnam/ns/doc/index.html
9. Wiethölter, S., Emmelmann, M., Hoene, C., Wolisz, A.: TKN EDCA Model for ns-2. Technical Report Technical Report TKN-06-003, Technische Universität Berlin (June 2006)

Accessing of Large Multimedia Content on Mobile Devices by Partial Prebuffering Techniques

Ondrej Krejcar, Dalibor Janckulik, and Leona Motalova

VSB Technical University of Ostrava, Center for Applied Cybernetics, Department of measurement and control, 17. Listopadu 15, 70833 Ostrava Poruba, Czech Republic
Ondrej.Krejcar@remoteworld.net, Dalibor.Janckulik@vsb.cz,
Leona.Motalova@vsb.cz

Abstract. New types of complex mobile devices can operate full scale applications with same comfort as their desktop equivalents only with several limitations. One of them is insufficient transfer speed on wireless connectivity in case of downloading the large multimedia files. Main area of paper is in a model of a system enhancement for locating and tracking users of a mobile information system. User location is used for data prebuffering and pushing information from server to user's PDA. All server data is saved as artifacts along with its position information in building or larger area environment. The new partial prebuffering techniques is designed and pretested. The accessing of prebuffered data on mobile device can highly improve response time needed to view large multimedia data. This fact can help with design of new full scale applications for mobile devices.

Keywords: Mobile Device, Localization, Prebuffering, Response Time, Area Definition.

1 Introduction

The usage of various mobile wireless technologies and mobile embedded devices has been increasing dramatically every year and would be growing in the following years. This will lead to the rise of new application domains in network-connected PDAs or XDAs (the "X" represents voice and information/data within one device; "Digital Assistant") that provide more or less the same functionality as their desktop application equivalents. The idea of full scale applications pursuable on mobile devices is based on current hi-tech devices with large scale display, large memory capabilities, and wide spectrum of network standards plus embedded GPS module (HTC Touch HD).

Users of these portable devices use them all time in context of their life (e.g. moving, searching, alerting, scheduling, writing, etc.). Context is relevant to the mobile user, because in a mobile environment the context is often very dynamic and the user interacts differently with the applications on his mobile device when the context is different [1].

My recent research of context-aware computing has been restricted to location-aware computing for mobile applications using a WiFi network (LBS Location Based Services). The information about basic concept and technologies of user localization

J. Wozniak et al. (Eds.): WMNC 2009, IFIP AICT 308, pp. 80–91, 2009.

(such as LBS, Searching for WiFi AP) can be found in my article [2]. On localization basis, we created a special framework called PDPT (Predictive Data Push Technology) which can improve a usage of large data artifacts of mobile devices [3]. We used continual user position information to determine a predictive user position. The data artifacts linked to user predicted position are prebuffered to user mobile device. When user arrives to position which was correctly determined by PDPT Core, the data artifacts are in local memory of PDA. The time to display the artifacts from local memory is much shorter than in case of remotely requested artifact.

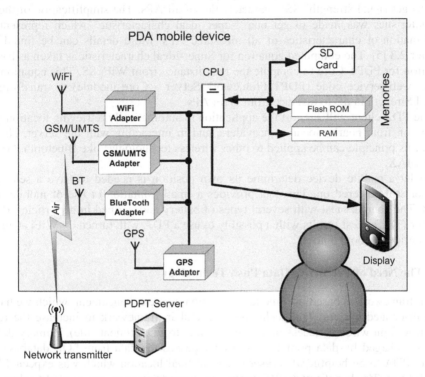

Fig. 1. Wireless networks and GPS sensor localization possibilities on mobile devices

The idea of prebuffering may not be only one application method for user position knowledge. As well as WiFi is not only one wireless network to use for localization of user device. WiFi has advantage in speed in indoor positioning therefore the GSM/UMTS can be used in outdoor (Fig. 1). The GPS sensor is also embedded in several types of current mobile devices, or it can be plugged by SDIO or BT interface.

We would like to describe a problem of long response in mobile device in the beginning of next chapter. Following subchapter will deal with optimal artifact size determination as well as new partial prebuffering techniques details. The position information obtaining from wireless networks (WiFi, BT, GSM, GPS) background will follow in next subchapter. The needed info about PDPT Framework design, area definition around the user and PDPT client application highlights are in the rest subchapters of chapter 2.

2   The PDPT Framework and PDPT Core

The general principle of my simple localization states that if a WiFi-enabled mobile device is close to such a stationary device – Access Point (AP) it may "ask" the provider's location position by setting up a WiFi connection. If position of the AP is known, the position of mobile device is within a range of this location provider. This range depends on type of WiFi AP. The Cisco APs are used in my test environment at Campus of Technical University of Ostrava. We performed measurements on these APs to get signal strength (SS) characteristics of all APs. The simplification of these characteristics was made to get one "super ideal characteristic" which represent a combination of characteristics of all measured APs. More details can be found in chapter 2.3 [5]. The computed equation for Super-Ideal characteristic is taken as basic equation for PDPT Core to compute the real distance from WiFi SS. This equation is in the web service code (PDPT Framework Server – Core module) to transform a Signal Strength in dB to distance from WiFi APs.

The PDA client will support the application in automatically retrieving location information from nearby location providers, and in interacting with the server. Naturally, this principle can be applied to other wireless technologies like Bluetooth, GSM or WiMAX.

To let a mobile device determine its own position is needed to have a selected adapter still powered on. This fact provides a small limitation of use of mobile devices. The complex test with several types of battery is described in my article [4] in chapter (3). The test results with a possibly to use a PDA with turned on WiFi adapter for a period of about 5 hours.

2.1   The Need of Predictive Data Push Technology

PDPT framework is based on a model of location-aware enhancement, which we have used in created system. This technique is useful in framework to increase the real dataflow from wireless access point (server side) to PDA (client side). Primary dataflow is enlarged by data prebuffering. PDPT pushes the data from SQL database to clients PDA to be helpful when user comes at final location which was expected by PDPT Core. The benefit of PDPT consists in time delay reducing needed to display desired artifacts requested by a user from PDA. This delay may vary from a few seconds to number of minutes. Theoretical background and tests were needed to determine an average artifact size for which the PDPT technique is useful. First of all the maximum response time of an application (PDPT Client) for user was needed to be specified.

Nielsen [6] specified the maximum response time for an application to 10 seconds [7]. During this time the user was focused on the application and was willing to wait for an answer. The book is over 20 years old (published in 1994). We suppose the modern user of mobile devices is more impatient so the stated value of 10 second will be shorter. This is for me even better, because my framework is more usable. We used this time period (10 second) to calculate the maximum possible data size of a file transferred from server to client (during this period). To define the amount of data is possible to download to mobile device; we executed a test of data transfer rate measurement of large data size artifacts throw the FTP protocol. We used as test devices four types of PDA (HTC Athena, HTC Universal, HTC Blueangel, HTC Roadster). Fist two devices are equipped with 802.11g standard while the rest two are only with 80211b standard of WiFi capability. These PDA devices were connected throw CISCO Wi-Fi AP. The FTP server holds 3 types of large artifacts (files) which were downloaded to internal PDA memory.

Table 1. Transmission speed on large files

	PDA device			
	Athena	Universal	Blueangel	Roadster
data size [MB]	Transfer Speed [kB/s]			
10	347	123	160	106
20	344	121	157	79
40	314	123	58	43

Unafraid the theoretical transfer rates (802.11g = 54 Mbit/s, 802.11b = 11Mbit/s) were not achieved (Table 1). The maximal transfer rate of 350 kB/s has HTC Athena, but this device is not a standard PDA device. Athena is a mini-notebook with windows mobile 6 operating system. All of others devices have only a quarter amount of such speed. It is much clear that the wireless connected mobile devices have not only a limitation in wireless network module HW, but it has a problem in a slow internal bus. Finally if transfers speed wary from 40 to 160 kB/s the result file size (file which can be downloaded in a defined time of 10 seconds) wary from 400 to 1600 kB.

2.2 Determination of Optimal Artifact Size

The next step is an average artifact size definition. We use a network architecture building plan as a sample database, which contained 100 files of average size of 470 kB. The client application can download during the 10 second period from 2 to 3 artifacts. The problem is the long time delay in displaying of artifacts in some original file types. It is needed to use only basic data formats, which can be displayed by PDA natively (bmp, jpg, wav, mpg, etc.) without any additional striking time consumption.

The abysmal difference of transfer speeds from previous table [Table 1] vamooses, when we use for transfer smaller data files (10KB – 150KB). The testing of data transfer throw the web services was executed on all of mentioned devices. Firstly the 50 kB and then the 150 kB data file were transferred. The response time for one access and then for 100 access and finally for 500 access were measured. The test results are in Table 2 and Table 3.

The third step to determine the optimal artifact size is testing of database response for buffering. The test was executed again on all mentioned devices. The information about SQL server response of SQL Server 3.5 Compact Edition was stored.

From the executed test over the web service is evident; the artifacts of size from 50 to 150 kB are more suitable for transfer. It is because the transfer speed of them is relatively affordable in compare to transferred data amount.

Table 2. Response time - 50KB artifacts [ms]

	Number of Artifacts		
	1	100	500
Type	**Transfer Speed [kB{s]**		
Mode	101.82	5208.62	16036.61
Median	101.86	5228.54	16024.24
Average	102.00	5229.81	16022.67

Table 3. Response time - 150KB artifacts [ms]

	Number of Artifacts		
	1	100	500
Type	**Transfer Speed [kB{s]**		
Mode	304.82	20208.93	61029.73
Median	301.85	20227.71	61026.08
Average	302.00	20229.99	61022.86

In this case is only a higher starter costs which going to fall after first executed query. The SQL server response results are better in case of 50 to 150 kB artifacts. The artefacts transfer is not striking burdened by dependence on real speed of connection between AP and mobile device. Difference will show after a huge amount of artifacts was transferred (hundreds). The buffer stream of such size is equivalent to download of minimally 14 presentation areas.

Table 4. Response time average- SQL query - Insert of 1 artifact into SQL CE database engine [ms]

Device Type	Artifact size[kB]		
	50	150	500
HP 614c	176.2	528.3	2377.3
HTC Roadster	222.4	667.6	3004.0
HTC Advantage	226.6	679.5	3057.6
HTC Blueangel	136.4	409.0	1840.4
Samsung Omnia	185.9	557.2	2507.3

Table 5. Response time average- SQL query - Select of 1 artifact into SQL CE database engine [ms]

Device Type	Artifact size[kB]		
	50	150	500
HP 614c	83.0	95.7	118.7
HTC Roadster	91.3	103.3	128.1
HTC Advantage	79.3	91.2	113.1
HTC Blueangel	79.1	91.1	112.9
Samsung Omnia	81.1	93.1	115.5

As will be shown from schema of presenting area (Fig. 3), the most of buffered parts will not be needed. The most part of the writing time of SQL buffer from web service is due to the response from SQL server. As it is resulted in tables, writing of tens 50 kB files to the database cost about a same time as writing of one 500 kB file. With smaller files we are able to be more effective to cover buffered area.

The large artifacts are better in compare to 1Byte transferred data, but the response time is not optimal.

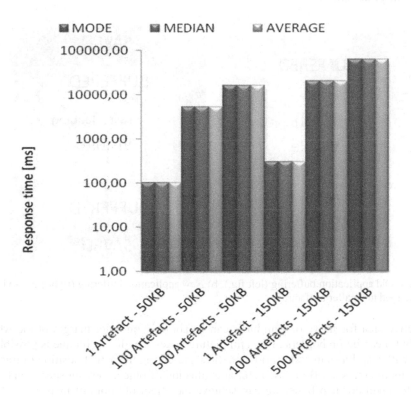

Fig. 2. Web service reaction times

Such large artifact size is not only with slow transfer speed, but they also allow move with steps, which are not affordable for quicker move or more accurately presentation of position in artifact. The most of tests and problems passes from the size of artifacts. Therefore we made a change to data artifact size to maximal size of 150 kB and we change also the access to these artifacts to new way.

PDPT framework design is based on the most commonly used server-client architecture. To process data the server has online connection to the information system. Technology data are continually saved to SQL Server database.

Previous version of PDPT framework buffered the data according to the predicted position of user. In case of unexpected change of direction, the buffer in most cases cannot act on this quick change.

The active presented area was divided to more partial artifacts (Fig. 3b). This new modified system is now implemented to our other projects, where the position of user is needed. One of these projects is a Guardian II. This project is for hospitals areas for patients and physicians monitoring. We have implemented the new possibilities of biomedical e-health systems are discovered for increasing of interactivity. Based on position of patient, the server can select the nearest physician or nurse to act on discovered problem. By this way the response on problem can be reduced and it can help to save more human life.

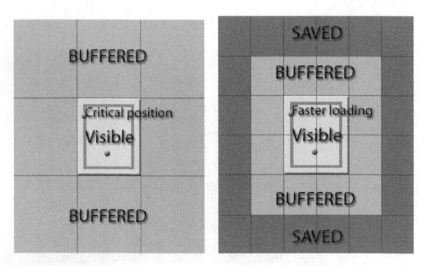

Fig. 3. a) Old application buffering (left fig.). b) New application buffering (right fig.) – visible parts merged to smaller artifacts.

Active area for move of map basis can be not only on outer margin of the whole area, but it can be on borders of several picture-boxes. By this technique is possible to move with map basis in softer grid and allow presenting the actual position of tracked object more precisely. By consequential evaluation of object moving speed, and with suitable modified map basis, we can achieve the effect of zoom of map basis. After application of such principle the system can be applicable for open space with sufficient signal for triangulation - WiFi, GSM or WiMAX. This part of framework is suitable for patient tracking in case of home care agencies. We can track the time of one attendance of nurse at the patient.

2.3 WiFi, BT, GSM and GPS Data Collection for Localization

When the use of PDPT technique is possible and it bears some advantages, the first key step of the PDPT is a data collection phase. We record information about the radio signals (WiFi, BT, GSM) as a function of a user's location. The signal information is used to construct and validate models for signal propagation.

If the mobile device knows the position of the stationary device (transmitter, BTS), it also knows that its own position is within a range of this location provider. The typical range wary from 30 to 100 m in WiFi case, respectively 50 m in BT case or 30 km for GSM. Granularity of location can be improved by triangulation of two or more visible APs (Access Points) or BTSs (Base Transceiver Stations).. The PDA client currently supports the application in automatically retrieving location information from nearby WiFi location providers, and in interacting with the PDPT server. Naturally, this principle can be applied to other wireless technologies like BT, GSM, UMTS or WiMAX. The application (locator) is implemented in C# language using the MS Visual Studio .NET with .NET Compact Framework and a special Open-NETCF library enhancement.

In previous research, we focused only to use of WiFi networks while the other wireless possibilities remained without a proper notice. Now we made an enhancement of Locator component of PDPT framework (Fig. 4) to allow operate with BT and GSM networks.

In BT network case, the position of BT APs must be known to allow the position determination. To collect BT APs position info in outdoor environment, the GPS can be used. For indoor area, the GIS (Geographic Information System) software with buildings map must be used to measure exact position of BT AP against to local environment. To manage with BT hardware of mobile device another library InTheHand 32Feet.NET is used.

GSM network is not local network but a cellular network. The problem is in position information of GSM BTSs. The operator doesn't provide any such information. One of possible solutions is based on unofficial BTSs lists which can be found on internet. The lists are typically available in HTML, TXT or CSV formats. The medium rate for BTSs with GPS position information is about 90 % of all BTSs in European countries. In case of PDPT Framework, the list must be converted to PDPT server database – GSM_BTS table.

In all three described cases of nearby BSs scanning, the data are saved to Locator Table in PDPT server DB. Data are processed from Locator Table throw the PDPT Core to Position Table. The processing techniques depend on selected wireless network. WiFi and BT network provide all visible APs nearby the user. From list of these APs is computed actual position (by triangulation).

Mobile devices with windows mobile operation system do not provide any GSM info to .NET Compact Framework. Even any special framework as in previous two cases is not known for me until now. Only possibility is in use of RIL (Radio Interface Layer) library. This library is divided into two separate components, a RIL Driver and a RIL Proxy. The RIL Driver processes radio commands and events. The RIL Proxy performs arbitration between multiple clients for access to the single RIL driver. When a module calls the RIL to get the signal strength, the function call immediately returns a response identifier. The RIL uses the function response callback to convey signal strength information to the module.

The GSM network provide only one BS info in each search cycle. This BS has the highest signal strength. The more BTSs info is collected by a several iteration cycles. During 10 cycles (per 10 seconds) the 4 BTS info is obtained on average.

The important info from BTSs is Signal Strength and Time Advance (TA). SS is refreshed every several seconds (in every scan) whereas TA is provided only during some type of communication with selected BTS (e.g. request to talk, move to another area - Location Area Code (LAC)). The list of these BTSs with info is further processed as in previous case for WiFi and BT networks. Only change is in usage of TA if it is accessible. Another possibility to get the user position in outdoor space is in GPS [8]. GPS provide a position by LONgitude and LATitude (X and Y). Only simple conversion is needed to transform a GPS coordinates to S-JTSK, which is used in PDPT Framework.

2.4   The PDPT Framework Design

The PDPT framework design is based on the server-client architecture. The PDPT framework server is created as a web service to act as a bridge between MS SQL Server (other database server eventually) and PDPT PDA Clients (Fig. 4).

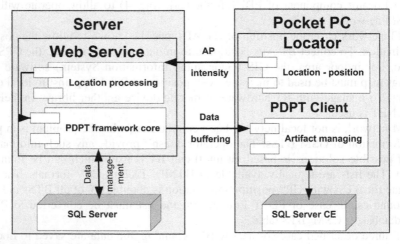

Fig. 4. PDPT architecture – Server has a web service with two modules (Locator processing and PDPT framework Core) and a connection to SQL Server. Mobile Client has a Locator module to send a signal info to sever and a PDPT Client module to manage downloaded artifacts to SQL Server CE mobile database.

Client PDA has location sensor component, which continuously sends the information about nearby AP's intensity to the PDPT Framework Core. This component processes the user's location information and it makes a decision to which part of MS SQL Server database needs to be replicated to client's SQL Server CE database [9][10]. The PDPT Core decisions constitute the most important part of PDPT framework, because the kernel must continually compute the position of the user and track, and predict his future movement. After doing this prediction the appropriate data are prebuffered to client's database for the future possible requirements. This data represent artifacts list of PDA buffer imaginary image (Fig. 5).

2.5   PDPT Core – Area Definition

The PDPT buffering and predictive PDPT buffering principle is shown in Fig. 5. Firstly the client must activate the PDPT on PDPT Client. This client creates a list of artifacts (PDA buffer image), which are contained in his mobile SQL Server CE database. Server create own list of artifacts (imaginary image of PDA buffer) based on area definition for actual user position and compare it with real PDA buffer image.

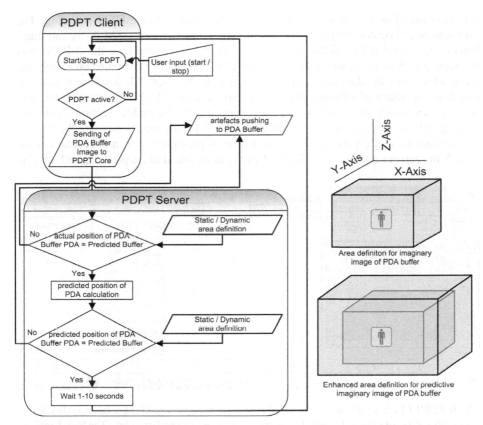

Fig. 5. Object diagram of PDPT prebuffering and predictive PDPT prebuffering. Right part shows the area definition for imaginary image of PDA buffer.

The area can be defined as an object where the user position is in the object centre. We are using the cuboid as the object in present time for initial PDPT buffering. This cuboid has static area definition with a size of 10 x 10 x 3 (high) meters. The PDPT Core will continue with comparing of both images. In case of some difference, the rest artifacts ale prebuffered to PDA buffer. When all artifacts for current user position are in PDA buffer, there is no difference between images. In such case the PDPT Core is going to make a predicted user position. On base of this new user position it makes a new predictive enlarged imaginary image of PDA buffer. The size of this new cuboid is statically defined area of size 20 x 20 x 6 meters. The new cuboid has a center in direction of predicted user moving and includes a cuboid area for current user position. The PDPT Core compares the both new images (imaginary and real PDA buffer) and it will continue with buffering of artifacts until they are same. In real case of usage is better to create an algorithm to dynamic area definition to adapt a system to user needs more flexible in real time. For additional info please refer to [11].

2.6 The PDPT Client Application

The PDPT Client application realizes thick client to the server side and an extension by PDPT and Locator modules. Figure 6 shows three screenshots from the mobile client.

The first one (Fig. 6a) shows the Locator module with selected GSM scanning. The info text box "Locator AP ret." Provide info about last founded GSM BTSs and number of recognized BTSs (BTSs with GPS position). In current case the 6 BTSs was founded and 5 of them was recognized by PDPT Framework. Figure 6b shows the classical view of the data artifact presentation from MS SQL CE database to user (in this case the image of Ethernet plan of the current area). The PDPT tab (Fig. 6c) presents a way to tune the settings of PDPT Framework. The middle section shows the logging info about the prebuffering process. The right side means measure the time of one artifact loading ("part time") and full time of prebuffering in millisecond resolution. More screens and details of PDPT Client can be found in chapters 2.7 and 2.8 [5].

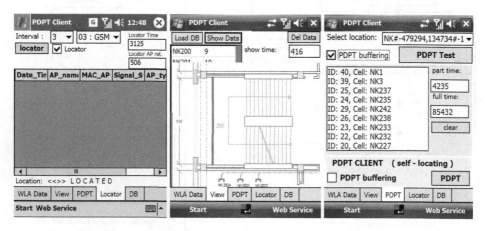

Fig. 6. PDPT Client – Left one figure 6a – Locator component with GSM scanning. Middle one figure 6b – View of classical client data representation. Right one figure 6c - The PDPT options screen allow to start and control the PDPT buffering.

3   Conclusions

We am focused on the real usage of the developed PDPT Framework on a wide range of wireless connected mobile devices and its main issue at increased data transfer. For testing purpose, five mobile devices were selected with different hardware and software capabilities. The high success rate found in the test data surpassed our expectations. This rate varies from 84 to 96 %. Please see the chapter 4 [5] for more info. The PDPT prebuffering techniques can improve the using of medium or large artifacts (from 50kB to 500 kB per artifact) on wireless mobile devices connected to information system. The localization part of PDPT framework is currently used in another project of biotelemetrical system for home care named Guardian II to make a patient's life safer [12]. Another utilization of PDPT consists in use of others wireless networks like BT, GSM/UMTS, WiMAX, or in GPS. This idea can be used inside the information systems like botanical or zoological gardens where the GPS navigation can be used in outdoor. Some improvements of Locator module are also described. The larger area of PDPT utilization can improve importance of PDPT Framework in wireless mobile systems.

Acknowledgment. This work was supported by the Ministry of Education of the Czech Republic under Project 1M0567.

References

1. Abowd, G., Dey, A., Brown, P., et al.: Towards a better understanding of context and context-awareness. In: Gellersen, H.-W. (ed.) HUC 1999. LNCS, vol. 1707, p. 304. Springer, Heidelberg (1999)
2. Krejcar, O.: User Localization for Intelligent Crisis Management. In: AIAI 2006, 3rd IFIP Conference on Artificial Intelligence Applications and Innovation, Boston, USA, pp. 221–227 (2006)
3. Krejcar, O., Cernohorsky, J.: Database Prebuffering as a Way to Create a Mobile Control and Information System with Better Response Time. In: Bubak, M., van Albada, G.D., Dongarra, J., Sloot, P.M.A. (eds.) ICCS 2008, Part I. LNCS, vol. 5101, pp. 489–498. Springer, Heidelberg (2008)
4. Krejcar, O.: PDPT Framework - Building Information System with Wireless Connected Mobile Devices. In: ICINCO 2006, 3rd International Conference on Informatics in Control, Automation and Robotics, Setubal, Portugal, August 01-05, pp. 162–167 (2006)
5. Krejcar, O., Cernohorsky, J.: New Possibilities of Intelligent Crisis Management by Large Multimedia Artifacts Prebuffering. In: I.T. Revolutions 2008, Venice, Italy, December 17-19, 2008. LNICST. Springer, Heidelberg (2008)
6. Nielsen, J.: Usability Engineering. Morgan Kaufmann, San Francisco (1994)
7. Haklay, M., Zafiri, A.: Usability engineering for GIS: learning from a screenshot. The Cartographic Journal 45(2), 87–97 (2008)
8. Evennou, F., Marx, F.: Advanced integration of WiFi and inertial navigation systems for indoor mobile positioning. Eurasip journal on applied signal processing (2006)
9. Arikan, E., Jenq, J.: Microsoft SQL Server interface for mobile devices. In: Proceedings of the 4th International Conference on Cybernetics and Information Technologies, Systems and Applications/5th Int. Conf. on Computing, Communications and Control Technologies, Orlando, FL, USA, July 12-15 (2007)
10. Jewett, M., Lasker, S., Swigart, S.: SQL server everywhere: Just another database? Developer focused from start to finish. DR DOBBS Journal 31(12) (2006)
11. Krejcar, O.: Utilization Possibilities of Area Definition in User Space for User-Centric Pervasive-Adaptive Systems. In: Hesselman, C., Giannelli, C. (eds.) Mobilware 2009 Workshops. LNICST 12, pp. 124–130. Springer, Heidelberg (2009)
12. Janckulik, D., Krejcar, O., Martinovic, J.: Personal Telemetric System – Guardian. In: Biodevices 2008, Setubal, Funchal, Portugal, pp. 170–173. Insticc Press (2008)

Performance Analysis of Multicast Video Streaming in IEEE 802.11 b/g/n Testbed Environment

Aleksander Kostuch[1], Krzysztof Gierłowski[2], and Jozef Wozniak[2]

[1] Sprint Sp. z o.o., Gdansk, Poland
[2] Gdansk University of Technology, Poland
Aleksander.Kostuch@sprint.pl, {gierk,jowoz}@eti.pg.gda.pl

Abstract. The aim of the work is to analyse capabilities and limitations of different IEEE 802.11 technologies (IEEE 802.11 b/g/n), utilized for both multicast and unicast video streaming transmissions directed to mobile devices. Our preliminary research showed that results obtained with currently popular simulation tools can be drastically different than these possible in real-world environment, so, in order to correctly evaluate performance of video streaming, a simple wireless test-bed infrastructure has been created. The results show a strong dependence of the quality of video streaming on the chosen transmission technology. At the same time there are significant differences in perception quality between multicast (1:n) and unicast (1:1) streams, and also between devices offered by different manufacturers. The overall results seem to demonstrate, that, while multicast support quality in different products is still varied and often requires additional configuration, it is possible to select a WiFi access point model and determine the best system parameters to ensure a good video transfer conditions in terms of acceptable QoP/E (Quality of Perception/Exellence).

Keywords: Performance, analysis, video, IEEE 802.11b/g/n, multicast, unicast, wifi, wireless, mos, qop, dcf, access point.

1 Introduction

Until recently, computer networks were mostly used for bulk data transfers (for example: digital files), generated by various applications. However, recently we have been witnessing a dramatic growth of interest in real-time multimedia transfers, especially with audio and video content. Successful transmission of such content over computer packet networks require specialised client-side mechanisms, in order to adjust transmission parameters and ensure appropriate transport and application protocol configuration to obtain required reception quality. Also, requirements regarding Quality of Service (QoS) in the network itself, necessary to support real-time streaming traffic, are much more restrictive and difficult to provide.

Obtaining such QoS support in case of classical, cable-based network technologies is a fairly well researched and documented task, comparatively easy due to high available bandwidth and stable nature of transmission medium.

Growing popularity of WiFi networks creates a natural need to provide the same services in wireless environment. Here the same task is much more difficult. The changing

J. Wozniak et al. (Eds.): WMNC 2009, IFIP AICT 308, pp. 92–105, 2009.
© IFIP International Federation for Information Processing 2009

and unpredictable transmission medium creates a very difficult environment for QoS-realted network mechanisms. Also, differences between particular implementations of network hardware tend to be much more prominent than in case of cable-based solutions, and can result in drastically different performance in similar conditions.

At the same time, the required high quality of video transmissions is dictated by the high expectations of end-users (clients), thus quality of network service available for audio/video streaming transmissions, delivered by any wireless network connectivity provider, becomes one of the crucial issues.

Recently some interesting studies were published concerning video and multicast over 802.11 networks. Research presented in [2] and [4] discussed the topic of maximizing number of users, balancing the load among APs and minimizing the load of APs. Authors based their research on multicast traffic in WLAN environment on simulations only. Different research presented in [8] indicated the influence of network streaming quality on MOS (Mean Opinion Score) in case of 25 fps (frame per second) video content. MOS was about 4,5 when movie was transmitted with a very small jitter. It dropped to about 3,5 when the transmission experienced a single long freeze/skip. MOS was still lower (2,5) when frequent short freezes/skips occurred. Moreover, the value of MOS has been proved to decrease when decrease frame rate – MOS was about 4 when movie streamed at 12,5 fps and 2,3 at 5 fps. In their research authors used H.264 codec (stream bandwidth about 1 Mbps) and WMV player version 9.

On the basis of these publications, we decided to concentrate our experiments on the popular (MPEG-4/H.263) codec and its well known usage scenarios, which are widely accepted in commercial networks.

Experiments similar to ours, but only in IEEE 802.11b environment are described in [9]. The measurements of network bandwidth show that such network is too slow for the transmission of high definition video streams which proved to be fully consistent with our own results. Authors in [5] focused their experiments on multicast streaming in 802.11g environment, but have not employed any subjective quality assessment methods.

In our paper we investigate different IEEE 802.11 technologies (IEEE 802.11 b/g/n), looking for the most beneficial operational parameters and configurations of WLAN networks for video streaming transfers. The variety of possible WLAN configurations and WiFi standards creates a need to determine their capabilities and to estimate the video stream transmission quality.

The paper is organized in the following way. In the following section (Digital Video Transmission) we present an overview of basic mechanisms of IP-based video transmissions, together with a short overview of MPEG standards. Differences in unicast (1:1) and multicast (1:n) streaming, are also briefly discussed. In the next section our test-bed environment and methods of quality analysis (in our case Quality of Perception / Excellence – QoP/E) are described, followed by Results section, containing discussion of obtained results, together with some comments and recommendations. The article is concluded with their summarization and plans of future research.

2 Digital Video Transmission

Parameters (such as resolution, number of frames displayed per second, color depth etc.) specified in current TV standards (see Table 1), starting from aged SECAM and

NTSC and ending with Full HD digital TV, make it very ineffective to even try to transmit over the network the video signal in its base, unmodified form. Bandwidth requirements would clearly be unacceptable and precise timing relations would be impossible to meet without allocation of a very large buffer space.

Table 1. Popular television standards

System	Resolution		Frequency of image fields changes	
	Number of lines	Numbers of points in line	Refresh rate [*Hz*]	Number of video frames per second
SECAM (*fr. Colour electronic system with memory*)	625	720	50	25
NTSC (*National Television System Committee*)	525	720 or 320 (VHS)	59,94	29,97
PAL (*Phase Alternating Line*)	625	720 or 320 (VHS)	50	25
CCTV (*Closed-Circuit TeleVision*)	288	360	60	30
Ready for HDTV	1280	720	50	25
Full HD	1920	1080	50	25

It is evident, that video information needs to be encoded in order to minimize the amount of traffic, to effectively transport it across a network system. Because of that requirement, development of efficient encoding techniques has an immense influence on the popularity of computers employed as audio-video systems [11]. Encoding mechanisms often include compression or reduction of primary video information [6].

Table 2. Characteristics of popular encoding standards

Codec name	Bandwidth necessary for PAL	Error sensitivity	Encoding computing power
MPEG-1	8 Mbps	Low	Small (x386)
MPEG-2	5 Mbps	Medium	Medium (x486)
MPEG-4	3-6 Mbps (variable)	Medium	High (x586)
H.264	1 Mbps	High	Very high (Dual core)

The reduction of transmitted information is usually based on unification of similar colours or not showing the details in similar colours. The most popular way of coding is the MPEG standard family, created by ISO organization (International Standards Organisation). Table 2 shows the basic characteristics of compression for particular versions of MPEG encoding and PAL television signal. All of MPEG standards are characterized by asymmetric computational requirements – the decoder part is much less complicated then the encoder, and requires only a small fraction of its computational requirements. This characteristic is one of main reasons of MPEG standards family practical popularity.

Table 3 shows a comparison of network bandwidth required in case of MPEG-2/H.262 (currently most popular solution) and MPEG-4/H.263 (which is rapidly gaining popularity) employed in case of high definition video: 4CIF resolution (704 x 576) – the highest employed in industrial monitoring. In our research we decided to concentrate on variable rate stream and high quality, as a most resource intensive and difficult to effectively transmit type of video traffic.

The MPEG-4/H.263 transmission rate is variable in time, as it depends on changes in motion. MPEG-2/H.262 bandwidth is mostly constant and it is currently the most popular compression method used for the digital television transmission and many new services, such as VoD (Video on Demand).

Another solution, namely H.264 is the most sophisticated way of video encoding available currently and it is gaining a great popularity in television transmission and video conferences through cell phones. Its downside is a very high computational power necessary for encoding and high sensitivity to transmission errors, which makes it poorly suited for employment in a wireless environment.

Apart from the smallest possible output bandwidth, a good codec intended for streaming transmission should also provide a decent resistance to stream errors, which can result from malformed, lost or reordered IP datagrams. It is especially important in case of wireless transmissions, where probability of such errors is much higher than in case of cable-based transport technologies.

Based on literature study and some preliminary experiments, we decided to choose MPEG-4/H.263 codec as the subject of our detailed research. It offers a good compression ratio, acceptable resistance to streaming errors and can be employed in case of variety of video signals – starting from low quality mobile-phone video, and ending with high resolution, full motion TV.

Table 3. Comparison of bandwidth (in kbps) required by MPEG-2 and MPEG-4 encoding in closed-circuit television for 4CIF resolution, 25 frames per second, without sound, based on [3]

4CIF (704x576 pixels) 25 frames per second	MPEG2/ H.262 Auto	X-series MPEG4/H.263		
		Highest compression	Medium compression	Highest quality
Many changes in image	5000	1700	2950	4200
Partially changing image	3500	950	1725	2500
No changes in image	2000	700	1550	2400

Another important element of real-time streaming system is data buffering. It allows the proper (constant and regular) timing of succeeding video frames display along with correct synchronization of sound. Experiences from practical usage of video coding and decoding applications show that the size of buffer is based on bandwidth of the network link and parameters of a video stream, such as its bandwidth and overall amount of data to be transmitted.

3 Network Transmission: Unicast (1:1) and Multicast (1:n)

The most popular transmission protocol utilized for video streaming in IP network is RTP (Real Time Protocol). It is an unreliable transport protocol based on well-known UDP (User Data Protocol), extended with a number of mechanisms designed specifically for real-time, inelastic data transmission.

RTP can be used in both unicast and multicast streaming – each of these approaches offers unique advantages, but also brings specific requirements.

The unicast traffic stream delivers information to one particular receiver (point to point). Every new unicast connection causes the increase of overall bandwidth usage, which is proportional to the number of unicast streams present. An advantage of unicast transmission is that it can be initiated on demand of the user – that means, that:

- the encoder can be idle if there are no requests,
- each user can negotiate different stream characteristics,
- each user can control the content of its own stream.

In contrast, a single multicast stream delivers information to the group of receivers at once. The encoder is operating constantly and interested users can joint multicast session to receive the content that is currently being transmitted. Multicast is an effective way of sending the data from a single source to many receivers in a network [20]. In order to set up multicast transfer it is necessary to fulfil the following criteria:

- the video transmitter must be able to send the multicast streams,
- the receiver should be able to receive the multicast transmission,
- in order to receive multicast transmission the recipient has to join the particular multicast group – Internet Group Management Protocol (IGMP) signalisation support is required [20],
- to properly route multicast traffic in complex network environment, dedicated multicast routing protocol (for example: DVMRP, MOSPF, PIM-DM, PIM-SM…) must be implemented in routers [20].

4 Test-Bed Environment

Today, many popular devices, such as notebooks, laptops, palmtops and smart-phones implement an IEEE 802.11 compliant interfaces.

Most often, popular wireless LAN installations employ IEEE 802.11b/g compliant hardware, and there is also a growing number of IEEE 802.11n Draft 2 compliant devices.

Devices operating according to the IEEE 802.11a standard are not very popular among Small Office Home Office (SOHO) users, but they occupy an important place as short range infrastructure links in more complex wireless networks. Because of similarities between 802.11a being and 802.11g standards, the empiric research was performed for the 802.11g standard instead and the obtained results should be valid in both cases.

Table 4. Features of particular 802.11 b/g/n standards

Parameters	Standard 802.11b	Standard 802.11g	Draft of standard 802.11n
Year of issue	1999	2003	Draft – version no. 3
Frequency range [GHz]	2,4000 ÷ 2,4835	2,400,0 ÷ 2,4835	2,4 ÷ 2.485; 5,150 ÷ 5,350; 5,470 ÷ 5,850
Channel bit rate [Mbps]	11	54	100 ÷ 300 ($MCS15$)
Modulation technique	DQPSK. DBPSK	BPSK, QPSK, 16-QAM, 64-QAM	Packet aggregation: A-MPDU, A-MSDU
Transmission technique	DSSS, CCK	OFDM	OFDM+MIMO 2x2, 2x3, 3x3 antennas

5 Hardware Configuration

Figure 1 presents a hardware configuration of our testbed. It consists of wired infrastructure and a stationary, IEEE 802.11 compliant, wireless access network working in an infrastructure mode.

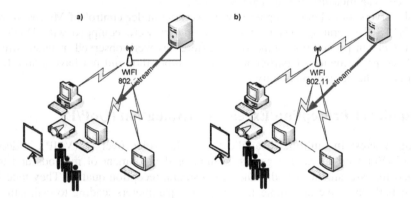

Fig. 1. WiFi network topology proposed with Access Point for empiric research. a) Cable to WiFi, b) WiFi to WiFi.

The measurements were conducted two configurations:

- The video streaming source was located in the cable network. Video stream was sent through a simple cable infrastructure (consisting of a single Fast-Ethernet point-to-point link) to an Wireless Access Point and then to wireless IEEE 802.11 compliant clients.
- The video streaming source is connected as a wireless network client. The stream of video data is transmitted to clients exclusively by IEEE 802.11 network operating in an infrastructure mode.

Different models of access points were tested. In all cases standard antennas built in the access point had been used:

- 802.11b – Linksys BEFW11S4v4,
- 802.11g – Linksys WAP54G-v3,
- 802.11n – Dlink DIR-655.

All clients were located in a room with size of about 25 square meters, lacking significant signal propagation barriers. Such environment provided good propagation conditions, characteristic to a well designed WLAN network in internal office spaces.

The radio channel has been verified as unused and lacking significant noise level during experiments. The transmission was conducted without encryption. Based on literature study and our earlier experiments, we can state that encrypting has a minimal influence on speed decrease in the transmission efficiency (about 5%) and as such it doesn't have a significant influence on a video transmission [18].

6 Software Configuration

Video LAN Connector application [22] in version 0.8.6f has been used as both streaming server and client. WireShark Network Protocol Analyzer [21] has been employed as an measurement and analysis tool.

Both servers and client computers were working under control of Microsoft Windows XP SP2 operating system. Client computers were equipped with DualCore, Athlon, Celeron processors. Some small differences were observed in video streaming-related performance of particular computers, but they did not have a significant influence on the perception quality.

7 Quality of Preception/Excellence Evaluation (QoP/E)

Growing interest in multimedia applications, like Voice over IP (VoIP), Video on Demand (VoD) or interactive games, stimulates development of methods and tools designed for assessment of audio and video stream reception quality. They take into account both objective and subjective metrics – parameters leading to estimation of the level of recipients satisfaction, so-called Quality of Excellence (QoE).

Objective methods for audio signals include, among others: PEAQ [10], PSQM [12] and PESQ [16] algorithms, proposed by ITU that take into account from 5 up to 11 measured parameters.

For video signals Peak Signal to Noise Ratio (PSNR), ANSI [1] and J144 [15] methods are recommended.

At the same time a variety of subjective methods are proposed. However, the most popular are still relatively simple methods. For both audio and video signals Absolute Category Rating (ACR) [14] or Degradation Category Rating (DCR) [13] are often employed. Both methods use 5-degree grading tables.

In our research we decided to employ a subjective assessment method, supplemented by statistical traffic analysis conducted with mechanisms available in Wireshark Network Analyzer (such as stream bandwidth, packet delta, jitter and packet loss rate).

A group of 10 test subjects has been selected. The scale of marks ranged from no degradation (mark 5) to very severe degradation (mark 1). The quality of received image has been measured according to values described in Table 5. Based on marks given by testers' so-called MOS (Mean Opinion Score) final parameter values were calculated. The average mark equal to 5 means excellent quality, while 1 – not acceptable. The observations of video quality were performed on client computers by users with no previous knowledge of the original material (so-called QoP method – Quality of Perception estimation). The one-stimulus (eyesight) method has been used in which a group of respondents estimated the succeeding videos [7].

Table 5. Subjective qualitative estimation of video stream

QoP	Levels of quality	Subjective impressions
5	Excellent	Imperceptible differences between local transmission and the one after transmission remote access to a video stream
4	Above Average	Perceptible delay of transmission, but not affecting the final reception
3	Average	Small errors, temporary image freezing up to 1 second
2	Below average	Plenty of errors, image freezes for more than 1 second
1	Fail	No image, or unrecognisable

The experiment result form (completed by test subjects) included the name of configuration scenario and video sample, time of measurement and the estimation. The estimations from all experiments of particular scenario were summarized thus receiving the average estimation of measurement. During the research, measurements of the transmission speed, jitter, packet lost and packet errors had also been conducted. The experiments were performed many times for the same scenario, in different rooms. The video material used during measurements was a PAL video recording encoded in the MPEG-4/H.263 standard where bandwidth of the stream was changing between 3 and 6 Mbps.

8 Results

In course of experiments conducted in our testbed installation, we were aiming to answer to the following basic questions:

1. How well various WiFi devices are prepared for transporting multicast and unicast real-time video streams?
2. What subjective quality of the video we can expect from a stream transported through the different WiFi network configurations and devices?

To assess the number of high definition (5 Mbps) MPEG4/H.263 video streams that a given 802.11 technology can support, we started with a simple network throughput assessment.

A 10 MB file has been transferred from the cable network to wireless client by unicast traffic, and transmission time, overall throughput and its stability has been measured – results are presented in Figure 2.

As we can see 802.11b does not provide a sustained throughput required for even one of such high-definition video streams. 802.11g provided stable bandwidth of about 23 Mbps which can be enough for roughly 4-5 streams. 802.11n effective bandwidth was much less stable than in case of previous technologies, but, with mean rate of over 60 Mbps, it has a potential ability to support a significant number of hi-def video streams.

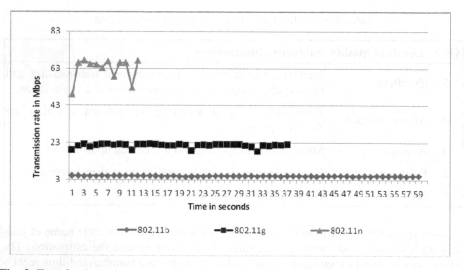

Fig. 2. Transfer time and transmission rate for 10MB file sent from cable network to a WiFi wireless client (802.11b/g/n standards)

Following this preliminary assessment, the main group of experiments has been conducted according to QoP evaluation rules described above. All popular WiFi base technologies were tested: IEEE 802.11 b, IEEE 802.11g (which earlier research describes as equivalent to IEEE 802.11a in our environment) and IEEE 802.11n Draft 2. Each technology has been tested for cable-wireless and wireless-wireless streaming. Both unicast and multicast tests were conducted.

The results of QoP evaluation are presented in Table 6.

The results of our experiments allow us to formulate the following direct conclusions:

1. IEEE 802.11b standard cannot support hi-def video streaming.
2. IEEE 802.11g based network supports up to 5 hi-def unicast video streams of acceptable quality.
3. IEEE 802.11n standard can support up to 5 independent hi-def video multicast streams.

One of the most significant results is a very profound difference in QoP scores between unicast and multicast streams for 802.11g standard.

Table 6. Summary of QoP evaluation

Number of concurrent MPEG4-compressed streams		The WiFi standard and configuration of settings in network of transmitters and receivers					
		802.11b WiFi to WiFi	802.11b Cable to WiFi	802.11g WiFi to WiFi	802.11g Cable to WiFi	802.11n WiFi to WiFi	802.11n Cable to WiFi
Multicast	1 stream	1,0	1,0	1,5	1,0	5,0	5,0
	2 streams					4,7	5,0
	3 streams					4,3	5,0
	4 streams					3,9	5,0
	5 streams					3,8	4,7
	6 streams					2,9	4,4
	7 streams						4,3
	8 streams						4,1
	9 streams						3,1
Average real speed in Mbps		2,76	5,41	13,34	19,45	20,52	62,9
Unicast	1 stream	1,8	4,2	5,0	5,0	5,0	5,0
	2 streams	1,0	3,6	5,0	5,0	5,0	5,0
	3 streams		1,7	5,0	5,0	5,0	5,0
	4 streams			5,0	5,0	5,0	5,0
	5 streams			4,0	4,9	5,0	5,0
	6 streams			2,3	4,7	4,9	5,0
	7 streams			1,3	4,7	4,7	5,0
	8 streams				4,5	4,4	5,0
	9 streams				4,3	4,3	5,0
	10 streams				3,9	3,8	5,0
Number of parallel video transmissions in PAL quality		0	1	3	4	5	12

By employing equipment from the year 2006, 5 parallel, unicast video steams were possible between wired LAN and WiFi.

This result is easy to predict theoretically, because by using the unicast transmission, we increase network load with each sender-recipient pair: 5 parallel transmissions, approximately 5 Mbps each, take about 25 Mbps of bandwidth, which is consistent with maximum bandwidth available for streaming with this technology. Above this limit, with each unicast stream created, the image quality was depreciating for all concurrent transmissions.

In case of multicast streams the situation is not obvious – it is notable, that even a single multicast stream transmitted through otherwise not utilized wireless network is of very bad quality.

Broadcast and multicast frames, which can also be referred to as group frames (because they are destined for more than one receiving station) are exchanged without confirming acknowledgements. The sender has no way of confirming the success of transmission and will not retransmit group addressed frame. I other words, broadcast and multicast frames are delivered without any reliability guarantees or even indications of failure.

Moreover, while unicast traffic is transmitted with transmission rates dependant on current radio conditions and ranging from 1 Mbps through 11 Mbps (802.11b) to 54 Mbps (802.11 a/g, and even faster for 802.11n), group addressed frames are transmitted with rates from a much smaller set.

Each BSS maintains a list of transmission rates, which are supported by both ALL devices in a given BSS – a Basic Rate Set (BRS). Control and mulitcast/broadcast data frames may be transmitted only with rates that are listed in BRS, as they must be received and understood by all (or at least by a significant group) of the stations in the BSS [17]. The BRS is broadcasted by an AP controlling a given BSS and only stations supporting all its rates are allowed to join the network.

The default list of basic rates depends on implementation, but they rarely exceed 11 Mbps.

Theoretically a station or AP can choose any of BRS rate to transmit a group addressed frame, but often devices choose the slowest one, to maximize chances of successful transmission in absence of acknowledgement mechanism.

For example an unmodified Linksys WAP54G chooses 1 Mbps rate. It is far too slow for streaming high-definition video. Moreover, we cannot change BRS set or rules for selecting transmission rates in a vast majority (probably about 90%) of home use devices.

Table 7 presents statistical information obtained with the Wireshark software at receiver station. The first stream was transmitted with multicast frames at the default 1Mbps. That speed is used in most of home use 802.11g APs for delivery of group addressed frames. In such transmission we lost 72% of transmitted multicast frames. Next we tested high level, professional AP from 3Com, model 8760 where we can select speed of multicast.

From our experiments, the highest rate we could choose in our environments, before transmission errors will offset throughput-based PoE advantages is 11 Mbps. We have only 2,1% of packet loss and mean jitter 2,1ms. [19]. That is a significantly

Table 7. Statistics of a single stream transmission in IEEE 802.11g standard. (3COM model 8760 AP).

Speed Transmission Type (for 802.11g)	Received Packets *Number*	Lost Packets *(percent)*	Max Delta *(ms)*	Max Jitter *(ms)*	Mean Jitter *(ms)*	QoP score
1Mbps Multicast	4694	12033 (71.9%)	94.90	88.74	5.33	1
11Mbps Multicast	16336	347 (2.1%)	200.27	41.78	2.91	3
54Mbps Unicast	16655	7 (0.0%)	194.94	24.18	4.23	5

better result and allows us to successfully conduct video streaming. In the comparison case of 802.11g unicast streams we have a more significant mean jitter, but just only 7 lost packets.

The new IEEE 802.11 standard extension – IEEE 802.11n in its current form of Draft 2 implementation, was first tested in an ad-hoc configuration (802.11n D-link external WiFi network adapters) and the maximum measured unicast transmission rate did not exceed 1 Mbps. The most probable reason for such a small bandwidth can be attributed to a very early firmware version present in that hardware. We hope that in next generation of firmware for 802.11n ad-hoc this will be corrected.

For the IEEE 802.11n network in infrastructure mode, the unicast transmission safely supported 9 parallel video streams between wireless stations. The overall consumed bandwidth reached 50 Mbps and still all transmissions were of a very good quality. The further increase of streams was stopped, due to the limited number of available receiver-sender hardware device pairs.

Overall results (see figure 3) show, that regardless of employed technology we can always transmit more independent unicast streams than similar multicast streams. On the other hand, number of simultaneous unicast receivers is limited to the number of unicast streams, while number of multicast receivers is limited only by overall AP client capacity. IEEE 802.11g (Modified) corresponds to a IEEE 802.11g technology, where BRS has been modified (extended) to better accommodate multicast video streaming.

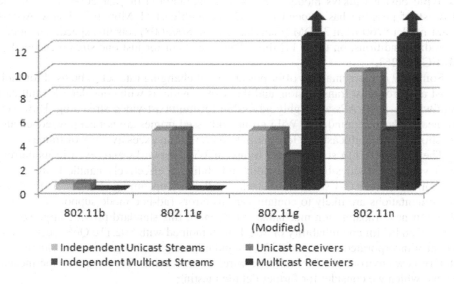

Fig. 3. Number of possible independent hi-def video streams and maximum number of receivers in case of various IEEE 802.11 technologies

9   Conclusions

The research proved that the real, effective rate of transmission of a given stream has a great influence on the quality of video material received from a WiFi network. The available network bandwidth should be enough to fully accommodate the needs of a video stream. Even slight deficiencies here result in drastic degradation of user experience (perception).

In case of unicast transmission this requirement is a straightforward one, despite the fact, that each client requires a separate point-to-point stream, so the number of simultaneous clients will be limited. Unicast WiFi traffic can easily utilize any transmission rate negotiated between AP and a client station, up to the maximum transmission rate. It makes it easy to assess the number of supported streams, and can be function on any popular WiFi hardware.

Multicast streaming has the undeniable advantage that the same material can be simultaneously provided to (theoretically) unlimited number of users which are in the range of AP's radio transmission. Unfortunately correct operation of this functionality depends on the particulars of AP implementation, and should not be considered advisable until it is verified for a particular model. That property results in serious differences in results obtained with use of simulation models and real life performance of particular devices. Implementation details of a particular device can cause differences in 0.3-2.5 times range (our estimation based on experimental research), between simulated performance and performance obtained in production environment.

IEEE 802.11 standard limits multicast transmission rates to a very limited rate set (rates < 11 Mbps), and practical implementations trend to choose the most stable (slowest) rate from this set resulting in 1Mbps multicast streaming bandwidth. For example device Linksys model WAP54G (manufactured in year 2006) is using the slowest of possible basic speed for 802.11g multicast (1 Mbps). The new Access Point model 8760 from 3COM (manufactured in 8/8/2008) sets this speed, depending on radio conditions, on up to 11Mbps. This is enough for just one stream encoded in MPEG4/H.263.

Some AP implementations offer possibility of changing rate set to be used for multicast and broadcast transmission, and that can provide us with considerable multicast bandwidth – such APs are highly advisable for multicast video streaming. However, we need to keep in mind, that WiFi group addressed frames are not acknowledged and we should keep multicast rate low enough to prevent an excessive loss of frames.

The IEEE 802.11n standard is a future of WiFi networks and, even in its current draft version, provides high bandwidth and ability to effectively handle both unicast and multicast MPEG-4 streams. There are still many upcoming changes, and today's implementations are likely to contain various errors (ad-hoc mode support, for example), but new transmission mechanisms present in this standard provide large advantages. Bandwidth and reliability of 802.11n, combined with 802.11e QoS mechanisms recently incorporated into main 802.11 standard, are going to provide us with a completely new environment for real-time wireless streaming. They are also the mechanisms, which we consider for further detailed testing.

References

1. ANSI T1.801.01-1996: Digital transport for video teleconferencing/ video telephony signals – Video test scenes for subjective and objective performance assessment, American National Standards Institute (1996)

2. Bejerano, Y., Dongwook, L., Sinha, P., Zhang, L.: Approximation Algorithms for Scheduling Real-Time Multicast Flows in Wireless LANs. In: The 27th Conference on Computer Communications (INFOCOM 2008), Ohio State University, April 2008, pp. 2092–2100 (2008)
3. Spread calculation used to estimate the bandwidth and space on disk used for planning infrastructure with the X-series devices coders made by the Bosh Corporation, Calculation made in July 22 (2005)
4. Chen, A., Dongwook, L., Sinha, P.: Optimizing Multicast Performance in Large-Scale WLANs, Ohio State University (2007)
5. Dujovne, D., Turletti, T.: Multicast in 802.11 WLANs: An Experimental Study, Project Planete INRIA, France (October 2006)
6. Fibush, D.K.: A Guide to Digital Television Systems and Measurments. Tektronix (1997)
7. Ghinea, G., Thomas, J.P., Fish, R.S.: Quality of perception to quality of service mapping using adynamically reconfigurable communication system. This paper appears in Conf. on Globecom 1999, Brazil, pp. 2061–2065 (1999)
8. Huynh-Thua, Q., Ghanbari, M.: Impact of Jitter and jerkiness on perceived video quality., Research made in University in Essex. Presented on Fourth International Workshop on Video Processing and Quality Metrics for Consumer Electronics (VPQM 2009), p. 308 (2006)
9. Ikkurthy, P.C.: Software testing testbed for mpeg-4 video traffic over ieee 802.11b wireless lans. University of South Florida (2003)
10. ITU-R Recommendation BS 1387:Method for Objective Measurements of Perceived Audio Quality (PEAQ) (1998)
11. ITU-R Recommendation BT.1208-1: Video coding for digital terrestrial television broadcasting (October 1997)
12. ITU-T P.861: Objective Quality Measurement of Telephone-Band Speech Codecs (February 1998)
13. ITU-T P.910: Subjective video quality assessment for multimedia applications (1996)
14. ITU-T Rec. P.800: Method for subjective determination of transmission quality (1996)
15. ITU-T Recommendation J.144: Objective perceptual video quality measurement techniques for digital cable television in the presence of a full reference (March 2004)
16. ITU-T Recommendation P.862: Perceptual evaluation of speech quality (PESQ), an objective method for end-to-end speech quality assessment of narrowband telephone networks and speech codecs (1998)
17. Gast, M.: 802.11 Wireless Networks The Definitive Guide. O'Reilly, Sebastopol (2005)
18. NetGear: Wireless Performance Study. LioNBridge (October 2006),
    http://www.veritest.com
19. Schulzrinne H., Casner S., Federick R., Jacobson V.: RTP: A Transport Protocol for Real-Time Applications, RFC 3550 (June 2003)
20. Wittmann, R., Zitterbart, M.: Multicast Communication: Protocols and Applications. Morgan Kaufmann Publisher, translated by Academic Press (2001)
21. http://www.wireshark.org/
22. http://www.videolan.org/vlc/

Optimization of Frame Structure for WiMAX Multi-hop Networks

Pavel Mach and Robert Bestak

Czech Technical University in Prague, Faculty of Electrical Engineering
Technicka 2, Prague, 16627, Czech Republic
{machp2,bestar1}@fel.cvut.cz

Abstract. To enhance the system throughput and to extend coverage of IEEE 802.16 networks, relay stations can be implemented. If a user station is attached to the base station (BS) through several relays stations (RS), a multi-hop communication occurs. To enable multi-hop communication, the IEEE 802.16j standard proposes two approaches how RSs can be implemented into the network. The first approach groups BSs and several RSs into a multi-frame with repetition of relay zones. In the second approach, the BS schedules several relay zones in one frame. While the first approach causes long packet delays, the second approach has high requirements on RS's processing capabilities. This paper proposes an optimized frame structure that allows using second approach whilst the requirements on RS's processing time are still kept in reasonable range. The obtained simulation results indicate that packet delays in downlink and uplink direction can be significantly reduced.

Keywords: Relay Station, frame structure, IEEE 802.16j, packet delay.

1 Introduction

Over the last years, broadband wireless systems established themselves as one of the fastest growing and developing area in the field of telecommunications. One of the most promising technologies represents WiMAX widely known as wireless networking standard that addresses interoperability across IEEE 802.16 standard-based products. So far, IEEE Std. 802.16-2004 [1] intended for fixed terminals was approved. In addition, amendment to the former standard labeled as IEEE 802.16e [2] was adopt as well. Its purpose is to enrich WiMAX by further features such as handover and power management modes to enable user's mobility.

To facilitate QoS management of individual users, the standard specifies five QoS scheduling services; i) Unsolicited Grant Service (UGS), ii) real time Polling Service (rtPS), iii) extended real time Polling Service (ertPS), iv) non real time Polling Service (nrtPS) and the last one v) Best Effort (BE).

In order to enhance the system throughput and extend Base Station's (BS) coverage, a Relay Station (RS) that enable multi-hop communication can be introduced into the network. The multi-hop communication occurs when data are transferred from the source to the destination node via one or more RSs. The implementation of relay stations into the WiMAX system is within the scope of IEEE 802.16j working group [3] that was established in 2006.

J. Wozniak et al. (Eds.): WMNC 2009, IFIP AICT 308, pp. 106–116, 2009.

According to [3], three types of RSs are specified; fixed, nomadic and mobile RSs. This paper considers only the fixed RSs, i.e. RSs that are permanently installed at the same location. The RSs are in most cases build in, owned and controlled by service provider. An RS is not directly connected to the wire infrastructure and has minimum functionalities to support the multi-hop communication. In addition, two types of RSs can be distinguished: transparent and non-transparent one. The transparent RS (T-RS) transmits neither long preamble at the beginning of frame nor broadcast MAC management messages such as DL/UL maps or DCD/UCD. Therefore, a Mobile Station (MS) has to be in the BS coverage area. The only aim of the T-RS is to enhance the throughput within the BS cell; a T-RS is used in cooperation scenarios when data are sent via several independent radio channels. The second category of RSs is known as non-transparent RS (NT-RS). In comparison with T-RS, NT-RSs transmit long preamble and broadcast MAC management messages. Thus, the NT-RSs are more suitable for a scenario where a MS is out of BS range, e.g. due to the shadowing effect or when the MS is actually to far-away to receive BS's signal with satisfactory quality. Nonetheless, a NT-RS can be also used to enhance the cell throughput.

Basically two types of NT-RSs can be considered: i) centrally controlled RS (CC-RS) and ii) de-centrally or distributed controlled RS (DC-RS). The former type of RS is completely controlled by the BS. This means that the BS handles and schedules all data and control transmissions between the RS and its own users. In case of DC-RS, the RS itself (without BS help) schedules all control and data transmissions of its users. As we consider exclusively multi-hop scenarios where a MS may be out of the BS range, only the NT-RSs will be taken into account.

The rest of the paper is organized as follows. In subsequent section several approaches of MAC frames for WiMAX system with relays are contemplated. The next section describes in more details the frame concepts based on IEEE 802.16 and proposal based on [4]. Section four investigates possible frame modifications in order to reduce packet delays in multi-hop scenarios. In section five, simulation scenario is depicted together with presentation of simulation results. The last section concludes our paper.

2 Related Work

If a RS is implemented into the WiMAX network, the original MAC frame needs to be updated in order to support multi-hop communication. So far, several research works had been done during the last few years. The main goal is to effectively integrate RSs to IEEE 802.16 standard while still keeping backward compatibility with the legacy MSs. In [5], authors proposed a simple and flexible frame structure based on IEEE 802.16e. Within the IEEE 802.16j working group, proposals for T-RS and NT-RS were introduced ([3], [6]). However, the NT-RS proposal supposes that a BS and its subordinate RSs simultaneously transmit during the broadcast part of frame and DL/UL access zones. In order to avoid mutual interference of BS and NT-RS transmissions, a multi-frame structure is proposed in [7]. Other interesting concept how to integrate NT-RSs into the existing WiMAX network and to avoid unwanted interferences is presented in [4]. The frame concept for more than two hops is defined in [8] where out-frame and in-frame multi-hop relaying is considered. By out-frame relaying is meant the situation when data from the BS to MS are forwarded subsequently in following frames. The second case reflects to the situation when data are sent between the BS and MS within one frame. The similar concept for multi-hop communication is also introduced in [3]. While the first method excessively prolong the packet delays, the second method interpose high requirement on RS's CPU processing capabilities. To that end we propose a modification to frame structure based on the second method which lower requirements on RS's CPU processing. Although the proposal is applied to CC-RS frame structure specified in [4], the whole idea can be extend to other frames concepts.

3 Frame Structure

The frame concept of NT-RS based on IEEE 802.16 is shown in Fig. 1. Both, BS and RS frames, begin with broadcasting part composed of long preamble, FCH (Frame Control Header) field and DL/UL MAPs. Subsequently, one or more DL access zones follow. The access zone is an interval in which the BS and RS send DL/UL data bursts to its subordinate stations. The rest of DL subframe is reserved to BS-RS transmissions specified by DL relay zone(s). Since the DL access zone precedes the DL relay zone, data are delivered to the MS in the second frame at best (for 2 hop scenario). In the first frame, the BS transmits within the DL relay zone data to the RS. The RS retransmits the data burst in a subsequent frame.

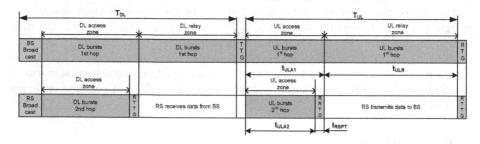

Fig. 1. Frame concept for NT-RS according IEEE 802.16j [3]

The UL subframe begins similarly as the DL subframe with the access zone during which MSs transmit data bursts to their super-ordinate stations. The MAC frame ends up by UL relay zone(s). Between the DL/UL subframes are included gaps to enable the BS antenna switch from the transmission to reception mode and vice versa. Furthermore, the RTTG (RS transmit/receive transition gap) and RRTG (RS receive/transmit transition gap) gaps are necessary to facilitate RS's antenna switching. The data can be sent from the MS to the BS in one frame if the RS has enough time for processing data received within UL access zone. In other words, the time which has the RS to process all received data (in Fig.1 specified as t_{RSPT}) must be greater or equal to t_{RSPC} as indicated in the following expression,

$$t_{RSPT} \geq t_{RSPC} . \tag{1}$$

where t_{RSPC} is exactly the time needed for processing of received data during t_{ULA2} which is directly proportional to that amount of data. The t_{RSPT} can be expressed as,

$$t_{RSPT} = T_{UL} - \left(t_{ULA2} + t_{ULR} \right).$$ (2)

where T_{UL} represents the duration of UL subframe, t_{ULA2} is the interval when MSs attached to the RS are transmitting and t_{ULR} corresponds to the length of UL relay zone when the RS is in transmitting mode. The length of T_{UL} depends on the current requirements and demands of individual users since the BS is able to dynamically change the DL and UL subframes duration. On the other hand, the length of t_{ULA2} and t_{ULR} generally depend on actual traffic load and could be computed as,

$$t_{ULA2} = \sum_{i=0}^{l} \left(\frac{D_{ULi}}{D_{bps_i}} * t_S \right), t_{ULR} = \sum_{j=0}^{m} \left(\frac{D_{ULj}}{D_{bps_j}} * t_s \right).$$ (3)

where D_{ULi} expresses the number of bits sent in the current frame by i-th MS, D_{bpsi} is the number of bits that could be transmitted per one OFDM(A) symbol by i-th MS, t_s represents the length of one symbol and l indicates the number of MSs active in the current frame and attached to the RS. The D_{bpsi} is derived from modulation and coding scheme applied for that particular transmission and dependents on received SNR (see [2]). Similarly, the D_{ULj} is the number of bits send by j-th RS, D_{bpsj} is the amount of bits transmitted per one OFDM(A) symbol by j-th RS and m is the number of active RS. Only at low traffic load, as frame is not fully occupied, the RS is able to re-transmit data within one frame. This is due to the fact that the BS is able to schedule individual transmission in such a way that the gap between the UL access zone on the 2nd hop and UL relay zone are far greater than t_{RSPC} (or at least of equal length as t_{RSPC}). With increasing traffic load, the condition indicated in expression (1) can not be satisfied and certain amounts of data have to be sent in consecutive frames.

The concept based on [4] assumes that RSs and BS share radio resources within every BS cell and utilize the same frequency channel. Consequently, dynamic sharing of radio resources according to current stations requirements may be used. The frame structure of CC-RS is illustrated in Fig 2. At the beginning of frame, the BS sends its control information to all subordinate stations. Note that BS transmits alone while other stations (both RS and MS) are in receiving modes; hence no intra-cell interference arises. Since the BS schedules all transmissions on any other subsequent hops, DL

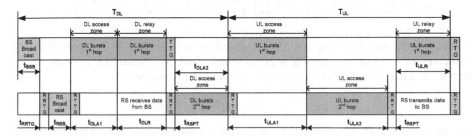

Fig. 2. Frame concept with CC-RSs [4]

and UL MAPs are considerably larger comparing to standard MAP messages in legacy WiMAX systems. After the BS's broadcast interval follows a RRTG gap which provides sufficient time to switch RS antenna from receiving to transmitting mode.

From the received BS broadcast information the RS extracts relevant parts and rebroadcast it to its subordinate stations (MSs or/and RSs). When the RS(s) broadcast transmission is over, its RS antenna needs to be switched to the receiving mode and waits for the DL access zone which ideally occurs at the end of DL subframe. At the same time, the BS starts to distribute data to the MSs on the first hop within the first DL access zone. While the beginning and end of DL subframe is dedicated to the DL access zones, the middle part is assigned to one or several DL relay zones. In order to enable the RS's antenna transition from receiving to transmitting mode, a RRTG gap is inserted. After the BS transmission, the RS itself sends data on the 2nd hop within the scheduled second DL access zone. Similarly as in case of DL subframe, the uplink subframe is composed of at least one UL access zone at the side of BS and one UL access zone at the RS's side. The end of UL subframes is assigned to the UL relay zone.

In comparison with the frame concept based on [3] data transmissions in both directions can be ideally accomplished in one frame. However, the same condition as in previous case must be fulfilled, i.e. the t_{RSPT} has to be greater or equal to t_{RSPC}. Note that the t_{RSPC} for the DL direction is proportional to the amount of data received during t_{DLR} (not to t_{ULA2} as in UL direction). In DL case, t_{RSPT} is computed as,

$$t_{RSPT} = T_{DL} - \left(t_{BSB} + t_{RRTG} + t_{RSB} + t_{DLA1} + t_{DLA2} + t_{DLR} \right) \qquad (4)$$

where t_{BSB} (t_{RSB}) is the length of BS (RS) broadcast interval, t_{RRTG} corresponds to the time dedicated for RS's antenna switch, $t_{DLA1,2}$ marks the duration of DL access zones and finally t_{DLR} represents the duration of DL relay zone. The time necessary for broadcasting part of the frame is further derived from the following expression,

$$t_{BSB,RSB} = t_{LP} + t_{FCH} + t_{BM} \qquad (5)$$

where t_{LP} is the length of long preamble (commonly 2 symbols), t_{FCH} corresponds to the duration of FCH field and t_{BM} is the time required for broadcasting of MAC management messages (e.g. DL and UL maps). The length of t_{DLA1}, t_{DLA2} and t_{DLR} can be evaluated by adopting the expression (3) as,

$$t_{DLA1} = \sum_{x=0}^{n} \left(\frac{D_{DLx}}{D_{bpsx}} * t_s \right), t_{DLA2} = \sum_{i=0}^{l} \left(\frac{D_{DLi}}{D_{bpsi}} * t_s \right), t_{DLR} = \sum_{j=0}^{m} \left(\frac{D_{DLj}}{D_{bps\,j}} * t_s \right) \qquad (6)$$

where n is the number of active MSs attached directly to the BS. The rest of the parameters have the same meaning as already described in expression (3) but are related to the DL direction. In UL direction, the t_{RSPT} may be expressed as,

$$t_{RSPT} = T_{UL} - \left(t_{ULA1} + t_{ULA2} + t_{ULR} \right) \qquad (7)$$

where t_{ULA1} corresponds to the time dedicated for data transmission between the MS and BS. The expressions (4) and (7) imply that requirements on RS' CPU are higher at heavy traffic load similarly as in the frame concept based on [3].

Up to now, only two hops between the BS and MS have been considered. According to [3] [5], two approaches are specified to support more than two hop communication. The first approach groups BS and several RSs into multi-frame with repetition of relay zones. For example, if the multi-frame is composed of two frames, the relay zone in the first frame is assigned to the RS on the first hop whereas in the second frame the relay zone is assigned to the RS on the second hop. Thus, data are sent in consecutive order between the BS and MS. The disadvantage of such principle is a significant increase of packet delay which is crucial for delay sensitive services such as VoIP. The solution is offered by the second approach as several relay zones are scheduled in one frame. Consequently, data are transferred within one frame which minimizes the packet delays. The drawback of this method is higher requirements on RSs processing capabilities. Nevertheless, to fully utilize the potential of second approach, the frame scheduling and planning of frame structure needs to be properly done. Our suggestion, how to schedule the frame is described in the next section.

4 Optimization of Frame Scheduling

In order to minimize the packet delay, individual parts of frame have to be scheduled in such manner that data in the UL and DL direction has to be sent as soon as possible. Generally, the packet delay depends on two parameters; i) the number of frames between packet arrival at the originator station and reception at the destination station and ii) the position of data burst in the frame. Especially the first parameter influences the overall packet delay. Concerning the position of data burst in the frame, the DL subframe always precedes subframe in the UL direction. Thus, in the DL direction are foreseen shorter packet delays. Decreasing of packet delay is achieved by successive filling of data burst from the left to the right side of frame. However, this method has only effect as long as the system load is low.

The optimized frame structure is presented in Fig. 3. In the same manner as described in Fig. 2, the individual DL/UL access and relay zones are allocated in such way that data can be ideally transmitted in one frame. In the DL direction, data are sent on the first hop between the BS-MS/RS prior to the transmission on the second or any other hop. On the other hand, in the UL direction is possible to change transmitting order as UL bursts on the higher hops are primarily scheduled.

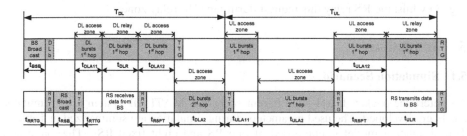

Fig. 3. Optimized CC-RS's frame structure based on [4]

In comparison with the original concept, two advantages are foreseen. Firstly, the radio resources are exploited in more efficient manner as all RRTG gaps are filled out with data transmissions. To be more specific, the RRTG gaps scheduled in the DL subframe can be used for DL transmissions between the BS and MS. The RRTG gap following the second UL access zone is utilized for transmission from the MS to BS. The second advantage is that the packet delays, both in DL and UL directions, are cut down to minimum.

While the beginning and end of BS's DL bursts are dedicated to DL access zones (in Fig. 3 the time intervals corresponds to the t_{DLA11} and t_{DLA12}), the middle part is assigned to DL relay zones. This way is ensured that a RS has time to switch to receiving mode during the first DL access zone and come over to transmitting mode within the second access zone. Additionally, the RSs have more time to process data which increase the probability that data are relayed in the same frame. In other words, the RS's processing time is prolonged exactly by the t_{DLA12}. Consequently, it is possible to schedule the DL relay zone is scheduled right after the RTTG gap in order to provide RS with sufficient processing time at heavy load. Note that in such case the t_{DLA11} is equal to t_{RTTG}. If this condition is satisfied, the t_{RSPT} in the DL direction could be formulated as,

$$t_{RSPT} = T_{DL} - \left(t_{BSB} + t_{RRTG} + t_{RSB} + t_{RTTG} + t_{DLR} + t_{DLA2}\right) \qquad (8)$$

In the UL direction, the UL access zone on the 1st hop is similarly split into two parts as in DL direction. The most significant gain is foreseen if the BS schedules UL bursts on the 2nd hop immediately after RTTG gap, i.e. duration of RTTG gap equals to t_{DLA11}. This will grant sufficient time (t_{RSPT}) to the RS to process all received data which could be expressed as,

$$t_{RSPT} = T_{UL} - \left(t_{RTTG} + t_{ULA2} + t_{ULR}\right) \qquad (9)$$

In the UL direction, the delay is further increased by requesting mechanism specified in WiMAX networks. If a MS needs to send data to a BS, a request has to be issued in the predefined time slot scheduled by the BS. When the MS is attached to the BS via one or more RSs, a RS retransmits this request on behalf of the MS. To deliver the request to the BS in the fast fashion, time slots allocated for that purpose need to be efficiently assigned. As an example, MS's request is sent at the beginning of UL bursts field on the second hop (which corresponds to the second UL access zone in Fig. 3) whilst the RS relay this request during the UL relay zone.

5 Simulations

5.1 Simulation Scenario

To determine packet delays for different scenarios, MATLAB system level simulator has been developed. The used parameters during simulation are summarized in Table 1. The simulation model is composed of one BS and eight fixed RSs. The maximum distance between the RS and BS is restricted to two hops. The RSs positions are chosen in such way that all MSs are always in a transmission range of at the least one station (BS or RS). The simulator works in the following way. In every simulation step (the length of one MAC frame) is generated certain traffic load. According to that traffic load, the BS is able to compute the duration of individual MAC frame's parts (i.e. duration of access and relay zones, etc.). Consequently, the BS can decide if the t_{RSPT} is greater than t_{RSPC} or not. If the former is true, all data are transmitted within one frame. Otherwise, some data must be sent in the consecutive frame.

There is implemented a mobility model for every MS. At the beginning of simulation, an initial position of each MS is randomly determined in such manner, that the MS is located within a defined range, i.e. between 0 to 800 m from the BS. Additionally, a velocity and random movement direction are determined for all individual MSs in the system. The MSs are moving along straight line until the distance from the BS is equal or larger than the defined BS cell area. In such circumstance, a new MS direction is established. This mechanism guarantees, that no MS moves out of the BS range during the simulation process.

The path between the BS and MS is determined according to the minimum Radio Resource Cost (RRC) metric (more details may be found in [9]). The RRC is measured in a number of OFDM symbols needed for transmission of certain amount of data burst (e.g., 1000 bits). To decide which point of attachment is the best from the point of system performance, the RRC compares all available routes from (to) the BS and determines how much system resources have to be allocated.

The traffic model assumed in simulation is based on VoIP with suppression of silence intervals as defined in [10]. The size of packets generated during active/inactive state is denoted in simulation as *AS/IS* (see Table 1). The packet size includes user's payload and protocol headers (RTP, UDP, IPv4, 802.16 generic MAC header and CRC).

Table 1. Simulation parameters

Parameter	Value
Frequency band [GHz]	3.5
Channel bandwidth [MHz]	20
Number of MS [-]	1-100
MS's velocity [m/s]	10-50
Frame duration [ms]	20
BS transmit power P_t [dBm]/height [m]	30/30
RS transmit power P_t [dBm]/height [m]	30/30
MS transmit power P_t [dBm]/height [m]	30/2
BS cell area [m]	800
Max number of hops between the BS and MS [-]	3
Channel model between BS-RS, RS-RS	LOS [10]
Channel model between BS-MS, RS-MS	NLOS [10]
Length of simulation [min.]	30
Noise [dBm]	-100.97
Packet size during active (AS)/inactive state (IS) [b]	696/456

The packet delay considered in simulation corresponds to a time interval between a packet arrival and reception at the station's MAC layer. Thus, delays introduced by higher protocol layers and the rest of network are not considered. Furthermore, the packet delay is evaluated under assumption that MSs use rtPS scheduling services in the uplink directions. Another factor which influences the packet delay is offered traffic load, i.e. number of active MSs in the system. The maximal considered offered traffic (MOT) corresponds to state when all MSs in the system are active and generates traffic which corresponds to two VoIP connections. Note that packets eventually discarded by system due to congestion are not considered. Thus, only packets successfully arriving at the destination are taken into account.

Three simulation scenarios are compared depending on RS's processing capabilities (i.e. different t_{RSPC}); i) RS needs from 0 to 5ms to process received data (in the next section depicted as Scenario A), ii) RS needs from 0 to 3ms to process received data (in the next section depicted as Scenario B) and iii) RS needs from 0 to 1ms to process received data (in the next section depicted as Scenario A). The bottom boundary corresponds to the situation when the RS has not received any data (no data are processed at all). The upper boundary is statistically derived from the average t_{RSPC} for 100% of MOT.

There are considered two cases for the DL direction in every scenario. The first one corresponds to non-optimized frame (see Fig. 2). The second scenario considers the optimized frame structure described in Fig. 3. In case of UL direction, three cases are taken into account. The first case reflects the situation when the RS cannot transmit/receive data during one frame. In addition, MS's requests are relayed to the BS in the following frame. The second case optimizes the requesting mechanism but data are relayed in similar manner as in the first case. Finally, the last case considers the fully optimized method as described in the previous section.

5.2   Simulation Results

Fig. 4 depicts the mean packet delay in the DL direction depending on the system traffic load and different RS's processing capabilities. If the system traffic load is low, all investigated scenario perform equally. The RSs have enough time to process received data and forward them to the MS/RS in the same frame. The reason is twofold,

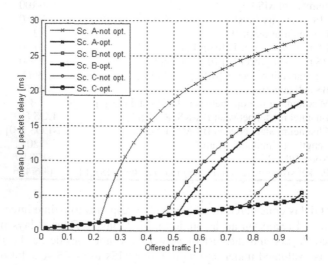

Fig. 4. Mean packet delay in DL direction

i) the whole frame is generally not occupied by data transmission and BS is able to include between DL relay and DL access zones sufficient time interval, ii) as small data burst are transmitted, the data are process faster. When increasing the traffic load, the packet delay roughly linearly increases. Nonetheless, from certain point the packet delay is significantly increased.

If we assume the frame structure according to Fig. 2, the system utilizing RSs equipped with long CPU's processing time (scenario A) shows sudden rise in packet DL approximately at 23% of MOT. From this point on, the RSs are not able to process all data burst in the same frame and have to wait for another frame to send the remaining data. With more powerful CPU's, the RSs are able to better cope with the heavy traffic load. To be more precise, RSs manage to relay data in one frame into the 49% of MOT for scenario B and into the 79% of MOT for scenario C. This situation is improved considerably if the proposed optimization of frame structure is applied. Even thought when the system is enhanced by RSs with low processing capability, the RSs are able to transfer all data within one frame up to 53% of MOT. For scenarios B and C, RSs actually manage to send all data bursts regardless the current system traffic load. The observed mean packet delays in the DL direction are up to 27ms depending on the current traffic load and processing capability of RS's CPU.

In the UL direction, the packet delays are generally longer (see Fig. 5). Firstly, this is due to the fact that the whole UL subframe follows the DL subframe. Secondly, the requested mechanism which allows MSs to ask for a transmission opportunity significantly increases packet delivery time. Note that the UL packet delay starts to substantially increase at the same value of MOT as in the DL case. The reason for it is that the BS's scheduler allocates the same guard intervals between the DL relay and DL access zones (respectively UL access and UL relay zone). If non-optimized frame is taken into account and RSs are not able to send MS's request in the same frame, the mean UL packet delay varies between 40 and 70ms. The mean packet delay can be considerably shortened by proper allocation of time slots for MSs (RSs) bandwidth requests. If the optimized frame is used, the packet delays further decrease, i.e. below 30ms.

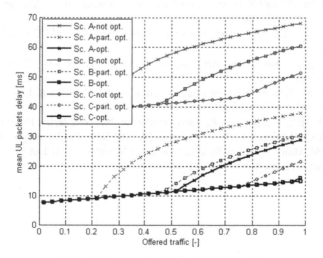

Fig. 5. Mean packet delay in UL direction

6 Conclusion

The paper focuses on frame structure of WiMAX networks and suggests an optimization for multi-hop communication in order to minimize the packet delay. A description of frame for CC-RS according to IEEE 802.16j and [4] has been introduced. To support the multi-hop communication, two approaches can be considered. The first one introduces multi frame and repetition of relay zones. However, the packet delay in multi-hop scenario is quite high. On the other hand, the second approach allows RS to receive (transmit) and transmit (receive) data bursts in one frame. The drawback of this method is high requirement on RSs processing capabilities. The simulation results show that if the proposed scheme of frame structure is applied, the RSs have more time to process data which decrease the overall packet delays as for DL as for UL directions.

Acknowledgments. This research work was supported by grant of Czech Ministry of Education, Youth and Sports No. MSM6840770014.

References

1. IEEE Std 802.16-2004, IEEE Standard for Local and metropolitan area networks, Part 16: Air Interface for Fixed Broadband Wireless Access Systems (2004)
2. IEEE Std 802.16e-2005, IEEE Standard for Local and metropolitan area networks, Part 16: Air Interface for Fixed and Mobile Broadband Wireless Access Systems (2006)
3. IEEE 802.16j, Baseline Document for Draft Standard for Local and Metropolitan Area Networks Part 16: Air Interface for Fixed and Mobile Broadband Wireless Access Systems (2007)
4. Hoymann, Ch., Klagges, K.: MAC Frame concepts to Support Multi-hop Communication in IEEE 802.16 Networks. Wireless Word Research Forum (2005)
5. Tao, Z., Li, A., Teo, K.H., Zhang, J.: Frame Structure Design for IEEE 802.16j Mobile Multihop Relay (MMR) Networks. In: Global Telecommunications Conference (GLOBE-COM), pp. 4301–4306 (2007)
6. Loa, K., et al.: In-Band Transparent Relay Frame Structure, IEEE C802.16j-07/179r6 (2007)
7. Ahn, D.H., et al.: Multiframe structure consistent to 802.16e for MR networks, IEEE C802.16-07/162r5 (2007)
8. Leng, X., et al.: Non-Transparent relay structure extension for multi-hop (>2 hops) support, IEEE C802.16j-07/145 (2007)
9. Mach, P., Bestak, R., Becvar, Z.: Optimization of Network Entry Procedure in Relay Based WiMAX Networks. In: Proceedings of IWSSIP 2008, Bratislava, pp. 61–64 (2008)
10. IEEE 802.16j, Multi-hop Relay System Evaluation Methodology (Channel Model and Performance Metric) paper No. 06/013r3 (2007)

Robust Transmission of H.264/AVC Video Using 64-QAM and Unequal Error Protection

Ahmed B. Abdurrhman, Michael E. Woodward, and Vasileios Theodorakopoulos

School of Informatics, Department of Computing, University of Bradford, U.K.
{A.Abdurrhman,M.E.Woodward}@bardford.ac.uk,
vtheodor@homtmail.com

Abstract. This paper presents a robust transmission of H264/AVC coded video using hierarchical quadrature amplitude modulation (HQAM), which takes into consideration the non-uniformly distributed importance of frames in a group of pictures (GOP) intracoded (I-frames) and predictive coded frames (P-frames) as well as the sensitivity of the coded bitstream against transmission errors. Unequal error protection is based on uniform and non-uniform HQAM constellations in conjunction with different scenarios of splitting the bits of transmitted symbol for protection of the more important information of the video content. The performance of the transmission system is evaluated under additive Gaussian noise (AWGN) conditions. The simulation results demonstrate that the proposed strategy produces a better quality of the reconstructed video data compared with a system that offers uniform protection.

Keywords: H.264/AV, video over wireless networks, hierarchical QAM, unequal error protection.

1 Introduction

With the development of multimedia communication services, robust video transmission in a wireless environment poses many challenges. The lossy wireless channel causes high transmission errors of the compressed video data and resulting in the deterioration of the video reconstruction quality. As the video compression standards have been developed for relatively error free environments they cannot be directly transferred to a hostile mobile environment due to the extensive employment of variable length coding techniques which are error sensitive since a single transmission error may result in an undecodable string of bits. Therefore an essential issue is how to protect highly error sensitive video information in hostile mobile environments [2], [4]. Several error resilient video coding techniques have been proposed, as seen in [1], in order to minimize the effects of the transmission errors on the reconstructed video image quality. Unequal error protection (UEP) of the coded video bit-stream is one of the most successful techniques. The main idea of (UEP) is based on the fact that the bits in a compressed video stream are not equally important. For example the motion vectors and picture header are much more

J. Wozniak et al. (Eds.): WMNC 2009, IFIP AICT 308, pp. 117–125, 2009.
© IFIP International Federation for Information Processing 2009

important than the texture video data [4]. The reconstructed video quality will be severely degraded if the visually important video data is affected by errors; these important bits should be given a higher protection order than the rest of the video bit-stream. Hierarchical modulation is another way of UEP, in which the high priority (HP) data bits of the coded video are mapped to the most significant bits (MSB) of the modulation constellation points while the low priority (LP) data bits of the coded video are mapped to the least significant bits (LSB) of the modulation constellation points [1], [3]. The overall video quality will be improved compared with non-hierarchical modulation at low channel signal to noise ratio (SNR) conditions since the highly sensitive HP data bits are mapped to the MSBs of the hierarchical QAM (HQAM) with low BER. The UEP employing the hierarchical quadrature amplitude modulation (HQAM) was proposed in [7] where NAL-A units constitute the HP bits while NAL-B and NAL-C units constitute the LP bits.

This paper presents the UEP of H.264/AVC coded video using hierarchical 64-QAM, which takes into consideration the non-uniformly distributed importance of I-frame and P-frame. The protection provided to the compressed video bits-stream is non–uniformly distributed between the video frames to minimize the picture quality degradation due to the transmission errors. The paper is organised as follows. An overview of 64 –HQAM is given in section 2. Section 3 shows the proposed UEP and results are presented in section 4. Conclusions and future work are given in section 5.

2   Hierarchical 64-QAM

To comply with the high bit rate requirement for video transmission, it is desirable to choose a bandwidth efficient modulation technique. For this reason a 64-QAM is selected mainly due to its superb spectral efficiency which can satisfy the high data rate required, by assigning more bits to each transmitted symbol. However, modulation schemes that allow a larger number of bits per symbol, have symbols closer to each other in the constellation diagram, and small errors can result in erroneous decoding. To overcome this problem, a twin-class 64-QAM is proposed, with uniform and non-uniform signal space constellations, to give different degrees of error protection. Each constellation point, in 64-level QAM systems, is represented by a unique 6-bit symbol $i1, q1, i2, q2, i3, q3$. The position of the bits in the 6-bit QAM symbol has an effect on their error probabilities and they can create various sub-channels with different integrities [4]. In the 64-QAM constellation diagram, the two MSBs have lower error probabilities than the four LSBs. Consequently, the bits $(i1, q1)$ and $(i2, q2, i3, q3)$ can be viewed as two sub- channels each having different integrities; this will be referred as (2 - 4) splitting. To increase the capacity of the better protected sub-channel, it is possible to use a (3 - 3) partitioning where HP is formed by the three MSBs $(i1, q1, i2)$ and LP is formed by the three LSBs $(q2, i3, q3)$. To improve the transmission efficiency of the system, higher error protection can be applied to the most important units of the coded video data by using non-uniform 64-QAM. In the 2-4 splitting case, to improve the performance of the HP, the constellation of Figure 1

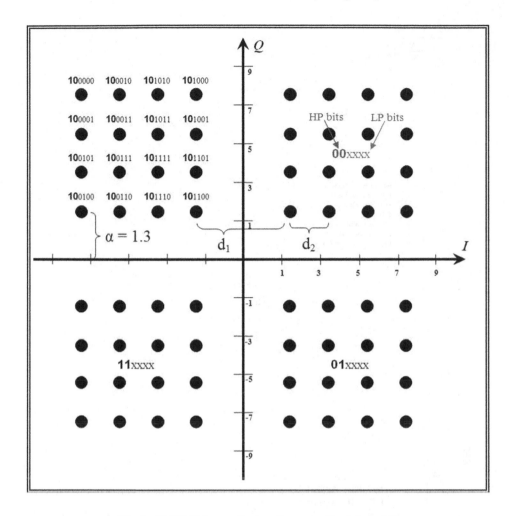

Fig. 1. 64-HQAM constellation diagram with alpha = 1.3

can be used with $\alpha > 1$, where alpha is the ratio of between the minimum distance between quadrants (d1) and minimum distance between points inside each quadrant (d2) and is given as $\alpha = d1/d2$ [7].

In this case the performance of the HP will be improved at the expense of LP. Consequently, Fig 2 shows the Symbol Error Rate of the high and low priority sub - channels in 64-HQAM for different values of α. For the sake of comparison the performance of the non-partitioned 64-QAM (STD) is also presented. For a Gaussian channel, the performance of the HP is seen to have only a small advantage over the LP sub-channel. However, by increasing the degree of non-uniformity, (d1 > d2), the improvement of the HP performance is significant at the expense of the LP sub-channel.

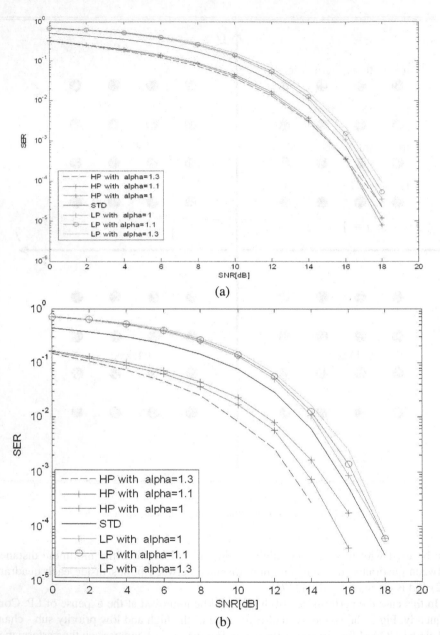

Fig. 2. (a) SER vs SNR for range of alpha for (3-3) splitting for 64 QAM (b) SER vs SNR for range of alpha for (2-4) splitting for 64 QAM

3   Proposed UEP Using 64-HQAM

The system block diagram of the proposed UEP scheme is shown in Fig 3. In this proposed scheme, the HP bits of the H.264/AVC coded video data are mapped to the

MSB's of the modulation constellation points and the LP bits are mapped to LSB's of the hierarchical 64-QAM. Compared to the uniform error protection, in which the amount of protection allocated to the coded video sequence is uniformly distributed, in the proposed UEP the compressed video bit-stream is divided into two classes of priority, namely HP data and LP data. HP data contains the part of the compressed video bit-stream which is most sensitive to transmission errors; if errors affect the HP data the reconstructed video image quality will be severely degraded and therefore a higher amount of protection is allocated to protect this data. On the other hand, errors that effect on LP data will not cause significant distortions on the reconstructed video image and therefore a lower amount of protection can be applied. The two MSBs of the constellation points in hierarchical 64-QAM have lower BER than the two LSBs therefore they are used to transmit HP data while the two LSBs are used to transmit LP data.

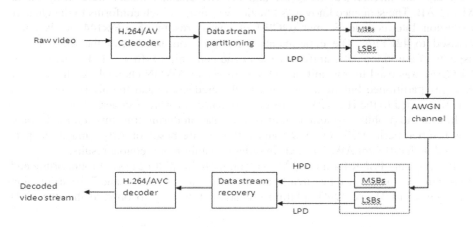

Fig. 3. Block diagram of the proposed UEP scheme

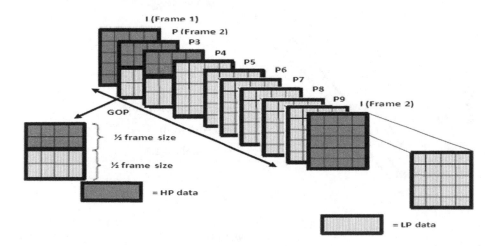

Fig. 4. Proposed UEP of the H.264/AVC

Fig 4 shows the group of pictures (GOP) of the video data. The first frame in a GOP, which is an-I frame, is classified as HP data, while the last six frames, which are P frames, are classified as LP data. The frames in the GOP have descending areas of important, so the first frame should have a higher protection. The transmission errors depend on the error position of the frame in the GOP, and when an error occurs in the beginning of a GOP the more frames are affected, while the errors in the last frames do not affect any other frames. The coded video data should be partitioned in a way that the first frame in a GOP is more protected than the last frame.

4 Simulation Results

For the purposes of this research work the H.264/AVC official reference software was used and the hierarchical 64-QAM and AWGN channel model were designed in MATLAB. The sequence known as "Suzie" was used, which conforms to the Quarter Common Intermediate Format (QCIF) of spatial resolution 176×144 pixels compressed to 42Kbit/s. The experiments were also concerned with the transmission aspects of the video data; uniform ($\alpha = 1$) and non-uniform ($\alpha = 1.1$, 1.3) partitioned 64-QAM was used to transmit the bit-stream via an AWGN channel. At the receiver end, the partitioned bit-stream was fed to the predecoder and the aligned bit-stream was forwarded to the H.264/AVC official reference software decoder.

In order to prohibit the temporal error propagation during transmission the I frame was inserted periodically every 9 frames. Results are based on fifty simulations performed with different AWGN seeds in order to obtain more reliable results.

The resulting PSNR for a 64-QAM system for both splitting cases for uniformly and non-uniformly partitioned bit-streams are presented in Figures 5, 6 and 7. By examining the uniform (3-3) and (2-4) splitting scenarios, it is easy to observe that the (2-4)

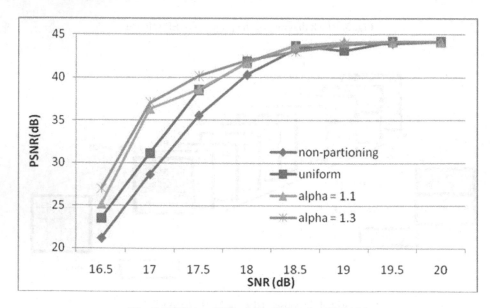

Fig. 5. PSNR vs. SNR , 3-3 splitting for 64-QAM

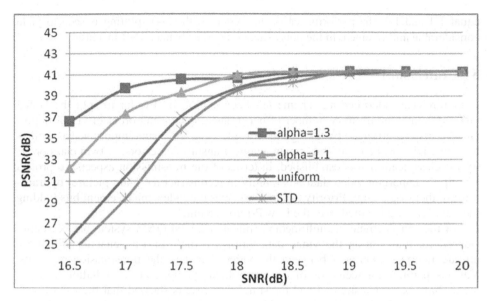

Fig. 6. PSNR vs. SNR, (2-4) splitting for 64-QAM

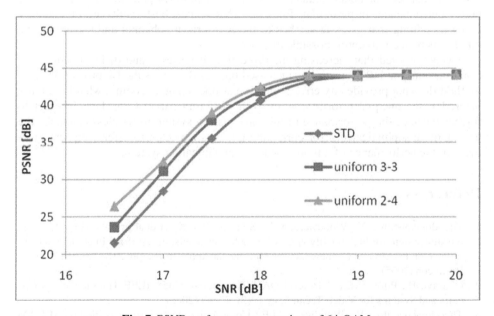

Fig. 7. PSNR performance comparison of 64-QAM

scenario leads to a better quality of reconstructed video signal. The (3-3) scenario offers a slightly worst performance than the (2-4) case because the HP data of the (3-3) scenario suffers from high BER. Both these scenarios have better performance to the non-partitioned bit-stream STD. In contrast, when applying non-uniform partitioning with α equal 1.1 and 1.3, the performance of the system with (2-4) splitting is seen to yield considerable improvement in the noisy situation as the channel SNR increases.

5 Conclusion

A hierarchical video coding scheme has been exploited in order to split the H.264 video data into an arbitrary number of partitions in decreasing order of visual importance. This was arranged by taking into consideration the visual importance as well as the sensitivity of the coded bit-stream against transmission errors. A twin class priority bit classification was established, in terms of the transmission aspects, to protect the vitally important video data bits (ability to reconstruct the main picture) by transmitting them via a High Priority channel, whereas the video enhancement bits (adding the detail) are transmitted over the Low Priority channel.

For the enhancement of multimedia applications, a 64-QAM system has been considered for transmitting the video data through a multi-priority channel utilizing a unique interface approach between the video data and the transmission elements. Various partitioning scenarios of the transmitted symbol and constellation arrangements have been investigated and the simulation results showed that the use of non-uniform constellations offer better video quality for error-prone channels. For the AWGN channel, for better channel conditions the uniform partitioned systems offer better performance. In particular, for small SNR values, best results are obtained by using a splitting that gives fewer bits to the high priority channel and separating the quadrants by the maximum possible distance.

It was observed that increasing the protection distances α and d_1 in the constellation diagrams the HP protection is increased but this decreases the LP protection. This method does not provide any error protection as does channel coding; what is actually offered is the unequal priority control to the different parts of the data. In order to further enhance the performance of the transmission system in wireless channels, the use of more sophisticated error correcting codes is expected to offer a big improvement in the performance of a mobile video communication system.

References

1. Theodorakopoulos, V., Woodward, M.E., Sotiropoulou, K.: A dual priority M-QAM transmission system for high quality video over mobile channels. In: 1st IEEE International Conference on Distributed Frameworks for Multimedia Applications, pp. 218–225. IEEE Press, Besancon (2005)
2. Gharavi, H.: Pilot- assisted 16-level QAM for wireless video. IEEE Transactions on Circuits and Systems for Video Technology 12(2), 77–89 (2002)
3. Theodorakopoulos, V., Woodward, M.E.: Comparative analysis of a twin-class M-QAM transmission system for wireless video applications. Journal of Multimedia Tools and Applications, Special Issue: Wireless Multimedia 28(1), 125–139 (2006)
4. Stedman, R., Gharavi, H., Hanzo, L., Steel, R.: Transmission of subband coded images via mobile channels. IEEE Transactions on Circuits and Systems for Video Technology 3(1), 15–26 (1993)

5. Wei, L.-F.: Coded modulation with unequal error protection. IEEE Transactions on Communications 41(10), 1439–1449 (1993)
6. O'Leary, S.: Hierarchical transmission and COFDM systems. IEEE Transactions on Broadcasting 33, 166–174 (1997)
7. Chang, Y.C., Lee, S.W., Komiya, R.: A Low-Complexity Unequal Error Protection of H.264/AVC Video Using Adaptive Hierarchical QAM. IEEE Transactions on Consumer Electronics 52(4), 1153–1158 (2006)
8. Abdurrhman, A.B., Woodward, M.E.: Unequal Error Protection For Data Transmission Using Adaptive Hierarchical QAM. In: 9th Annual Postgraduate Symposium on The Convergence of Telecommunications, Networking and Broadcasting, pp. 1153–1158 (2008)

An Optimization Approach to Coexistence of Bluetooth and Wi-Fi Networks Operating in ISM Environment

Tomasz Klajbor, Jacek Rak, and Jozef Wozniak

Gdansk University of Technology,
Faculty of Electronics, Telecommunications and Informatics,
Narutowicza 11/12, Pl-80-233 Gdansk, Poland
{klajbor,jrak,jowoz}@pg.gda.pl

Abstract. Unlicensed ISM band is used by various wireless technologies. Therefore, issues related to ensuring the required efficiency and quality of operation of coexisting networks become essential. The paper addresses the problem of mutual interferences between IEEE 802.11b transmitters (commercially named Wi-Fi) and Bluetooth (BT) devices. An optimization approach to modeling the topology of BT scatternets is introduced, resulting in more efficient utilization of ISM environment consisting of BT and Wi-Fi networks. To achieve it, the Integer Linear Programming approach has been proposed. Example results presented in the paper illustrate significant benefits of using the proposed modeling strategy.

Keywords: Bluetooth, IEEE 802.11b, interference, coexistence, ILP optimization, minimization of interferences.

1 Introduction

We have been witnessing a very fast development of various wireless technologies and network devices making use of the common unlicensed ISM (Industrial, Scientific and Medical) band, e.g. Bluetooth (BT) [3], IEEE 802.11b (Wi-Fi) [8] or IEEE 802.11g. As a result, it becomes more and more difficult to provide transmission parameters that can guarantee the quality of services required by coexisting networks. This is especially true of network devices operating in a close vicinity around other devices belonging to different independent networks, very often based on different technologies and functional solutions.

In order to provide for higher operational efficiency of diverse technological solutions working within the same area, coexistence mechanisms have been proposed. Such mechanisms can be divided into two groups [9]:

- collaborative mechanisms, requiring information exchange between IEEE 802.11b and Bluetooth devices and,
- non-collaborative mechanisms, which can be adapted separately by 802.11b and/ or Bluetooth devices without (like e.g. AFH [4] - Adaptive Frequency Hopping algorithm).

J. Wozniak et al. (Eds.): WMNC 2009, IFIP AICT 308, pp. 126–139, 2009.

Apart from the mechanisms presented in [9], examples of collaborative algorithms can be found in the literature. Some solutions, facilitating the coexistence of various technologies, are based on prediction of propagation conditions. In [4], the *Interference aware BLUEtooth Segmentation mechanism (IBLUES)*, with a dynamic BT frame selection (depending on the propagation conditions), has been presented. This method relies upon the theoretical determination of the probability of successful transmission of frames and the queuing tasks analysis. Based on such information, IBLUES takes decisions concerning the choice of a frame format between those defined in specification [3], according to which the data will be transferred (e.g. DM1, DM3, DM5).

In [10], a new coexistence mechanism, related to the management of a Bluetooth network topology has been proposed. This mechanism, named *Interference Aware BLUEtooth Scatternet (RE)configuration Algorithm* (IBLUEREA), is based upon the idea of the switching functionality built in BT devices. Generally, BT devices more frequently operating in the ISM band, should be moved further from IEEE 802.11b devices and other BT piconets as well as from each other. It can be achieved via changing functionalities of piconet/scatternet devices (from master to slave or vice versa). The IBLUEREA algorithm leads to electing a BT device operating as a master device (in a given piconet) when such a device causes, and is exposed to, little interference, compared to other BT piconet devices.

In this paper, we propose a more systematic approach to the problem of coexistence of different BT and Wi-Fi devices, operating in a certain area and utilizing ISM resources. The paper extends the idea presented in [10], and formulates a more general optimization problem of coexistence of BT and Wi-Fi networks. To strengthen the concept of a coexistence mechanism, the Integer Linear Programming approach has been introduced. The solutions lead to optimal topology, in terms of location of BT devices in a complex ISM environment. The issue is to minimize the mutual interferences in order to maximize the transmission efficiency and quality.

The paper is organized as follows. Section 2 illustrates the idea of modeling the complex ISM environments. Section 3 formulates the optimization problem. Section 4 describes the simulation environment, and Section 5 presents the benefits of using the optimization strategy, leading to minimization of interference. Some exemplary topology scenarios are investigated via simulation experiments. Conclusions and remarks complete the paper.

2 Modeling the ISM Environment

While analysing the efficiency of utilization of a given ISM environment, a lot of factors, like the number of coexisting various technology devices, their parameters, propagation conditions, etc., need to be taken into consideration. In order to estimate with a required accuracy the efficiency of a given ISM environment, the influence of all interfering devices on prospective receivers located within their range need to be accounted for.

Bluetooth specification [3] precisely determines the rules of formation and maintenance of piconets as well as the rules of joining and leaving the active network by a device. Devices within one piconet may take on the superior as well as the inferior role. Superior status – master – is assigned to the device that initiated the process of piconet creation. Switching (changing) of master and slave functionalities is possible with regard to the BT network devices.

A master device manages the transmission within a piconet. On the other hand, a slave device wishing to communicate with another station, sends data first to the master device that redirects it to an appropriate recipient. More complex BT networks – called scatternets – can be also created in which a number of piconets are connected.

Some useful mathematical tools have been invented that can describe those network structures. In [5], the general principles of scatternet description (in the form of matrices) have been presented. The authors also proposed metrices making the aggregated (and standardized) link capacity determination in scatternet possible. The metrices are of little significance when tackling the interference issue, which has only been mentioned in this article. Moreover, those metrices do not enable the analyses of interference coming from other systems. In [7], an original methodology and key metrices necessary for determination of interaction between Bluetooth scatternets and IEEE 802.11b network devices have been presented.

Fig. 1 depicts an example of an ISM environment, with possible mutual interference areas of Bluetooth piconets and coexisting IEEE 802.11b network devices. It has been assumed that the mutual interference areas (marked by dotted lines) are those where a given technology has a substantial negative impact on the receivers of other BT piconets or 802.11b network devices (for example can cause the frame error rate to exceed a given threshold value).

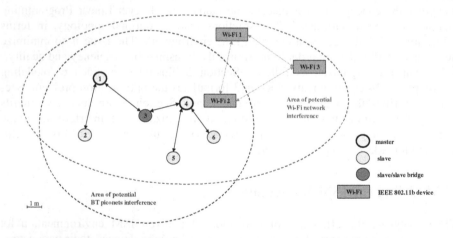

Fig. 1. Example of Bluetooth and Wi-Fi mutual interference areas

In accordance with the above illustration:

– BT devices can potentially interfere with the transmission of Wi-Fi device 2,
– IEEE 802.11b network interferes with the transmissions of all BT scatternet nodes,
– Wi-Fi nodes 1 and 3 are beyond the interference range of the BT scatternet.

The coexistence mechanism, proposed in [10], allows for an efficient and effective reconfiguration of a Bluetooth network topology. This mechanism, named *Interference Aware BLUEtooth Scatternet (RE)configuration Algorithm* (IBLUEREA), is based upon the idea of performing master functions by those BT devices more frequently using the ISM band and closer to receivers/transmitters of other wireless technology devices (e.g. 802.11b) and BT piconets. IBLUEREA accepts switching (changing) of device functionalities from master to slave, and vice versa and operating as a master (in a given piconet) is only permitted to a device that causes, and is exposed to, little interference, compared to other BT piconet devices and networks using other technologies. In order to ensure minimum interactions between BT and Wi-Fi devices, IBLUEREA uses a new approach for modeling the complex ISM environments.

In this paper, a formal optimization problem of coexisting BT and Wi-Fi networks is formulated.

3 Optimization Problem

We assume the coexistence of M networks with static placement of nodes (i.e. no mobility is considered). In each network, we need to choose one node as a master node. However, each choice has a certain impact on the performance of other networks.

In this section, we propose the Integer Linear Programming model to determine the optimal placement of master nodes in coexisting networks with the objective to minimize the mutual interference between networks i.e., maximize the transmission efficiency and quality. A set of the following constants and variables is introduced:

constants

$d_{a_m, a'_{m'}}$ - measure of a negative influence between networks m and m' in the case where node a is a master node in network m and node a' is a master node in network m'

V_m - number of nodes in the network m

M - number of potentially interfering Bluetooth networks (piconets)

variables

$J_{a_m, a'_{m'}}$ - equals 1, if node a is a master mode in network m and node a' is a master node in network m'; 0 otherwise

Our primary goal is to find the optimal placement of master nodes in BT networks such that it minimizes the total mutual negative influence of coexisting networks, defined as:

$$F = \sum_m \sum_{m'} \sum_a \sum_{a'} \left(d_{a_m, a'_{m'}} \cdot J_{a_m, a'_{m'}} \right) \tag{1}$$

subject to

constraints assuring that there is only one master node in each BT network

$$\sum_a \sum_{a'} \left(J_{a_m, a'_{m'}} \right) = 1 \qquad (2)$$

$$m = 1, 2, \ldots, M; \quad m' = 1, 2, \ldots, M; \quad m \neq m'$$

where:

$$J_{a_m, a'_{m'}} \in \{0, 1\}.$$

We may optimize the network configuration assuming various efficiency parameters, e.g. the average number of frames received with errors at all the nodes in a network S. In the remaining part of the paper, we will assume that the values of $d_{a_m, a'_{m'}}$ are calculated as:

$$d_{a_m, a'_{m'}} = \frac{1}{n^S} \sum_{x \in N^S} f(x) \qquad (3)$$

where $f(x)$ is the sum of probability values of receiving frames with errors at a node $x \in N^S$ (estimated with regard to all other nodes interfering with k), calculated as:

$$f(x) = \sum_{\substack{k=1 \\ k \neq x}}^{N} f^k(x); \quad x, k \in N^S \qquad (4)$$

Formula (3) expresses the average frame error rate calculated for all coexisting n^S nodes belonging to the network S. Next, we will try to minimize the value defined by (3).

We are interested in finding the representation of $Y_r \in \Re$ for which the sum of average probabilities of receiving frames with errors at all the nodes from a network S_r is minimized. Therefore, the optimal placement of master nodes may be calculated as follows:

$$K^*(Y_r) = \min_{Y_r \in \Re} K(Y_r) \qquad (5)$$

Taking into account the previously assumed definitions, formula (5) may be rewritten as:

$$K(Y_r) = \min \frac{1}{n^S} \sum_{x \in N^S} \left(f(x) \right) \qquad (6)$$

The coefficients $f^k(x)$ in formula (4), depend on the frame error rate at nodes belonging to S and are defined as:

$$f^k(x) = (1 - P_{S(k,x)}) \cdot v_{k,x} \qquad (7)$$

where:

$P_{S(k,x)}$ – conditional probability of successfully receiving a frame at node x assuming the occurrence of interference from node k,

$v_{k,x}$ – reflects the intensity of interferences between nodes x and k as well as the frequency of using the ISM band by node x.

The probability $P_{S(k,x)}$ can be written as follows:

$$P_{S(k,x)} = \sum_N P_S(P_E \mid n) \cdot P_C(n, N)$$

(8)

where:

$P_C(n,N)$ probability of a collision of a given technology frame with n other technology frames out of N possible collisions (frequency analysis),

$P_S(P_E \mid n)$ the probability of a successful reception of a IEEE 802.11b frame (Bluetooth), which was subject to a collision (time analysis).

In order to calculate the probability $P_S(P_E \mid n)$, it is important to determine the bit error rate P_E, in two cases: with and without collisions. Formulas describing the bit error rates have been presented in [9].

It is easy to see that any node interfering with node k uses the ISM band only at certain timeslots. Similarly, any node x being interfered by other nodes also transmits data only at certain timeslots. While calculating the values of $v_{k,x}$, we will thus determine the respective upper bounds for certain collisions (assuming that all the networks are mutually independent).

We will now investigate in detail all possible scenarios of collisions, in which nodes k and x may be the master and slave devices, respectively, in a BT as well as in a Wi-Fi network. Assuming that node k is a master node ($k \in M^S$) and interferes with a given node $x \in S^x$ ($x \in M^S$), we may assume that it transmits once every two frames. Therefore, the element $v_{k,x}$ (for scenarios where a master node interferes with other master node) takes the form:

$$\forall_{a_{k,x} \neq 0} \quad \forall_{k \in M^S} \quad x \notin S^k, x \in M^S \Rightarrow v_{k,x} = \frac{a_{k,x}}{2} \cdot \frac{a_{x,k}}{2} = \frac{1}{4}$$

(9)

However, if we investigate the scenario in which a master node $k \in M^S$ interferes with other slave node $x \in S^x$ ($x \notin M^S$), then $v_{k,x}$ may be given as:

$$\forall_{a_{k,x} \neq 0} \quad \forall_{k \in M^S} \quad x \notin S^k, x \notin M^S \Rightarrow v_{k,x} = \frac{a_{k,x}}{2} \cdot \frac{a_{x,k}}{2 \cdot (n^x - 1)} = \frac{1}{4 \cdot (n^x - 1)}$$

(10)

where n^x denotes the number of nodes in S^x i.e., the BT scatternet including node x.

We will now investigate the scenario in which the interfering node k is a slave node ($k \in S^k$, $k \notin M^S$), i.e. it may interfere with other nodes $x \in S^x$ ($k \neq x$), according to its frequency of transmitting the data in scatternet S^k. In this case, $v_{k,x}$ may be written as:

$$\forall_{a_{k,x} \neq 0} \quad \forall_{k \notin M^S} \quad x \notin S^k, x \in M^S \Rightarrow v_{k,x} = \frac{a_{k,x}}{2 \cdot (n^k - 1)} \cdot \frac{a_{x,k}}{2} = \frac{1}{4 \cdot (n^k - 1)}$$

(11)

In the case where node k interferes with another slave node, $v_{k,x}$ takes the form:

$$\forall_{a_{k,x} \neq 0} \quad \forall_{k \notin M^S} \quad x \notin S^k, x \notin M^S \Rightarrow v_{k,x} = \frac{a_{k,x}}{2 \cdot (n^k - 1)} \cdot \frac{a_{x,k}}{2 \cdot (n^x - 1)} = \frac{1}{4 \cdot (n^k - 1) \cdot (n^x - 1)}$$

(12)

where n^k and n^x denote the numbers of nodes in networks S^k and S^x respectively.

In the case of nodes that are used to connect the Bluetooth network bridges, we assume the formulas (11)-(12), taking into consideration that any bridge device may interfere with all the networks it connects, even though it is not exchanging data with these particular networks.

We will denote by N^{Wi-Fi} the set of Wi-Fi nodes in network S. Assuming that Wi-Fi nodes use the CSMA/CA technique, the impact of any 802.11 device $k \in N^{Wi-Fi}$ on other technology devices $x \notin N^{Wi-Fi}$ (BT here) may be described by the following formulas:

$$\underset{k \in N^{Wi-Fi}}{\forall} x \notin N^{Wi-Fi}, x \in M^{S} \Rightarrow v_{k,x} = \frac{a_{k,x} \cdot a_{x,k}}{2 \cdot n^{k}} = \frac{1}{2 \cdot n^{k}} \tag{13}$$

$$\underset{k \in N^{Wi-Fi}}{\forall} x \notin N^{Wi-Fi}, x \notin M^{S} \Rightarrow v_{k,x} = \frac{a_{k,x} \cdot a_{x,k}}{2 \cdot (n^{x}-1) \cdot n^{k}} = \frac{1}{2 \cdot (n^{x}-1) \cdot n^{k}} \tag{14}$$

where: n^{k} is number of IEEE802.11b devices in a Wi-Fi subnetwork that contain the interfering node k.

In order to analyze the influence of BT devices on other technology devices (here Wi-Fi), we modify the formulas (13)-(14) by a pair-wise exchange of k and x indices. The set of possible values of parameter $v_{k,x}$ is illustrated in Table 1.

Table 1. Possible settings of parameter $v_{k,x}$

Type of interfering node k	Type of a node x being interfered	$v_{k,x}$
master BT	master BT	$\dfrac{1}{4}$
master BT	slave BT	$\dfrac{1}{4 \cdot (n^{x}-1)}$
master BT	Wi-Fi	$\dfrac{1}{2 \cdot n^{x}}$
slave BT	master BT	$\dfrac{1}{4 \cdot (n^{k}-1)}$
slave BT	slave BT	$\dfrac{1}{4 \cdot (n^{k}-1) \cdot (n^{x}-1)}$
slave BT	Wi-Fi	$\dfrac{1}{2 \cdot (n^{k}-1) \cdot n^{x}}$
Wi-Fi	master BT	$\dfrac{1}{2 \cdot n^{k}}$
Wi-Fi	slave BT	$\dfrac{1}{2 \cdot (n^{x}-1) \cdot n^{k}}$

n^{k} i n^{x} - the numbers of nodes in networks S^{k} and S^{x}, respectively, i.e. in scatternets containing nodes k and x
n^{k} - number of nodes IEEE802.11b of a Wi-Fi subnet, such that includes the node k.

4 Simulations Environment

To illustrate the benefits of the optimum location of BT scatternet devices co-existing with Wi-Fi devices, a dedicated network simulator was used. In this simulator, channel models with either additive white Gaussian noise (AWGN) or with fading, have been implemented. In the second case, the Rice model [12], with small-scale fading, and no obstacles between transmitter and receiver was applied. The simulator uses a simplified implementation of a non-zero mean value compound Gaussian process with parameters measured in the real office environment [1].

Due to the fact that the main sources of disturbances in investigated networking environments were the transmitters of other systems (BT and Wi-Fi, respectively), the Gaussian noise was accepted at a relatively low (usually treated as standard) level of −95 dBm, according to commonly accepted characteristics of "good" channels [13].

There are many models of indoor signal attenuation, both statistical as well as taking into account the geometrical shapes of rooms. For the simulation purposes, a statistical model described in [11] has been selected. According to this model, signal attenuation PL [dB] over a distance d [m], can be expressed as follows:

$$PL(d) = \begin{cases} 40.0 + 20 \lg(d), & d \leq 8\,[m] \\ 58.5 + 33 \lg(d/8), & d > 8\,[m] \end{cases} \quad (15)$$

The network simulator used in modeling included the basic functionality of physical and MAC layers (IEEE 802.11 standard), as well as the physical (RADIO) layer of the Bluetooth standard. The implementation was performed subject to certain simplifications. The most important elements of the implemented environment are described later in this section. In particular, the structure of the analyzed transmission environment (802.11b) is presented in Fig. 2.

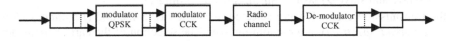

Fig. 2. Functional scheme of implemented transmission environment (IEEE 802.11b). QPSK – *Quadrature Phase-Shift Keying* modulation, CCK – *Complementary Code Keying* modulation used in IEEE 802.11b standard, S/P – serial/parallel converter, P/S – parallel/serial converter.

Data incoming from the MAC layer (at 11 Mbps) are grouped in the S/P converter into 8-bit sequences. The pairs of bits are then input to QPSK modulation. Sets of four QPSK symbols are next concatenated appropriately in the CCK modulator to obtain a codeword consisting of 8 symbols of a signal transmitted into the radio channel at 11 Mbps. The receiver utilizes the same simplified algorithm to obtain the 8-bit sequence as was used to generate the received CCK codeword. Decoded sets of bits are then utilized to form a frame passed to the upper layers at the receiving node.

On the other hand, in the implementation of 802.11b receivers, a simplified algorithm proposed in [2], has been used for decoding the CCK signals (*Complementary Code Keying*).

In the case of IEEE 802.11b environment, we assumed that in a each simulation scenario:

– the network nodes did not change their location,
– the parameters of transmission channel were constant. In particular, the bandwidth of a single radio channel was set to 22 MHz.

Additionally, in simulation experiments it was assumed that network devices generate and constantly exchange CBR (*Constant Bit Rate*) type of traffic. CBR is a stream of constant-length frames being generated by the source node, using the constant interspaces (i.e. at constant transmission rate). Parameters of the used CBR model included:

– maximum bit rate: 11Mb/s (IEEE 802.11b standard),
– frame length: 1500B.

The data and ACK frames transmitted with errors were retransmitted using the ARQ *Stop and Wait* algorithm. The maximum power of IEEE 802.11b transmitter was set to 100 mW.

When analyzing the interferences from a Bluetooth system, it was necessary to determine the probability of time-frequency collisions and the relative power levels for both the signals at the input of the 802.11 receiver. For the sake of simplicity, the GFSK modulation scheme was replaced by BPSK, since the latter has lower computational complexity (the same approach has been used in [6]).

Regarding the BT networks, the following assumptions were made:

- data transmission was performed with the use of collisionless TDD (*Time Division Duplex*) scheme. BT frames were transmitted between master and slave devices according to the assumed timeslot pattern (i.e. even and odd timeslots for master→slave and slave→master transmission, accordingly),
- the transmission channels were selected using the FHSS method (*Frequency Hopping Spread Spectrum*) in a random manner. The selection of the carrier frequency was based on the formula: $f = 2.402 GHz + k$ MHz, where $k = 0, 1, 2, ..., 78$,
- the data of each transmitting node fit in a timeslot of a length 625 µs (1/1600 s). Transmission breaks (i.e. periods with no data transmission) were used to stabilize the frequency after selecting a next transmission channel. Therefore, the real time of channel utilization was equal to $\tau_{BT} = 366$ µs,
- the bandwidth occupation of a single channel was set to 1 MHz,
- data transmission was performed by means of asynchronous connections without retransmission, using DH1 frames (each carrying 366 bits of data). A single DH1 link used a single channel of a piconet (CBR source),
- in each piconet, data transmission was performed between the master node and one of the slave nodes,
- Class 3 devices were in use (the with maximum power of 1 mW; 0 dBm).

Simulations were performed for the case of several randomly generated network to-pologies. Simulation scenarios were prepared to analyze the typical network configu-rations met in real environments. In all cases, the reconfiguration scenarios were to ensure the optimal location of BT master nodes, minimizing the value of the objective function (1), formulated in Section 3. The results of simulation experiments, pre-sented below, were obtained on the basis of 30 simulation runs lasting 10s each (warm-up periods were disregarded).

5 Example Results

Taking into account the assumptions from Section 3, optimum network topologies were sought, leading to minimization of mutual interferences (at the same time mini-mizing the average FER value).

The analyzed network topologies are depicted in Fig. 3. Fig. 3a shows an example BT scatternet, which may be referred to as the initial solution Y_1. The second one (Fig. 3b) presents the result of making use of the Bluetooth network structure recon-figuration algorithm (the network configuration with optimum placement of master BT nodes according to the assumed criterion (1)).

In the analysed case, 1500B-long 802.11b frames and DH1 BT frames (generated by CBR sources) were transmitted. The respective distances between network devices are determined by the distance scale presented in the bottom left corner of Fig. 3.

a)   b)

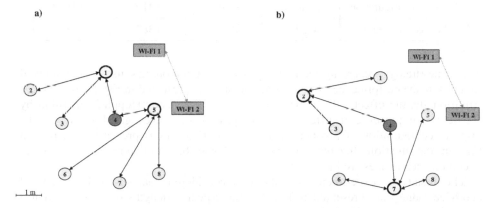

Fig. 3. Example illustration of IBLUEREA algorithm: network topology (a) before and (b) after reconfiguration

The influence of the AFH algorithm on the network performance was also investi-gated. Example results are presented in Fig. 4. The respective lengths of 95% confi-dence intervals are shown in Table 2. The modeling performed for the fading channel (the Rice model) for the initial configuration from Fig. 3a proved that the proposed algorithm ensures:

– 57% improvement of the average FER value - regarding the Wi-Fi network transmission,
– 15% improvement of FER - regarding the AFH mode of BT devices and Wi-Fi networks.

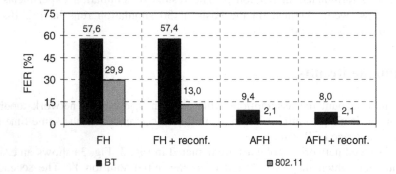

Fig. 4. Average Frame Error Rates (FER) for Bluetooth (class II devices) and 802.11b nodes (example 1–Fig. 3; AWGN channel with Rice fading)

Table 2. Lengths of the 95% confidence intervals for the average FER values

	FH BT	*FH 802.11*	*AFH BT*	*AFH 802.11*
before reconfiguration	0,46%	0,70%	0,41%	0,31%
after reconfiguration	0,54%	0,38%	0,38%	0,23%

The same strategy, regarding the topology (re)configuration, has also been employed to measure the performance of coexisting piconets, as presented in Fig. 5 (example 2). Additionally, the effects of utilizing the Adaptive Frequency Hopping algorithm by BT devices in presence of different channel models were investigated. In Fig. 5a, the original configuration of piconets with three Wi-Fi nodes is presented, whereas in Fig. 5b, the situation after reconfiguration is shown. In both diagrams, distances between piconet devices are indicated.

The results of experiments for the two considered channel models (AWGN and Rice fading) are presented in Fig. 6. The respective lengths of 95% confidence intervals are shown in Table 3.

The second example scenario actually describes and investigates the problem of mutual interferences between BT scatternets. As shown in Fig. 6, the use of the reconfiguration algorithm and the strategy to change the role of nodes may also have a very positive impact on the efficiency of ISM band utilization by devices employing the AFH algorithm (a 71% improvement in BT frame loss was observed). It may be of great importance especially in the case of sensor networks formed by large sets of Bluetooth devices. For modeling scenarios with AFH mechanism, the Wi-Fi technology does not benefit by using the algorithm. This is reasonable, since BT devices do not utilize the Wi-Fi band due to the AFH algorithm.

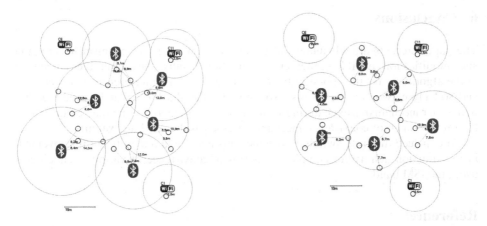

Fig. 5. Example illustration of the proposed the coexistence algorithm minimizing the mutual interferences: a) topology before reconfiguration, b) topology after executing the algorithm

The results of experiments for the two considered channel models (AWGN and Rice fading) are presented in Fig. 6. The respective lengths of 95% confidence intervals are shown in Table 3.

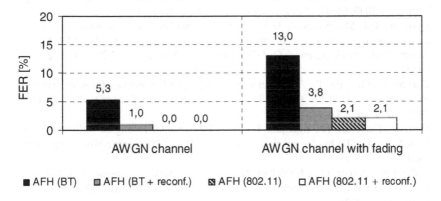

Fig. 6. Average FER values for Bluetooth (class II devices) and 802.11b network devices (example 2 – Fig. 5)

Table 3. Lengths of the 95% confidence intervals for the average FER values

	AFH BT	*AFH BT+ reconf.*	*AFH 802.11*	*AFH 802.11+reconf.*
AWGN channel	0,18%	0,10%	-	-
AWGN channel with fading	0,43%	0,16%	0,18%	0,29%

6 Conclusions

The paper presents the advantages of using a role-switching mechanism for creation and modification of BT scatternets. Simulation results show that the advantage of the algorithm is the possibility to reduce mutual interference between BT and 802.11b networks. The results also show that the algorithm may be used together with other mechanisms even during formation of BT scatternets and may contribute to the improvement of coexistence mechanisms (such as the AFH algorithm).

To conclude, the role-switching mechanisms may yield a significant improvement in transmission quality for complex scenarios of coexistence of various technologies using the ISM band.

References

1. Andersen, J.B., Rappaport, T.S., Yoshida, S.: Propagation Measurements and Models for Wireless Communications Channels. IEEE Commun. Mag. 33(1), 42–49 (1995)
2. Batabyal, S., Sarmah, S.J.: A Computationally Efficient Algorithm for Code Decision in CCK Based High Data Rate Wireless Communications. In: Proc. of IEEE International Conference on Personal Wireless Communications (ICPWC 2002), Delhi, pp. 143–146 (2002)
3. Bluetooth SIG, Inc., Specification of the Bluetooth System, Covered Core Package version: 2.0 + EDR (2004), http://www.bluetooth.org
4. Cordeiro, C.D.M., Agrawal, D.P.: Employing dynamic segmentation for effective colocated coexistence between Bluetooth and IEEE 802.11 WLANs. In: IEEE Global Telecommunications Conference, GLOBECOM 2002, November 17-21, vol. 1, pp. 195–200 (2002)
5. Cuomo, F., Melodia, T.: A general methodology and key metrics for scatternet formation in Bluetooth. In: IEEE Global Telecommunications Conference, 2002. GLOBECOM 2002, vol. 1, pp. 941–945 (2002)
6. Galli, S., Famolari, D., Kodama, T.: Bluetooth: channel coding considerations. In: IEEE 59th Vehicular Technology Conference, 2004. VTC 2004-Spring, vol. 5, pp. 2605–2609 (2004)
7. Golmie, N., Rebala, O., Chevrollier, N.: Bluetooth adaptive frequency hopping and scheduling. In: IEEE Military Communications Conference, 2003. MILCOM 2003, vol. 2, pp. 1138–1142 (2003)
8. IEEE, IEEE Std 802.11b-1999 (R2003), Supplement to IEEE Standard for Information technology – Telecommunications and information exchange between systems – Local and metropolitan area networks – Specific requirements – Part 11: Wireless LAN Medium Access Control (MAC) and Physical Layer (PHY) specifications: Higher-Speed Physical Layer Extension in the 2.4 GHz Band, New York (2003)
9. IEEE, IEEE Std 802.15.2-2003, IEEE Recommended Practice for Information technology - Telecommunications and information exchange between systems - Local and metropolitan area networks - Specific requirements. Part 15.2: Coexistence of Wireless Personal Area Networks with Other Wireless Devices Operating in Unlicensed Frequency Bands (2003)
10. Klajbor, T., Woźniak, J.: Interference Aware Bluetooth Scatternet (Re)configuration Algorithm IBLUEREA. In: Telecommunication Systems (2008), doi:10.1007/s11235-008-9092-2, ISSN 1018-4864

11. Lansford, J., Stephens, A., Nevo, R.: Wi-Fi (802.11b) and Bluetooth: Enabling Coexistence. IEEE Network 15(5), 20–27 (2001)
12. Sklar, B.: Rayleigh Fading Channels in Mobile Digital Communication Systems Part I: Characterization. IEEE Communications Magazine 35(7), 90–100 (1997)
13. Song, C., Hamid, S.: Improving goodput in IEEE 802.11 wireless LANs by using variable size and variable rate (VSVR) schemes: Research Articles. Wireless Communications and Mobile Computing 5(3), 329–342 (2005)

Performance Analysis of IEEE 802.15.3 MAC Protocol with Different ACK Polices

S. Mehta and K.S. Kwak

Wireless Communications Research Center, Inha University, Korea
suryanand.m@gmail.com

Abstract. The wireless personal area network (WPAN) is an emerging wireless technology for future short range indoor and outdoor communication applications. The IEEE 802.15.3 medium access control (MAC) is proposed, specially, for short range high data rates applications, to coordinate the access to the wireless medium among the competing devices. This paper uses analytical model to study the performance analysis of WPAN (IEEE 802.15.3) MAC in terms of throughput, efficient bandwidth utilization, and delay with various acknowledgment schemes under different parameters. Also, some important observations are obtained, which can be very useful to the protocol architectures. Finally, we come up with some important research issues to further investigate the possible improvements in the WPAN MAC.

Keywords: Analytical Modeling, Performance Analysis, MAC Protocol, IEEE 802.15.3.

1 Introduction

The IEEE standard 802.15.3 MAC layer [1] is based on a centralized, connection oriented topology which divides a large network into several smaller ones termed "piconets". A piconet consists of a Piconet Network Controller (PNC) and DEVs (DEVices). The DEV is designed to be low power and low cost. One DEV is required to perform the role of PNC (Piconet Coordinator), which provides the basic timing for the piconet as well as other piconet management functions, such as power management, Quality of Service (QoS) scheduling, security, and so on. The standard also allows for the formation of child piconets and neighbor piconets. The original piconet is called the parent piconet and the child/neighbor piconets are called the dependent piconets. These piconets differ in the way they associate themselves to the parent piconet. IEEE 802.15.3 standard supports multiple power saving modes and multiple acknowledgement (ACK) policies (NO ACK, Imm-ACK, and Dly-ACK,). It is very robust and supports coexistence with the other WLAN technologies such as IEEE 802.11. In IEEE 802.15.3 MAC protocol, although communications are connection–based under the control of the PNC, connections and data transfer can be made with peer to peer connections. In IEEE 802.15.3 MAC protocol, the channel time is divided into superframes, which each superframe beginning with a beacon. The superframe is composed of the three major parts: the beacon, the optional contention access

J. Wozniak et al. (Eds.): WMNC 2009, IFIP AICT 308, pp. 140–151, 2009.

period (CAP), and the channel time allocation period (CTAP) or channel allocation time (CTA).

Wireless channel is usually vulnerable to errors. Error control mechanisms should be designed in MAC protocol to provide a cretin level of reliability for the higher network layers. In accordance with that, IEEE 802.15.3 defines three types of acknowledgment mechanisms for CTAs: the No-ACK, Imm-ACK, and Dly-ACK mechanisms. For the Imm-ACK and Dly-ACK mechanisms, retransmission is adopted to recover corrupted frames in previous transmissions. For the No-ACK mechanism, ACK is not sent after a reception. In the Imm-ACK mechanism, each frame is individually ACKed following the reception of the frame. The Dly-ACK mechanism allows the source to send multiple frames without waiting for individual ACKs. Instead, the ACKs of the individual frames are grouped into a single response frame to be sent to the source DEV. In [2] and some other literature proposed implied-acknowledgment (Imp-ACK) for bidirectional communication. Implied acknowledgement (Imp-ACK) permits a CTA to be used bi-directionally within a limited scope. Implied ACK policy is not allowed in the CAP to avoid ambiguities between a frame that is transmitted in response to a frame with an implied ACK request and a frame that is transmitted independently when the original frame was missed. To reduce the overhead of the IEEE 802.15.3 MAC, we combine the frame aggregation concept [3] and Dly-ACK mechanism with different burst sizes, and we define this new mechanism as Dly-ACK-AGG. The idea of frame aggregation is to aggregate multiple MAC frames into a single (or approximately single) transmission.

All these ACK policies have a large impact on the throughput, delay, and channel utilization of the network and required a detailed study to find overall performance of the network. In this paper, we present the performance analysis of IEEE 802.15.3 from protocol architecture's point of view. Furthermore, we show the effect of aggregation with Dly-ACK, i.e. Dly-ACK-AGG, on the network performance.

2 Related Works

To the best of our knowledge, there is little work on the performance or channel analysis of IEEE 802.15.3 MAC with respect to different ACK policies, under different parameters. However, a large amount of literature is available on IEEE 802.15.3 MAC scheduling, optimization of superframe size, various traffic analyses and so on. Some of the important related works are as follow.

In [4] authors, presents the implementation of IEEE 802.15.3 module in ns-2 and discuses various experimental scenario results including various scheduling techniques. Specially, to investigate the performance of real-time and best-effort traffic with various super frame lengths and different ACK policies. In [5] authors, presents a novel CTA sharing protocol, called VBR-MCTA that enables the sharing of CTAs belonging to streams with the same group identity. This feature allows the proposed protocol to exploit the statistical characteristics of variable bit rate (VBR) streams by giving unused time units to a flow that requires peak rate allocation. And they presents two optimizations to VBR-MCTA, namely VBR-Blind and VBR-TokenBus, as well. In [6, 7], the authors proposed an adaptive Dly-ACK scheme for both TCP and UDP traffic. The first one is to request the Dly-ACK frame adaptively or change the burst size of Dly-ACK according to the transmitter queue status. The second is a retransmission counter to enable the destination DEV to deliver the MAC data frames to upper layer timely and orderly. While later is more focused on optimization of channel capacity. Both papers laid a good foundation in simulation and analytical works of IEEE 802.15.3 MAC protocol. Similarly, in [8] authors, formulate a throughput optimization problem under error channel condition and derive a closed form solution for the optimal throughput. The work presented in [8] is close to our work but their analysis scope is limited only in terms of throughput analysis. While our work span covers the delay, throughput, and channel utilization with different ACK policies under frame aggregation and error channel condition.

The paper is outlined as follows. In section 3 we present the performance analysis from protocol architecture's point of view and finally, conclusions and future work are drawn in section 4.

3   Performance Analysis of IEEE 802.15.3 MAC Protocol

In this section, we present the performance analysis of IEEE 802.15.3 MAC to answer several questions like optimization of payload, optimization of ACK policies, effect of aggregation etc., under various parameters.

3.1   Analytical Model

We use ground works of Bianchis's model [9] and [8] to present our analysis work. We divide our analysis in two parts; CTA analysis and CAP analysis. Table 1 shows the notations that we used for the analytical model.

Table 1. Parameters notations

T_{SIFS}	Short Inter Frame Space (SIFS) time
T_{DIFS}	Distributed Coordinate Function Inter Frame Space (DIFS) time
T_{MIFS}	Minimum Inter Frame Space (MIFS) time
CW_{min}	Minimum back-off window size
$T_{pre.}$	Transmission time of the physical preamble
T_{PHY}	Transmission time of the PHY header
L_{MAC-H}	MAC overhead in bytes
L_{ACK}	ACK size in bytes
L_{Data}	Payload size in bytes
T_{MAC-H}	Transmission time of MAC overhead
T_{ACK}	ACK transmission time
T_{Data}	Transmission time for the payload
T_{f-CAP}	The time for a transmission considered failed during CAP
T_{s-CAP}	The time for a transmission considered successful during CAP
T_{f-CTA}	The time for a transmission considered failed during CAT
T_{s-CTA}	The time for a transmission considered successful during CAT
T_{ACK-TO}	The time-out value waiting for an ACK
Th_{CAP}	Normalized throughput during a CAP time
Th_{CTA}	Normalized throughput during a CTA time
n_{burst}	Burst size in packets

The throughput is given by

$$Th = \frac{Transmitted\ Data}{Transmission\ Cycle\ Duration} \tag{1}$$

We assume a Gaussian wireless channel model. The channel bit error rate (BER), denoted as p_e $(0 < p_e < 1)$, can be calculated via previous frames or some other mechanism. How to obtain p_e is way out of the scope of this paper. From [9], a frame with a length L in bits, the probability that the frame is successfully transmitted can be calculated as

$$p_s = (1 - p_e)^L \tag{2}$$

Here, for the simplicity we assume that a data frame is considered to be successfully transmitted if both data frame and ACK are successfully transmitted in different ACK mechanism policies. We use Imm_ACK, No_ACK, D_ACK, and D_AGG_ACK to denote the immediate acknowledgement, No acknowledgement, delay acknowledgement, and delay acknowledgement with aggregation, respectively. Then we can define p_s for different ACK mechanisms as follows

$$p_s, \text{Imm\_ACK} = (1 - p_e)^{(L_{Data} + L_{MAC-H} + L_{ACK-Imm})*8}$$

$$p_s, \text{No\_ACK} = (1 - p_e)^{(L_{Data} + L_{MAC-H})*8}$$

$$\tag{3}$$

$$p_s, \text{Dly\_ACK} = (1 - p_e)^{(L_{Data} + L_{MAC-H} + L_{ACK-Dly})*8}$$

$$p_s, \text{Dly\_ACK-AGG} = (1 - p_e)^{(L_{Data} + L_{MAC-H} + L_{MAC-Ih} + L_{ACK-Dly})*8}$$

A successful transmission time during a CTA is given by[1]

$$T_{s-CTA} = \begin{cases} \left(T_{MIFS} + T_{Data} + T_{MAC-H} + T_{pre} + T_{PHY}\right) & for\ No-ACK \\ \left(2*T_{SIFS} + T_{Data} + T_{MAC-H} + 2*T_{pre} + T_{PHY} + T_{MAC-ACK} + T_{PHY-ACK} + T_{Imm-ACK}\right) & for\ Imm-ACK \\ \left(K(T_{MIFS} + T_{Data} + T_{MAC-H} + T_{pre} + T_{PHY}) + T_{SIFS} + T_{MAC-ACK} + T_{PHY-ACK} + T_{Dly-ACK} + T_{pre}\right) & for\ Dly-ACK \\ \left(K*T_{Data} + T_{MAC-H} + T_{MAC-IIs} + 2*T_{pre} + T_{PHY} + T_{SIFS} + T_{MAC-ACK} + T_{PHY-ACK} + T_{Dly-ACK-AGG}\right) & for\ Dly-ACK-AGG \end{cases} \tag{4}$$

---

[1] Readers are advice to have a look at table 1 while referring equations as we couldn't explain every parameter due to space limitation.

from (2), (3) and (4) the normalized throughput during a CTA is given by

$$
Th_{CTA} = \begin{cases} \dfrac{p_s, No\_ACKL_{Data}*8}{T_{s-CTA}} & for\ No-ACK \\[2mm] \dfrac{p_s, Imm\_ACKL_{Data}*8}{T_{s-CTA}} & for\ Imm-ACK \\[2mm] \dfrac{p_s, Dly\_ACK\ KL_{Data}*8}{T_{s-CTA}} & for\ Dly-ACK \\[2mm] \dfrac{p_s, Dly\_ACK-AGG\ KL_{Data}*8}{T_{s-CTA}} & for\ Dly-ACK-AGG \end{cases}
\tag{5}
$$

To demonstrate the effect of n-Dly-ACK and n-Dly-ACK-AGG on bandwidth utilization, we define a metric named maximum effective bandwidth (MEB) based on [10], which is a fraction of time the channel is used to successfully transmit data frames versus the total channel time. The maximum effective bandwidth utilization during a CTA/CAP slot is given by

$$
MEB_{CTA} = \begin{cases} n_{burst}.\dfrac{L_{Data}p_s, Dly\_ACK}{T_{s-CTA}} & for\ Dly-ACK \\[2mm] n_{burst}.\dfrac{L_{Data}p_s, Dly\_ACK-AGG}{T_{s-CTA}} & for\ Dly-ACK-AGG \end{cases}
\tag{6}
$$

$$
MEB_{CAP} = \begin{cases} n_{burst}.\dfrac{L_{Data}n\psi(1-\psi)^{n-1}p_s, Dly\_ACK,}{T_{s-CAP}} & for\ Dly-ACK \\[2mm] n_{burst}.\dfrac{L_{Data}n\psi(1-\psi)^{n-1}p_s, Dly\_ACK-AGG}{T_{s-CAP}} & for\ Dly-ACK-AGG \end{cases}
$$

During the CAP, we use the analytical model similar with CSMA/CA mechanism of IEEE 802.11. Also, we study how to optimize the channel throughput using different ACK policies under error channel condition. When the Imm-ACK mechanism is used in CAP, each node adopts CSMA/CA with binary exponential backoff. When NO-ACK mechanism is used, each node will use a fixed backoff window as it has no knowledge whether or not the transmitted data frame is successfully received. If the Dly-ACK mechanism is used, as long as the backoff timer of a node reaches zero, the node will first send a number (K) of data frames each separated by an MIFS and a delay-request frame separated by an MIFS, and wait for the ACK. If a burst transmission (of K data frames) is considered successful, the sender will reset the backoff window to the initial value; otherwise, the backoff window will be doubled. Dly-ACK-AGG follows the same backoff procedure as Dly-ACK. From [9], the failure probability of a transmission during a CAP is given by

$$
p_f = \begin{cases} 1-(1-p)p_s, Imm\_ACK & for\ Imm\_ACK \\ 1-(1-p)p_s, Dly\_ACK, & for\mathrm{Dly\_ACK} \\ 1-(1-p)p_s, Dly\_ACK-AGG, & for\ Dly\_ACK-AGG \end{cases} \tag{7}
$$

For n number of stations, the probability of a transmitted frame collision is given by

$$
p = 1-(1-\psi)^{n-1} \tag{8}
$$

ψ, probability of a station to transmit during a generic slot time is also depends on number of retry limit. Then, the probability of the busy channel is given by

$$
p_b = 1-(1-\psi)^n \tag{9}
$$

From (8) and (9), the probability of a success transmission occurs in a slot time is given by

$$
p_S = \begin{cases} n\psi(1-\psi)^{n-1}p_s, No\_ACK, & for\ No-ACK \\ n\psi(1-\psi)^{n-1}p_s, Imm\_ACK, & for\ Imm\_ACK \\ n\psi(1-\psi)^{n-1}p_s, Dly\_ACK, & for\ Dly\_ACK \\ n\psi(1-\psi)^{n-1}p_s, Dly\_ACK-AGG, & for\ Dly\_ACK-AGG \end{cases} \tag{10}
$$

A successful transmission time during a CAP is given by

$$
T_{s-CAP} = \begin{cases} \left(\overline{CW}+T_{MFS}+T_{Data}+T_{MAC-H}+T_{pre}+T_{PHY}\right) & for\ No-ACK \\ \left(\overline{CW}+2*T_{SIFS}+T_{Data}+T_{MAC-H}+2*T_{pre}+T_{PHY}+T_{MAC-ACK}+T_{PHY-ACK}+T_{Imm-ACK}\right) & for\ Imm-ACK \\ \left(\overline{CW}+K(T_{MFS}+T_{Data}+T_{MAC-H}+T_{pre}+T_{PHY})+T_{SIFS}+T_{MAC-ACK}+T_{PHY-ACK}+T_{Dy-ACK}+T_{pre}\right) & for\ Dly-ACK \\ \left(\overline{CW}+K*T_{Data}+T_{MAC-H}+T_{MAC-HS}+2*T_{pre}+T_{PHY}+T_{SIFS}+T_{MAC-ACK}+T_{PHY-ACK}+T_{Dy-ACK-AGG}\right) & for\ Dly-ACK-AGG \end{cases} \tag{11}
$$

where $\overline{CW}$ represents the average back-off time. The average back-off defines the back-off duration for "light loaded networks", i.e. when each station has access to the channel after the first back-off attempt and is given by

$$
\overline{CW} = \frac{CW_{min}.T_{slot}}{2} \tag{12}
$$

A failure transmission time during a CTA is given by

$$
T_{f-CTA} = \begin{cases}
\left(T_{MIFS} + T_{Data} + T_{MAC-H} + T_{pre} + T_{PHY} \right) & \text{for } No-ACK \\
\left(T_{SIFS} + T_{Data} + T_{MAC-H} + T_{pre} + T_{PHY} + T_{ACK-To} \right) & \text{for } Imm-ACK \\
\left(K(T_{MIFS} + T_{Data} + T_{MAC-H} + T_{pre} + T_{PHY}) + T_{ACK-To} + T_{SIFS} \right) & \text{for } Dly-ACK \\
\left(K*T_{Data} + T_{MAC-H} + T_{MAC-HS} + T_{pre} + T_{PHY} + T_{ACK-To} + T_{SIFS} \right) & \text{for } Dly-ACK-AGG
\end{cases}
\tag{13}
$$

from (10), (11), and (13), the normalized throughput during a CAP is given by

$$
Th_{CAP} = \begin{cases}
\dfrac{P_S L_{Data} *8}{(1-p_b)\delta + P_S T_{s-CAP} + (p_b - P_S)T_{f-CAP}} & \text{for } No-ACK \\[3mm]
\dfrac{P_S L_{Data} *8}{(1-p_b)\delta + P_S T_{s-CAP} + (p_b - P_S)T_{f-CAP}} & \text{for } No-ACK \\[3mm]
\dfrac{P_S KL_{Data} *8}{(1-p_b)\delta + P_S T_{s-CAP} + (p_b - P_S)T_{f-CAP}} & \text{for } Dly-ACK \\[3mm]
\dfrac{P_S KL_{Data} *8}{(1-p_b)\delta + P_S T_{s-CAP} + (p_b - P_S)T_{f-CAP}} & \text{for } Dly-ACK-AGG
\end{cases}
\tag{14}
$$

from (1), we can also calculate the average access delay during a CTA/CAP.

3.2 Performance Evaluation

For the performance evaluation, we adopt the following parameters from [11] as shown in table 2.

Table 2. Parameters values

Parameters	Values
SIFS	2.5 usec
MIFS	1 usec
Preamble and PLCP Header	9 usec
CW_{min}	8
Payload Size	1~5 KB
ACK Policy	3 basic +Dly-ACK-AGG policies
Data Rate	1~2 Gbps
Control Signal Rate	48 Mbps

3.2.1 CTA Analysis

Figures 1 and 2 show the normalized throughput for different payload sizes with different ACK polices with and without aggregation, respectively. We assume an ideal channel conditions for these results. Here, we can observe that No-ACK gives the superior results as most of the CTA time is utilized for data transfer. However, No-ACK policy is not suitable for every application and channel condition. The Dly-ACK-AGG policy can achieve somewhat close results to No-ACK policy, as it reduces the unnecessary inter-frame time as well as the header size. Figures 1 and 2 give us the normalized value of throughput at different payload sizes but can't answer for optimum payload size. So, we produce the same results with Gaussian wireless channel model with a given BER rate.

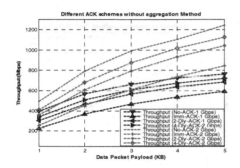

Fig. 1. Throughput versus payload size with different ACK policies

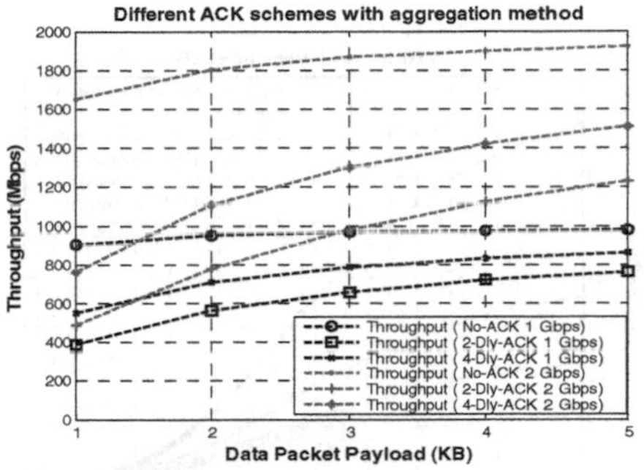

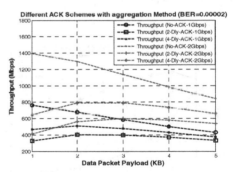

Fig. 2. Throughput versus payload size with different ACK policies

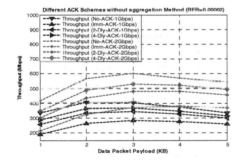

Fig. 3. Throughput versus payload size with different ACK policies

Fig. 4. Throughput versus payload size with different ACK policies

Figures 3 and 4 show the throughput for different payload sizes under a given BER value. It can be seen that an optimal payload size exists for a given BER value. As shown in the mentioned figures the throughput first increases, and then decreases with increasing payload size (even with the aggregation) in error prone channels. This is because without the protection of FCS in individual payload frame, a single bit error may corrupt the whole frame which will waste lots of medium time usage and counteract the efficiency produced by an increased payload size. Figure 5 shows the normalized throughput for different BER values with different ACK policies when payload size is set to 1KB. As the BER value increases the throughput decreases. The No-ACK policy with aggregation has larger throughput over large range of BER values than any other ACK policies. To find the effect of n-Dly-ACK on bandwidth utilization as well as to find the optimal value of burst size for n-Dly-ACK policy, we define the MEB metric in (6). Figures 6 and 7 show the MEB with different burst values under a given BER value. From these figure we can observe that burst size = 4 gives good results in fairly all given payload values. Figure 8 shows the access delay performance for different burst sizes with the aggregation method. Here, we only analyzed n-Dly-ACK-AGG policy to get the upper bound on the delay performance.

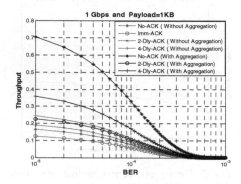

Fig. 5. Throughput versus BER value with different ACK policies

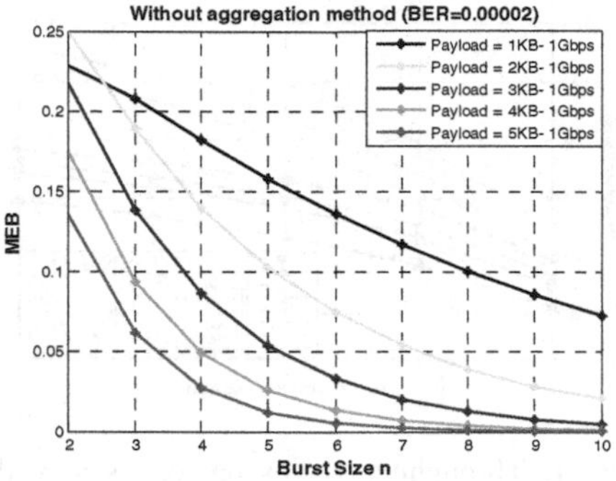

Fig. 6. MEB versus burst size

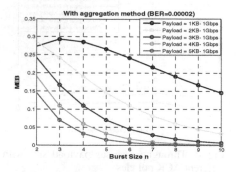

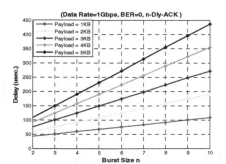

Fig. 7. MEB versus burst size

Fig. 8. Access delay versus burst size

3.2.2 CAP Analysis

Figures 9 and 10 show the normalized throughput over different payload sizes with different ACK policies. Here, we assume no competition between active nodes, as our main focus is to get maximum normalized throughput for each payload size. Figures 11 and 12 show the normalized throughput for different payload sizes with different ACK policies under a given BER value. For each ACK policies, with the increase of the payload size, the throughput first increases, then decreases after the maximal point. This can be explained as follows. In CAP, the time to transmit the payload is only a small portion of the total time used. Therefore, when the payload size increases, the transmission efficiency can be increase, but the error probability also increases. The increase of the curves in figures 11 and 12 is because the effect of increased transmission efficiency is more significant than the effect of increased frame error probability, and the decrease of the curves is due to dominant effect of increased frame error probability when payload size further increases. From the above mentioned figures we can find out the optimum payload size value for a given BER value. Here, it is worth to note that normalized throughput performance depends on the number of active stations and backoff window size during a CAP time. Figure 13 shows the normalized throughput for different BER values when payload size is set to 1KB.

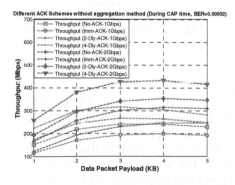

Fig. 9. Throughput versus payload size

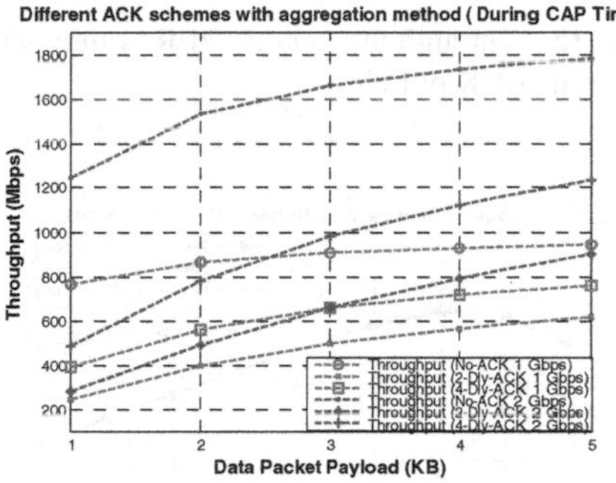

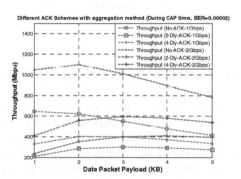

Fig. 10. Throughput versus payload size

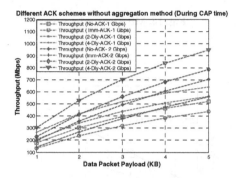

Fig. 11. Throughput versus payload size

Fig. 12. Throughput versus payload size

Similar to CTA duration analysis, Figure 14 and 15 shows the MEB for different burst values under a given BER value. Figure 16 shows the access delay performance for different burst sizes under a given BER value. The access delay performance with different ACK policies is very sensitive to backoff window size and number of active nodes. In this paper, we focused only on n-Dly-ACK-AGG policy to get the upper bound on the delay performance.

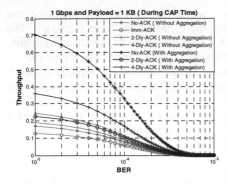

Fig. 13. Throughput versus BER value with different ACK policies

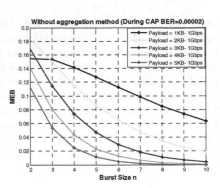

Fig. 14. MEB versus burst size

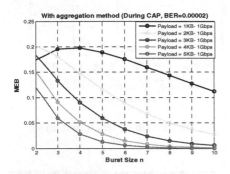

Fig. 15. MEB versus burst size

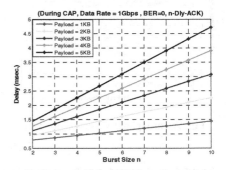

Fig. 16. Access delay versus burst size

4 Conclusions and Future Work

In this paper, we presented a performance analysis of WPAN (IEEE 802.15.3) MAC for wireless sensor networks. We have extensively studied the different ACK policies in CTA and CAP under a given BER value. The optimal payload size as well as optimal burst size can be determined analytically from the presented analysis. For future work, we want to study the impact of backoff window size on channel utilization and access delay. We hope that the results of this paper will help sensor network designers to easily and correctly provision systems based on IEEE 802.15.3 MAC technology.

References

1. P802.15.3/D17: (C/LM) Standard for Telecommunications and Information Exchange Between Systems - LAN/MAN Specific Requirements - Part 15.3: Wireless Medium Access Control (MAC) and Physical Layer (PHY) Specifications for High Rate Wireless Personal Area Networks (February 2003)

2. Part 15.3: Wireless Medium Access Control (MAC) and Physical Layer (PHY) Specifications for High Rate Wireless Personal Area Networks. Amendment 1: MAC Sub layer, IEEE Std 802.15.3b (February 2005)
3. Xiao, Y.: IEEE 802.11n: Enhancements for higher throughput in wireless LANs. IEEE Wireless Communications, 82–91 (December 2005)
4. Chin, K.-W., Lowe, D.: Simulation study of the IEEE 802.15.3 MAC. In: Proc., Australian Telecommunications and Network Applications Conference (ATNAC), Sydney, Australia (2004)
5. Chin, K.W., Lowe, D.: A Novel IEEE 802.15.3 CTA Sharing Protocol for Supporting VBR Streams. In: Proc. of IEEE ICCCN (2005)
6. Chen, H., Guo, Z., Yao, R., Li, Y.: Improved Performance with Adaptive Dly-ACK for IEEE 802.15.3 WPAN over UWB PHY. IEICE Trans. Fundamentals E88-A(9) (September 2005)
7. Tseng, Y.H., Wu, E.H., Chen, G.H.: Maximum Traffic Scheduling and Capacity Analysis for IEEE 802.15.3 High Data Rate MAC Protocol. In: Proc. of IEEE VTC (2003)
8. Xiao, Y., Shen, X., Jiang, H.: Optimal ACK Mechanisms of the IEEE 802.15.3 MAC for Ultra-Wideband Systems. IEEE JSAC 24(4) (April 2006)
9. Bianchi, G.: Performance analysis of the IEEE 802.11 distributed coordination function. IEEE JSAC 18(3), 535–547 (2000)
10. Chen, H., Guo, Z., Yao, R.Y., Shen, X., Li, Y.: Perfoamnce Analyis of delyed Acknowledgemnet Scheme in UWB-Based High-Rate WPAN. IEEE transaction on vehicular technology 55(2) (March 2006)
11. IEEE P802.15 Working Group for Wireless Personal Area Networks (WPANs).: Unified and flexible millimeter wave WPAN systems supported by common mode. DOC: IEEE 802.15-07-0761-05-003c (July 2009)

Efficient Sensor Data Gathering and Resilient Communication for Rescue Scenarios

Daniele Munaretto[1], Chunlei An[2], Joerg Widmer[1], and Andreas Timm-Giel[2]

[1] DOCOMO Euro-Labs, Munich, Germany
lastname@docomolab-euro.com
[2] University of Bremen, TZI ikom/ComNets, Germany
{atg,chunlei}@comnets.uni-bremen.de

Abstract. Wireless Sensor Networks (WSNs) have been used mainly to collect environmental data and send it to a base station. Routing protocols are needed to efficiently direct the information flows to the base station. Since sensor nodes have strict energy constraints, data gathering and communication schemes for WSNs need to be designed for an efficient utilization of the available resources. An emergency management scenario is investigated, where a sensor network is deployed as virtual lifeline when entering a building. In addition to navigation support, the virtual lifeline is also used for two purposes. Firstly, to exchange short voice messages between fire fighter and command post. For the communication between command post and fire fighter a fast and reliable routing protocol (EMRO) has been developed based on a broadcasting scheme. Secondly, for data gathering a network coding based algorithm has been designed. The feasibility of simultaneously using this virtual lifeline for data gathering and communications is investigated in this paper by means of simulation and real experiments. The resilience to packet loss and node failure, as well as the transmission delay are investigated by means of short voice messages for the communication part and temperature readings for data gathering.

1 Introduction

Sensor networks play an increasingly important role in emergency and rescue scenarios. For example, fire fighters today take a physical lifeline (rope) with them into a burning building, to be able to find the way back in poor visibility (smoke). However, it may happen that fire fighters on their way back run into fire and risk their life. Motivated by casualties in France, it was investigated in the wearIT@work project if an electronic, virtual lifeline consisting of sensor nodes can replace the physical lifeline. The virtual line has several advantages. First, it helps a fire fighter to orient himself without the hassle of being attached to a physical rope. Second, it allows to monitor the environment along the lifeline on the return path, and third it can be used for limited communication between the fire fighter and the group leader at the command post outside the building.

For indoor navigation and in particular to guide the fire fighter back out of the building, Pedestrian Dead Reckoning (PDR) is proposed as core technology. Based on accelerometers, the track into the building and thus a relative position can be estimated. However, the accuracy of the heading information is limited and the distance estimation suffers from irregular movement (e.g., crawling) of fire fighters in zero visibility

J. Wozniak et al. (Eds.): WMNC 2009, IFIP AICT 308, pp. 152–163, 2009.

environments. Over the distance the error accumulates significantly. Therefore hybrid technology is investigated including PDR and distance measurements to sensor nodes (RSSI-based or ultra-sound) of the lifeline [1, 2]. For real deployment of nodes a node dispenser is being developed in the wearIT@work project that can be mounted, e.g., at the back of a fire fighter oxygen bottle (see Fig. 1).

Fig. 1. Prototype of sensor dispenser for automated hands-free node deployment. Photo: BIBA, Germany.

From a communication technology point of view, fire fighting is a very challenging application scenario. Fire fighters enter buildings with unknown communication infrastructure. The availability of cellular (e.g. UMTS, GSM) or wireless networks cannot be relied upon. Radio communication inside buildings with fire and smoke is possible, but degraded in vapor [3].

In this paper, the feasibility of using this virtual lifeline for communication is investigated through experiments and simulations. We focus on the performance of the sensor lifeline for sensing the environment and for the communication between command post and fire fighter. We specifically investigate the *joint* simultaneous use of the lifeline for direct communication and for the distribution of sensor data. For the distribution of sensed data, a network coding approach is implemented, which provides high loss resilience at a comparatively low overhead. For the data communication, a broadcast based scheme is developed, which is robust and yet efficient in topologies given by linear deployment of sensor nodes. Both approaches are described in detail in Sections 2 and 3. Finally, we want to clarify that communication through a virtual lifeline is designed just as a supplement to normal communication methods, e.g. GSM or WLAN.

2   Routing

Multi-hop routing protocols for ad hoc sensor networks [4] have been extensively studied in the past few years due to the difficulty to provide features such as self-organization and robust multi-hop routing [5]. Using a multi-hop routing protocol in WSNs, it is possible to gather data over a wide area with only one sink and to arbitrarily modify that monitored area by moving/adding/removing stations whenever needed. Since the stations are always monitoring their network neighborhood, these changes are quickly and automatically taken into account without the need to reconfigure the network. A station may also fail, for instance due to lack of energy, without impacting the data gathering. If that station was a part of a route to the sink, a new route will automatically be created and used to replace the deprecated one. The multi-hop routing schemes proposed for WSNs are to a great extend based on ad-hoc routing schemes as discussed in the IETF working group MANET [6].

In the specific case of a lifeline, sensor nodes are dropped sequentially by the sensor dispenser as the fire fighter moves. This sequence can be utilized for the routing protocol and allows for a much simpler and more robust protocol design. Node IDs are assigned when a node is being placed and as long as the fire fighter moves further into the building, nodes with lower node ID are closer to the command post. We propose a simple protocol EMRO (EMergency ROuting) based on broadcasting information in two directions: if a packet is sent by the fire fighter, it has to be forwarded to lower node IDs (towards the command post); if a packet is sent by the command post, it has to be forwarded to higher node IDs (towards the fire fighter). Each packet has a unique sequence number. Every time an intermediate node receives a packet, it will first compare this packet sequence number with the one previously stored in its memory. If the incoming packet has a higher sequence number, it will be forwarded and the intermediate node will update its status with the new sequence number. If a packet with lower sequence number is received, the packet is discarded. When the fire fighter sends data to the command post, any node receiving a message will rebroadcast it, provided that the message is received from a node with a higher node ID. In case the command post sends a packet to the fire fighter, intermediate nodes receive the message and wait for an ACK message from the fire fighter for a specified amount of time. If an intermediate node receives this ACK, it drops the current data message, since it has already reached its destination. If an intermediate node does not get any ACK before the timer expires, it retransmits the message. This helps to reduce redundant traffic inside the network. We also define several emergency messages for special alerts with high priority. If one of those messages is transmitted from the command post to a fire fighter, the intermediate nodes will forward it immediately. Because of the nature of broadcasting, the network is very robust to the addition/removal of nodes to/from the network. For a comparison, a conventional RSSI-based routing protocol is used. The protocol is a modified version of the routing algorithm used in the Sensor Scope project [7]. Each node sniffs the traffic of the neighborhood and in case the RSSI of the incoming packet from a neighbor exceeds a given threshold, that node is stored in a neighborhood list and a specific field called *cost to destination* is filled with 1 in case the fire fighter is reached in 1 hop, 2 in 2 hops and so on. This field is updated hop by hop by the visited nodes. The same applies to the command post, using a second field. For routing, nodes choose the neighbor with lowest *cost to destination* for routing packets to the fire fighter or command post.

3   Sensor Data Gathering

3.1   Network Coding

Network Coding (NC) [8] allows nodes to transmit packets that are combinations of multiple original packets, instead of simply forwarding the packets they receive or that originate at the node. For practical reasons, random linear network coding is often used. Coding operations involve addition and multiplication over a finite field. An outgoing packet Y_{out} is a random linear combination of the m coded or uncoded packets Y_{in}^i available at a node $Y_{out} = \sum_{i=1}^{m} k_i Y_{in}^i$. The packet can thus be written as a random linear combination of the original data packets $X^1, ..., X^n$ to obtain $Y_{out} = \sum_{i=1}^{n} g_i X^i$. At the destination, decoding requires knowledge of the coding coefficients g_i, which can be transported, for example, in the packet header [9]. A node that has received a sufficient number of linear equations can decode by inverting the matrix of coding coefficients G to retrieve the original X.

3.2   Data Gathering in the Lifeline

Each node in a lifeline senses the surrounding environment and is responsible for disseminating these measurements, but may also forward other data such as voice messages. In this section we consider the problem of designing a coding algorithm which locally sets a suitable transmission schedule with respect to the sensing scope and communication range of the sensor nodes. Sensor readings are taken at regular time intervals and a node broadcasts them to its neighbors. When a node receives readings from its neighbors, it stores and combines the data received before broadcasting a new coded packet. Additional messages may be sent in case a dangerous situation is detected (e.g., high temperature, CO_2, smoke, etc.). Hence, the goal of the algorithm is to find the minimum number of transmissions required by the nodes to spread such readings throughout the lifeline so that each node can recover the sensed information. In particular, the fire fighter may pass by that node at a later time and may require these readings to determine if it is safe to proceed along the line. Furthermore, it may be useful for nodes to record the sensing history in their flash memory for longer term data gathering (with a low priority).

   Two main issues need to be addressed: first, each node has a time varying transmit and receive range, due to changes in the transmit power and changes in the radio propagation environment; second, nodes may fail due to the limited battery life or to external causes (e.g., melting). The data gathering algorithm has to be designed to cope with these dynamics. We first address the issue of achieving a low energy consumption in the sensor network by maintaining a low number of transmissions (the largest source of power consumption). In [10], the authors consider the problem of finding the minimum number of transmissions needed in a ad-hoc network of N nodes for all-to-all broadcasting. For a range of different settings, they calculate the node's optimum number of transmissions C. However, this study is limited to the case of homogeneous settings where each node communicates only with its 1-hop neighbors. For a realistic network, it is necessary to extend the analysis to the case of heterogeneous node densities and radio ranges. We first consider a line topology where nodes are able to communicate with the $1, 2, .., i$-hop neighbors, $i = 1, .., N$, in homogeneous settings. Then, we further extend the analysis to the heterogeneous case.

Homogeneous Settings. According to [10], we model the network as an undirected graph $G = (V, E)$ with $|V| = N$ vertices, where the number of transmissions assigned to each node has to fulfill the cut conditions of the underlying graph as given in [10]:

$$\min \sum_{i \in V} C_i \tag{1}$$

$$\sum_{j \in N_S} C_j \geq |S|, \text{ with } 0 < |S| < N, \forall S \subset V \tag{2}$$

$$C_j \geq 0 \tag{3}$$

i.e., the number of transmissions over any cut of G, which partitions V into two sets S and $\overline{S}$, has to be at least as large as the number of nodes (and thus the number of information units) contained in the cut set[1]. N_S is the set of nodes that have an edge from S to $\overline{S}$. In case nodes communicate only with their 1-hop neighbors, the optimal number of transmissions C_i for node i in a line of N nodes is

$$C_i = \begin{cases} 1, & \text{for } i = 1, N \\ N - i + 1, & \text{for } i = 2, ..., (N-1)/2 \\ i, & \text{for } i = (N+1)/2, ..., N-1 \end{cases} \tag{4}$$

We adapt this analysis to settings where each node communicates with its r-hop neighbors, $r = 1, ..., N$. Due to space constraints, we report only the case of an odd number of nodes; the analysis for an even number of nodes is a simple modification. We first consider the case $r = 2$, where each node may communicate with its 1- and 2-hop neighbors. Due to the doubled transmit range, each node increases the redundancy of information spread in its neighborhood by a factor of 2. As a first step towards the computation of the optimal transmission rates, we reduce the values in Eq. 4 by half, rounding up to the next integer. At the edge nodes, we leave the number of transmissions unchanged at 1, since only have their own reading to spread in the line. Modeling the network as for the 1-hop case, the new transmission numbers have to fulfill the same cut conditions of the underlying graph as the original optimization problem given in [10].

Using half the transmissions of Eq. 4 and rounding up gives a feasible but not optimal solution. The optimum number of transmissions can be found through the following simple algorithm:

1. The edge node asks its 1-hop neighbor to reduce its transmissions by one. To fulfill the conditions in Eq. 2, the 2-hop neighbor increases its transmissions by one. The ripple effect propagates through the line, with each even neighbor reducing its number of transmissions by one and each odd neighbor increasing it by one. Since the edge nodes do not modify their transmissions, the number of modified even nodes is larger by one than the number of odd nodes, and thus the total number of transmissions is reduced by 1.

---

[1] For simplicity, we assume that a coding is only performed over a specific set of packets (called *generation*), and that a generation of packets is composed of one reading from each sensor at roughly the same time instant.

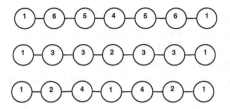

Fig. 2. Optimum transmission rates for 7 nodes on a line, starting from the 1-hop optimal values (top), reducing such values by half (middle), then after performing the iterations to find the optimum assignment (bottom)

2. This process is repeated until the assignment of transmissions does not change any further.

Fig. 2 gives the steps to compute the optimal transmissions for an example topology with 7 nodes, where each node communicates with its 1- and 2-hop neighbors. The extension of the algorithm (including the initial division of the number of transmissions by the number of nodes covered by the transmit range in each direction) for settings with transmit ranges of $r = 3, ..., N$ is straight-forward.

Heterogeneous Settings. We now consider the case of time-varying transmit ranges. Let Z_i be the set of nodes that is covered by a node i with transmit range r_i. Assume node j has a large transmit range. Clearly, if $Z_i \subseteq Z_j$ for any $i \in Z_j$, node j does not benefit from transmissions of i. In order to reduce the total number of transmissions in a lifeline, one should increase the number of transmissions of nodes with large transmit range and reduce it at nodes with smaller transmit range. The change of the number of transmissions is computed locally by each node based on a piggy-backed *feedback vector* sent with the data packets. Once a node j receives a packet from a neighbor i, it fills a vector at position i with value 1 in case the packet is non-innovative, 2 if the packet is innovative but cannot be decoded, and 3 if the packet is innovative and lets the node decode a new symbol. By default, the value of each entry is set to 0, which corresponds to the node not having received any packet from this neighbor. Once a transmission opportunity occurs, node j broadcasts a data packet, piggy-backing the feedback vector. Conversely, whenever node j receives a packet, it checks the value at position j of the incoming feedback vector. This provides the node with information about the previous packet it had sent. The information given by the feedback is used to assess the innovativeness of the node's own transmissions.

We now discuss how the algorithm works in practice. When a new generation of data (i.e., sensor readings from the environment) starts, each node broadcasts its own reading as uncoded information packet. After this first round, each node broadcasts packets coded over its own reading and the packets received from other nodes. Since feedback information about packet transmissions is obtained only in the following round, we ensure that feedback for all packets is given by sending a further packet after completing the assigned number of transmissions as dictated by the schedule. If the last packet sent is acknowledged as innovative, the node increases the transmission rate by one packet per generation, for the next generation of packets. The transmissions are reduced by one whenever the number of non-innovative packets sent is larger than one. We keep a safety margin of one additional packet to ensure that there is feedback for all required packets and as a precaution in case the topology changes or nodes fail. Reducing or increasing the transmissions only by one unit per generation caters to the time-varying nature of the transmit range. A gradual adaptation of the number of transmissions avoids that generations may frequently not be fully decodable when the radio environment is very dynamic.

Each generation involves the sensor readings of all the nodes generated within a certain time window. Thus, a new generation is started at the nodes at approximately the same time, which causes the nodes to broadcast their first uncoded packet. (This only requires a very loose synchronization among nodes.) To avoid collisions, packet transmit times are delayed by a small random offset. The following transmissions will occur periodically every $\frac{G_{time}}{C_i-1}$, where G_{time} is the expected recovery time of a generation. This leads to an even spacing of the C_i transmissions of a node. G_{time} can either be fixed before the deployment of the network or can be dynamically adapted by the nodes generation by generation. With this algorithm, some packets sent with a high transmission rate are not acknowledged by neighbors with fewer transmissions. In this case, nodes with lower transmission rate send feedback regarding the most recent packet received by their neighbor.

The feedback vector is also important for checking the connectivity of the network. When a node does not receive any packet from neighbors for a whole generation, it increases the transmit range by a fixed amount (see [11]). The same applies when the feedback vectors from neighbors have all entries set to 0, which means that the node can receive but it cannot reach the neighbors.

In summary, each node locally learns how to gradually adapt its transmission rate and transmit range. The scheme is decentralized, takes into account the time-varying nature of the node's transmit and receive range, ensures connectivity between remaining nodes in case of failures, and increases the node's life time by avoiding the transmission of non-innovative packets.

4   Simulations and Experimental Results

In this section we discuss our experimental and simulation results regarding the integration of the network coding based data gathering (Sec. 3) module with a broadcasting based routing protocol (Sec. 2). Experiments are performed on a lifeline composed of seven Tmote Sky sensor nodes [12]. They feature the Chipcon CC2420 radio chip [11] for radio communication (2.4 GHz, 250 Kbps). The radio chip is controlled by the TI MSP430 microcontroller (8 MHz, 10K RAM, 48K Flash). TinyOS [13] is used as sensor operating system. Due to the limited number of nodes available, we run simulations in TOSSIM, the TinyOS simulator [14], to evaluate the performance of our integrated protocols for "X and Y" topologies (Fig. 5), where two lifelines cross each other (X) or merge at a point (Y). Concerning the data gathering protocol, we show how the algorithm dynamically adapts the nodes' transmission rates and transmit ranges according to the actual network topology. For evaluating the broadcasting-based routing protocol, we compare it against a RSSI-based routing protocol. We analyze the robustness of the integrated protocols for this set of settings in terms of resilience to packet loss and node failure.

Fig. 3. Experiment Lifeline Setup

4.1 Line Scenario

We present the experimental results for a lifeline of 7 nodes, where 5 out of them are the sensing and forwarding nodes and the remaining two at the edges are the Fire Fighter (FF) at one side and the Command Post (CP) at the opposite side (Fig. 3). Each of the forwarding node runs the data gathering algorithm with a data transmission rate according to Sec. 3. FF and CP exchange short voice messages. When using a code rate of 20Kbps, a short voice message containing 7 words can be compressed to around 7 KB. The maximum size of a packet is 128 bytes and we use 110 bytes out of them for the data payload. Thus, we need 68 packets in total per voice message. The packets are transmitted at a rate of 5 packets per second. Setting a higher rate is detrimental to performance due to the higher number of collisions. The FF and CP use wearable computers (represented by laptops in the experiments) for displaying or playing out messages which are sent or received over the lifeline through a java interface.

The experiment evolves in 5 different steps. In step 1, all nodes communicate with their 1-hop neighbors and most of the time even with their 2-hop neighbors (we properly set the initial transmit ranges to ensure full connectivity between 1-hop neighbors w.h.p.). After letting the protocols run and reach a steady state, node 3 fails (step 2). At this stage we analyze the protocol's resilience to node failure. At step 3, a further node (number 5) fails. Up to now, the lifeline is still able to maintain end-to-end connectivity for most of the time. But when node 2 fails at step 4, the protocol reacts by increasing the transmission power of the isolated nodes to reestablish connectivity. The same applies at step 5 when a further node (number 6) fails. At this last stage, only one forwarding node is alive and lets the FF and the CP communicate.

We define the transmission time as the time between the first packet being sent from the source and the last packet being received at the destination. The transmission rates are the rates for generating network coding packets (Sec. 3), which is simply background traffic from the point of view of the voice message transmissions. Fig. 4 shows how the algorithm adapts the number of transmissions that occur at each of the aforementioned steps. Note that we only show the final steady state of each step. As we can see from the graph, the algorithm well adapts the transmission rates after each failure. We compare the performance of EMRO against the RSSI-based multihop routing protocol, which is a modified version of the one used by the Sensor Scope project [7]. Results in Fig. 4 show that the performance of EMRO is improved as far as delay is concerned. Regarding packet loss rate, the performances of these two protocols are comparable. We observe a slightly higher packet loss rate (overall packet loss rate inside the network) for EMRO, which is due to the redundancy inherent in the broadcasting. It is also related to the low buffer size in TinyOS 1. With the RSSI-based multi-hop routing, the performance in terms of transmission time and packet loss rate improves when more nodes fail, since the contention on the medium due to the multi-hop routing is reduced. Furthermore, the background traffic from network coding is lower as fewer nodes contribute to the sensor measurements.

4.2 X and Y Scenarios

The packet transmission rates used in the experiments must be changed in the simulator according to the new bit rates of 10 Kbps for the start symbols and of 40 Kbps for the payload featured by TOSSIM. Hence, sending a packet of, say, 128 bytes takes about 25.6 ms, which is substantially longer than the 4.1 ms required by real motes (without taking into account the channel busy time). As TOSSIM simulates a processor with a frequency of up to 4 MHz [15], which is much lower than that of a real TelosB mote, a node, in TOSSIM, is not capable to send packets as fast as in the real experimental environment.

Starting with the X setting, there are 2 FFs (node at the top-left as FF1 and node at the top-right as FF2) and accordingly 2 CPs (node at the bottom-right as CP1 and node at the bottom-left as CP2). In absence of intersections between these 2 lifelines, FF1 sends packets to CP1, and FF2 communicates with CP2 respectively. Each single life-line needs 34 seconds (Table 1) for transmitting a short voice message, which consists of 68 TinyOS packages, as aforementioned. In the setup shown in Fig. 5, the node in the middle is shared by both lifelines, therefore it should have double the amount of traffic. However, we decided to decrease the distance between the middle node and its 4 neighbors in the X topology and its 3 neighbors in the Y topology, so that the data load of the two lines can be shared by all these "core" nodes, avoiding loss of connectivity in case the middle node fails (which can be the result of the huge amount of data to be processed by only one node). Without such accommodation at the "core", the reaction

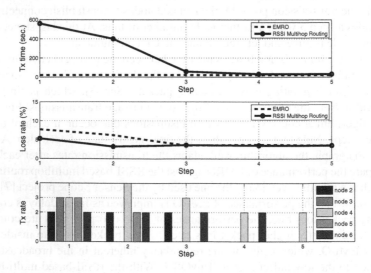

Fig. 4. Transmission rate adaptation, as well as loss rates and total transmission time for each step of the experiment. (a) and (b) for voice communication, (c) for data gathering.

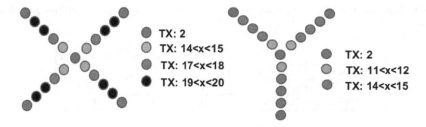

Fig. 5. Simulation results: transmission rate adaptation in X and Y topologies (1-hop communications): respectively, 2 crossing lifelines and 2 merging lifelines. At each node corresponds a color reporting the number of packets to be transmitted.

time required to adapt the transmit range in order to recover from connectivity losses could be too long[2].

As a remark, the 2 FFs are scheduled to send packets in an interleaving manner to avoid collisions at the intersection node. With the aforementioned accommodations, each lifeline has an increased transmit duration of 68 seconds (Table 1), which is almost twice the time needed for single lifeline transmission.

Given the X setting in Fig. 5, both CPs are able to receive packets from both FFs. This kind of information redundancy can be utilized to increase the reliability of connections between FFs and CPs. Furthermore, due to the NC based data gathering spreading data along both the lines, the robustness of the X network is increased as well as data persistence. Even if one of the two crossing lines (X topology) fails, the other line can still combine (through coding) in its packets the information which came from the other line before it failed and decode it at the other CP.

Without loss of generality, the Y type scenario shown in Fig. 5 can be considered as an extension of the X type scenario. If one FF loses connection to a CP, it still has a chance to get contact to the same CP (or to another CP) by merging its lifeline into another one.

Simulation results are shown in Fig. 5 and Table 1. For each scenario the simulation is repeated several times, and the average values are reported. Absolute values of the transmission delay are not of interest, but they indicate the increment of time needed for transmitting a short voice message, when 2 FFs are trying to send messages at the same time. Furthermore, since the frequencies at which packets are sent out are chosen purposely to avoid channel and nodes' overloading, no packet erasures are experienced.

In Fig. 5 simulation results show that nodes' transmission rates fit well with the theory (even when increasing the transmit range to consider also 2- and 3-hop communication models, but results are not shown due to the space constraints). The nodes at the edges as well as the node in the middle, for both X and Y settings, just need to send the minimum amount of packets per data generation. The nodes around the middle node for both topologies share the load of the coded data coming from the lines so that, in case of node failure in the middle of the network ("core", i.e. middle node and its 1-hop neighbors), the connectivity is maintained.

---

[2] In reality, sensor nodes should also be deployed in this manner, to ensure sufficient resilience against node failure.

Table 1. Simulation results: short voice message transmission in X and Y settings and single lifeline

	X Setting	Y Setting	Single Lifeline
Packet Sending Frequency (Hz)	1	1	2
Tx Delay (s)	68.137	68.465	34.765
Packet Loss Rate (%)	0	0	0

5 Conclusions

In contrast to conventional WSN routing protocols, the proposed broadcasting based protocol has the advantage of adapting to changes in the network topology very quickly. This feature is beneficial in scenarios with mobile fire fighters and frequent node failures. In the X and Y topologies, 2 lifelines can coexist with each other well, and a fire fighter can immediately recover from the lack of connectivity with the command post by merging his lifeline with another one. The data gathering protocol running in the background has no negative impact on the transmission of short voice messages or alerts. Like the routing protocol, it is very well suited for the dynamics of our settings, and nevertheless operates at a comparatively low overhead.

References

1. Klann, M., Riedel, T., Gellersen, H., Fischer, C., Oppenheim, M., Lukowicz, P., Pirkl, G., Kunze, K., Beuster, M., Beigl, M., Visser, O., Gerling, M.: LifeNet: an Ad-hoc Sensor Network and Wearable System to Provide Firefighters with Navigation Support. In: Ubicomp (September 2007)
2. Fischer, C., Muthukrishnan, K., Hazas, M., Gellersen, H.: Ultrasound-aided pedestrian dead reckoning for indoor navigation. In: First ACM International Workshop on Mobile Entity Localization and Tracking in Gps-Less Environments, San Francisco, California, USA (September 2008)
3. Hofmann, P., Kuladinithi, K., Timm-Giel, A., Goerg, C., Bettstetter, C., Capman, F., Toulsaly, C.: Are IEEE 802 Wireless Technologies Suited for Fire Fighters? In: 12th European Wireless 2006, Athens, Greece (April 2006)
4. Al-Karaki, J.N., Kamal, A.E.: Routing Techniques in Wireless Sensor Networks: A Survey. IEEE Wireless Communications (2004)
5. Becker, M., Schaust, S., Wittmann, E.: Performance of Routing Protocols for Real Wireless Sensor Networks. In: International Symposium on Performance Evaluation of Computer and Telecommunication Systems (SPECTS 2007), San Diego, USA (2007)
6. Mobile Ad-hoc Networks,
   http://www.ietf.org/html.charters/manet-charter.html
7. SensorScope website, http://sensorscope.epfl.ch/
8. Ahlswede, R., Cai, N., Li, S.Y.R., Yeung, R.W.: Network information flow. IEEE Transactions on Information Theory 46(4), 1204–1216 (2000)
9. Chou, P.A., Wu, T., Jain, K.: Practical network coding. In: 41st Allerton Conf. Communication, Control and Computing, Monticello, IL, US (October 2003)
10. Loyola, L., De Souza, T., Widmer, J., Fragouli, C.: Network-Coded Broadcast: from Canonical Networks to Random Topologies. In: NetCod 2008: Fourth Workshop on Network Coding, Theory, and Applications, Hong Kong, China (January 2008)

11. CC2420 data sheet, `http://www.ti.com/`
12. Tmote Sky data sheet, `http://www.sentilla.com/moteiv-endoflife.html`
13. TinyOS home page, `http://www.tinyos.net`
14. Levis, P., Lee, N.: TOSSIM: A Simulator for TinyOS Networks, Version 1.0. (June 2003)
15. TinyOS lesson - using TOSSIM,
    `http://www.tinyos.net/nest/doc/tutorial/tossim-lesson.html`

Self-Organizing Wireless Monitoring System for Containers

Ryszard J. Katulski, Jacek Stefański, Jarosław Sadowski,
Sławomir J. Ambroziak, and Bożena Miszewska

Department of Radiocommunication Systems and Networks,
Gdansk University of Technology
11/12 Gabriela Narutowicza Street
80-952 Gdańsk
rjkat@eti.pg.gda.pl, jstef@eti.pg.gda.pl,
Jaroslaw.Sadowski@eti.pg.gda.pl, sj_ambroziak@eti.pg.gda.pl,
bozena.miszewska@yahoo.pl

Abstract. This paper presents a description of new global monitoring system for containers, with its layer-modular structure, as a solution for enhance security and efficiency of container transport with particular emphasis on the practical implementation of that system for maritime container terminals. Especially the Smart Container Module (SCM) architecture and its operation as a part of the Self-Organizing Container Monitoring Network is presented.

1 Introduction

The first ship with containers, *SS Ideal X*, sailed from Newark for Houston, on 26th April 1956 and opened a new age in transportation. Currently container transport includes more than 90% of the global trade and takes place along trade lanes, both marine and terrestrial. However, despite all the advantages of this mode of transport, it carries a lot of risk, as a result of the exploitation of containers. First of all, containers are closed during the transportation, so the load on their interior remains beyond any control. For this reason, in peer-reviewed literature on the safety of transport, container is known as the *Trojan Horse of twenty-first century*. In this state of affairs it is particularly important issue of controlling and monitoring of cargo container, both during transportation and storage.

This issue became a subject of research at the Department of Radiocommunication Systems and Networks in the Gdansk University of Technology, where the Group of Self-Organizing Ad-Hoc Wireless Sensor Networks was founded. This Group developed the original concept of a global wireless system for monitoring of containers cargo, with particular emphasis on the practical implementation of that system for maritime container terminals [1, 2].

2 General Characteristics of the System

2.1 A Global Grasp of the System

The proposed new monitoring system solution for containers has a layer-modular structure (Fig.1). The basic layer of the system is the *Smart Container Module (SCM)*,

J. Wozniak et al. (Eds.): WMNC 2009, IFIP AICT 308, pp. 164–172, 2009.

designed for installation on containers and eventually should be an integral part of each container. Another, higher layer of the system is the Ship Subsystem - the wireless self-organize network, based on the *802.11* standard, working on the container ship. Due to the fact that the typical container ship can carry up to dozen thousand containers, there is a necessity for the location a special database on the ship. Data about all containers are regularly sent to the ship network controller. Information can also be sent on request.

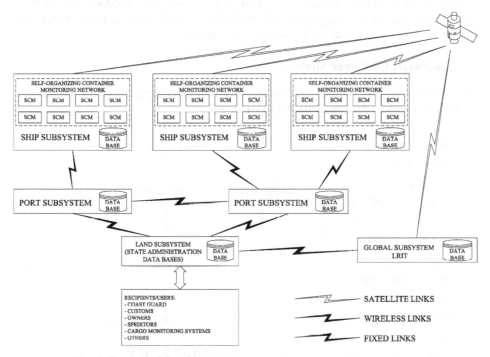

Fig. 1. Concept of the global wireless monitoring system for containers

Container terminals and sea ports form third layer of the system - the Port Subsystem with the Port Administration Data Base, equipped with wireless access points located at entrances to ports and wireless radio networks covering container depot. All of the Port Subsystems located inside the country are connected to each other and provide all data to the Land Subsystem - The State Administration Data Base. The Land Subsystem is the fourth layer of the monitoring system.

Transmission of information between each part of the system is in accordance with certain rules, using an appropriate transmission medium. In the case of a container ship, the period of reporting depends on its position, i.e. on distance to the mainland, the home port and/or destination port. Frequency of sending the relevant data increases with the approach to the mainland.

The Global Subsystem is the highest layer of the monitoring system, which can use the database of a new *LRIT* (*Long Range Identification and Tracking*) system for long-range ship monitoring. Data are transferred to the global database via satellite, when the vessel is on the open sea. Approaching the mainland and entering the harbour, information are communicated to the Port Subsystem that supports appropriate area. Transmission of this information is carried on using the *GPRS* system. It is also necessary to realize co-operation and data exchange between local and global databases, and also co-operation between databases inside the country. It can be realized by means of fixed lines, e.g. optical fiber.

The recipients (users) form a separate layer of the system. We can distinguish two classes of users: the services responsible for cargo security and the carriers. The first group is formed by the relevant services, such as Coast Guard, Customs and Police. The owners and the shipping companies belong to second group of users, which access to the system is carried out by authorized employees.

2.2 The Smart Container Module

The Smart Container Module is one of the most important elements of the monitoring system [1]. Figure 2 depicts its functional diagram.

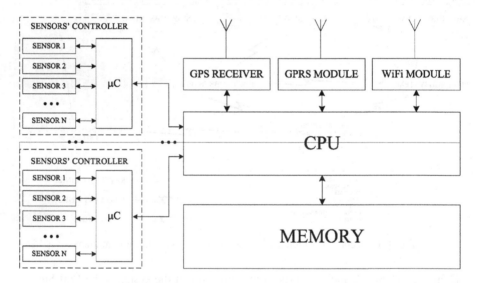

Fig. 2. Functional diagram of the Smart Container Module

The versatility of the SCM stems from the possibility of parameters measuring and monitoring of state within the container. To this end, the module is equipped with various sensors placed inside the container [3]. This sensor network enables detection of movement, temperature changes, pressure, humidity, presence of gases or radioactive substances, etc. An important aspect is also the location of the container, so each SCM is equipped with the *GPS* receiver. The GPRS module and the *WiFi* module are components of SCM. This parts of container module enable communication with port database (GPRS) or the local wireless network working on the container ship (WiFi).

Transportation documents concerns shipment (such as contents of the container, a shipper, a point of shipment, a point of destination and other information) are stored in SCM's memory. This data are sent to the appropriate database, whenever necessary.

In case of lack of connection with the radio resources of the system, it is necessary to store data about status of container in the module's internal memory and send them shortly after the SCM gets in a range of network.

The container module is characterized by very low power consumption and thus it enables unattended work for a long time (from several months to even few years). Equally important is compact construction of the SCM and to minimize its dimensions. The module's casing is matched to construction of the container, to damage prevent during transshipment. It is resistant to the weather conditions and it also has got a protection against removal or modification of its interior, with automatic alert generating in these cases.

During the research on the SCM, it was decided to use two types of industrial computers: the *Axiomtek SBC84710* and the *Advantech ARK1380*. Three prototypes of the SCM have been produced basing on these computers. The Smart Container Module prototype using the ARK1380 is shown in Figure 3.

Fig. 3. A prototype of the Smart Container Module

The module is enclosed in a sealed casing and there are various connectors available, as following:

- *COM1, COM2* - used to connect Sensors' Controllers in various configurations,
- *USB1, USB2, USB3, USB4* - two of them are used to connect external GPRS and GPS modules, next two are designed for future purposes,
- *LAN* - used to connect the SCM with the ship/port local network, in case of usage the SCM as a gateway,
- *DC IN* - used to connect the supply voltage range from 9V to 35V, but because of the usage of this voltage by Sensors' Controllers, it is recommended not to exceed a value of 15V,
- *PCMCIA* - used to connect the WiFi module,
- *VGA* - used to connect the monitor,
- *PS2* - used to connect the keyboard,
- *AUDIO* - not used, designed for future purposes,
- *LVDS* - not used, designed for future purposes.

Additionally, there are LEDs placed on the SCM's casing, which indicate supply voltage connection and work of the hard disk. There is also the on/off button placed on the casing.

The internal construction of the Smart Container Module is presented on an example of the SCM prototype based on SBC84710. The top view of internal construction of the SCM, with marked available interfaces, connectors, power supply voltage and connected the WiFi module is shown in Figure 4.

Fig. 4. Top view of the internal SCM construction

The bottom view of the internal SCM construction, with marked the RAM chip and the Compact Flash Card, which stores the operating system and all of the SCM's data is shown in Figure 5.

The container module is powered by two batteries thus it enables unattended work for a long time, which depends on batteries capacity. The first of batteries has a smaller capacity and it is enclosed in the SCM's casing. The second of batteries could have as large capacity as necessary. It is placed inside the container and connected to SCM. The SCM is equipped with a voltage control circuit, whereby it is possible to distinguish SCM turning off for lack of power against other cases.

The Sensors' Controller have been designed in a manner making construction of the measuring module independent from a type of using computer model and to enable free deployment of reconfiguring sensors inside the container. The exemplary Sensors' Controller board is shown in Figure 6.

Composition of this circuit includes three parts: the microcontroller, the RS485 converter with auxiliary circuits and the set of sensors dedicated to the specific cargo container. Sensors, which can be connected the Sensors' Controller with:

- *sabotage* - this sensor allows to detect a sabotage, e.g. SCM's casing tamper or cutting off any of Sensors' Controllers,
- *door switch* - this sensor indicates whether or not the door of the container is open or closed,
- *motion* - this sensor allows to detect movement inside the container,
- *smoke* - this sensor allows to detect smoke inside the container,
- *temperature* - this sensor measures ambient temperature inside the cargo container,
- *humidity* - this sensor measures relative humidity inside the cargo container,
- *acceleration* - this sensor measures acceleration on three axes (range to 10g or to 50g).

Fig. 5. Bottom view of the internal SCM construction

Usage of Sensors' Controllers allows connect many sensors to the SCM simultaneously. Additionally it allows to unrestricted changes types of sensors and their deployment inside the container.

3 Self-Organizing Network of SCM Modules

Because of the vertical way of containers storage on container ships, there is a problem of access to the on-board wireless network for Smart Container Modules mounted on the containers at the lowest level. Additionally, GPS satellites are beyond the sight of view of the GPS receiver, installed in the adversely located SCM modules, thus it is impossible to read the geographical position. In order to solve this problem SCM modules have been programmed in a manner that allows them to self-organize in ad-hoc network. During the research a new algorithm for data transmission control over a multi-hop ad-hoc network (assuming slow-moving nodes) has been developed.

Fig. 6. The sensors' controller board

Fig. 7. The SCF self-organizing ad-hoc network tests

By dint of this solution it is possible to transfer data about cargo to the Ship Sub-system's database, even from the Smart Container Module installed on the lowest container. The data about the cargo are transferred to adjacent modules by the WiFi radio link. If module has a connection to the on-board wireless network, it will forward data to the database. If this module is beyond the reach of the on-board wireless network, it will forward the data to the next SCM, etc. Additionally, the self-organizing ad-hoc network enables SCM modules beyond the reach of GPS system to read and save the geographical position on the basis of data from modules at the top of containers' stack. The ability to self-organize modules in the ad-hoc network can also be useful in harbours or container terminals, where direct access to the Port Subsystem's wireless network or receiving signals from GPS satellites can be impossible. Figure 7 depicts the laboratory for testing the Self-Organizing Container Monitoring Network composed of three Smart Container Modules.

The results of these tests have proved the ability of testing network to self-organize in real time, especially to choose SCM module to communicate with the database.

4   User Interface

In the course of the research, the user interface including a database storing all data about the system has been created. Depending on the granted access privileges, user may get the following information: state of the system, defined types of containers, registered containers (with the assigned sensor network), ongoing and completed shipments and communication history. In addition, there is the ability to users manage (adding/removing user, changing password) only for users with administrator privileges.

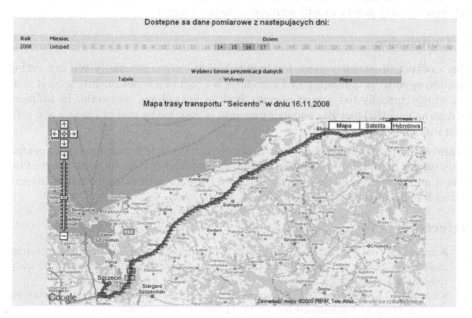

Fig. 8. Sample view of the user interface with a map of the transport route

Each transport can be remotely monitored. Cargo information (parameters from the sensor network, current position and route of transport) are available in three forms: a table, graphs and a map. Figure 8 depicts a map of portion of the transport route. This function uses the Google Maps available for free by Google.

Data can also be presented in a transparent form of graphs, which are created on the basis of data from different sensors. Sample record of the temperature inside the cargo is presented in Figure 9.

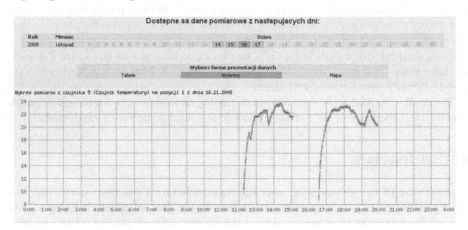

Fig. 9. Sample view of the user interface in the control of temperature scope

Using the interface is intuitive and user-friendly, which greatly facilitates access to the system at any point on earth with access to the Internet.

5 Conclusion

This concept of global system for monitoring of cargo, especially self-organizing monitoring network consisting SCM modules is a part of a wider security subject matter, which currently has a high priority, both in Poland and the world. In addition, increasing the efficiency of container transport is possible by dint of the safe and efficient user interface.

Acknowledgments. This new solution of system for wireless monitoring of containers is funded by the Polish Ministry of Science and Higher Education, as a part of research and development project No R02 012 01. The authors express their sincere thanks for allocated funds for this purpose.

References

[1] Katulski, R., Niski, R., Stefański, J., Żurek, J.: Concept of the container monitoring system in Polish harbours. In: 2006 IEEE Conference on Technologies for Homeland Security, Enhancing Transportation Security and Efficiency, Boston, USA, June 7 (2006)

[2] Katulski, R.J., Ambroziak, S.J., Sadowski, J., Stefański, J.: Containers wireless monitoring systems, Przegląd Telekomunikacyjny, No 12/2008, p. 1065 (2008)

[3] Karl, H., Willig, A.: Protocols and Architecture for Wireless Sensor Networks. Wiley & Sons, Chichester (2006)

Enabling SAML for Dynamic Identity Federation Management

Patricia Arias Cabarcos, Florina Almenárez Mendoza, Andrés Marín-López, and Daniel Díaz-Sánchez

University Carlos III of Madrid, Telematic Engineering Department, Avda. Universidad 30, 28911 Leganés (Madrid), Spain
{ariasp,florina,amarin,dds}@it.uc3m.es

Abstract. Federation in identity management has emerged as a key concept for reducing complexity in the companies and offering an improved user experience when accessing services. In this sense, the process of trust establishment is fundamental to allow rapid and seamless interaction between different trust domains. However, the problem of establishing identity federations in dynamic and open environments that form part of Next Generation Networks (NGNs), where it is desirable to speed up the processes of service provisioning and deprovisioning, has not been fully addressed. This paper analyzes the underlying trust mechanisms of the existing frameworks for federated identity management and its suitability to be applied in the mentioned environments. This analysis is mainly focused on the Single Sign On (SSO) profile. We propose a generic extension for the SAML standard in order to facilitate the creation of federation relationships in a dynamic way between prior unknown parties. Finally, we give some details of implementation and compatibility issues.

1 Introduction

Federation has emerged as a key concept for identity management. Its main goal is to share and distribute attributes and identity information across different trust domains according to certain established policies. The federation model enables users of one domain to securely access resources of another domain seamlessly, and without the need for redundant user login processes. Particularly, the most popular use case is Single Sign On (SSO), which allows users to authenticate at a single site and gain access to multiple sites without providing any additional information. Thus, separating identity management tasks from service provisioning is possible in order to reduce complexity in service providers. So service providers can concentrate on their core business and also improve user experience when interacting with various administrative domains.

The main actors in a federation scenario, as depicted in Fig. 1, are:

- *Service Providers (SPs)*, entities which consume identity data, they rely on user authentication made by a third party; SPs are also called *Relying Parties* (RPs).

J. Wozniak et al. (Eds.): WMNC 2009, IFIP AICT 308, pp. 173–184, 2009.
© IFIP International Federation for Information Processing 2009

- *Identity Providers(IdPs)*, entities that assert information about a subject; IdPs are also called *Asserting Parties* (APs). IdPs focus on authentication of users and management of identity information, which can be shared with various SPs.
- *Users*, which interact (usually via a user agent, e.g. web browser) with SPs. They are the subject of the assertions.

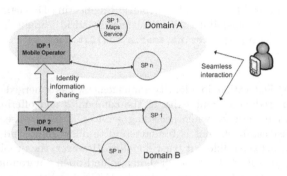

Fig. 1. Identity federation across two different administrative domains

In this example, sharing identity information between the Mobile Operator (IdP1) and the Travel Agency (IdP2) allows Bob to log in only once and gain seamless access to services and applications offered in different domains. By entering the Maps Service web page, a location-based services offered by the Mobile Operator, and present his credentials (e.g. username and password), Bob can follow a link to the Travel Agency web page, and access resources such as accommodation or restaurant information without re-authentication.

Identity federation implies many advantages, including typical use cases like cross-domain SSO, user account provisioning, entitlement management and user attribute exchange. However, current frameworks for identity federation have not been designed taking into account the requirements of dynamic and open environments. In these scenarios, companies should have an easy and agile way to provide services. But establishing relationships between entities is usually hard to scale, because trust must be preconfigured before any interaction between parties takes place.

Thus, trust between parties involved in a federation process should be managed in a dynamic fashion, avoiding or minimizing dependence on central administration and reducing preconfiguration needs. As a direct consequence, service provisioning and user interaction would be easier and more flexible, facilitating composition, enrichment and customization.

The remainder of this document is concerned with showing the main challenges in dynamic identity federation and explaining a generic extension to a widely known identity standard in order to overcome these challenges. Section 2 reviews current technologies for identity federation and the related work, Section 3 provides a comparative analysis of the trust mechanisms underlying these technologies and Section 4 describes our solution proposal. Finally, Section 5 explains some implementation issues and Section 6 contains conclusions and future work.

2   Background

Identity federation can be accomplished by means of formal Internet standards, such as the OASIS SAML specification [15], or using open source technologies and other openly published specifications, like the Liberty Alliance Identity Federation Framework (ID-FF) [13], Shibboleth [9], OpenID [17] or WS-Federation [21]. Next, the main technologies for identity federation are introduced with the aim to provide a background knowledge to the reader. Also, we include a summary of the current related work regarding dynamic identity federation.

2.1   Reviewing the Current Solutions for Identity Federation

- **Security Assertion Markup Language (SAML)** defines an XML based framework to allow the exchange of security assertions between entities. Basically, SAML specifies four different elements: *Assertions*, which are statements related to authentication, attribute, or authorization about a Principal, issued by an IdP; *Protocols*, which define how and which *Assertions* are requested; *Bindings*, which define the lower-level communication or messaging protocols (such as HTTP or SOAP) that the SAML *Protocols* can be transported over; and *Profiles*, which are combinations of SAML *Protocols* and *Bindings*, together with the structure of *Assertions* to cover specific use-cases.

  SAML is highly flexible, in fact, all of the above components can be extended. Furthermore, there is another extension point: the *Metadata*. The *Metadata* can be used to specify how configuration information is shared between two communicating entities. For instance, these data can include an entity's support for given SAML *Bindings*, identifier information, and Public Key Infrastructure (PKI) [5] information.

- **OpenID** is defined as an open, decentralized, and free framework for user-centric digital identity. It is based on well-known existing internet technologies (URI, HTTP, SSL, Diffie-Hellman), and it is clearly oriented to be used in web scenarios.

  A major feature of OpenID is user-centricity, which means that users can decide which IdP they trust the most to authenticate them. In fact, users can also become their own IdP without the need of registration or authorization from a third party. Thus, the OpenID protocol does not rely on a central authority to authenticate a user's identity.

  But OpenID is mainly an authentication protocol and federation is achieved with extensions, such as [7] that allows some attribute exchange. Thus, the OpenID specification is rigidly defined and it only covers a narrow range of Web SSO use cases. Despite having great benefits, such simplicity and no preconfiguration requirements before interactions, the main limitations are that trust, security and privacy are still not thoroughly examined.

- **Liberty Alliance (LA)** was formed with the aim to establish open standards to easily conduct online transactions while protecting the privacy and security of identity information. The Liberty specifications, built on top of SAML, enable identity federation and management through features such as identity/account linkage, single sign on, and simple session management. The Liberty Alliance contributed its federation specification, ID-FF, to OASIS, forming the foundation for SAML 2.0, the converged federation specification that Liberty now recognizes.
- **Shibboleth** is a project of the Internet2 Middleware Initiative that has developed an architecture and open-source implementation for identity federation based on SAML specifications. It is mainly intended to be used in educational environments (e.g. to help students and faculty in accessing online shared resources from federated Universities). The Shibboleth software provides a federated single sign-on and attribute exchange framework, also including extended privacy functionality and allowing the user's browser and their home site to control the attributes released to each application.
- **WS-Federation** defines mechanisms to allow different security realms to federate, such that authorized access to resources managed in one realm can be provided to security principals whose identities and attributes are managed in other realms. This includes mechanisms for brokering of identity, attribute, authentication and authorization assertions between realms, and privacy of federated claims. The specification is part of the larger Web Services Security framework (WS-*) and describes how to use WS-Trust [23], WS-Security [22] and WS-Policy [3] all together in order to provide federation between security domains.

2.2   Related Work

As will be shown in the next section, none of the above solutions define a suitable trust model to allow dynamic federation establishment. In this sense, there are various parties working with SAML to allow easy deployment and minimize mutual beforehand configuration steps.

The Internet2 group, Ping Identity and Stockholm University are working in Distributed Dynamic SAML [10] [6] to deal with challenges regarding deployment, scalability and interoperability. They aim to achieve: 1) distribution, in the sense of changing the operations of typical multi-party SAML federations to be less dependent on central administration; and 2) dynamism, which implies various means to support discovery and autoconfiguration instead of static prearrangement between parties. Thus, the group is developing proposals [19] [4] to be promoted in various communities, including potential submissions to the OASIS Security Services Technical Committee for consideration as standards.

The main important aspects of their contribution are that the partner keys used to sign and validate SAML SSO messages are included in the SAML metadata document, and trust in these keys is derived from the established trust in the metadata document itself. Also, the metadata document must be signed and the X.509 certificate chain used to validate the signature is included in the document. Thus, each partners trust anchor list just contains the root CA certificates.

But the idea of relying on signed metadata is quite similar to just relying on X.509. There are only two ways to establish trust in the metadata signatures: based on metadata signing certificates together with a traditional PKI or using out-of-band certificates as a form of pre-shared keys for signature validation. The proposal is focused on reducing the manual steps but it does not address dynamism in the sense of trust establishment and evolution. Although the process is lighter and federations are established more rapidly, trust continues to lie in pre-established arrangements, with no evolution over time, and entities cannot take autonomous decisions without some preconfigured information. Furthermore, the proposal is tied to certificate-based trust decisions and it is focused on the web SSO profile, but we think that a more general solution is needed that can be applied to a broader range of federation use cases.

3   Underlying Trust Models: A Comparative Analysis

Comparative studies of different identity management approaches [8] [14] show the main commonalities and differences regarding many different aspects: design centers, terminology, specification set contents and scope, user identifier treatment, security, IdP discovery mechanisms, key agreement approaches, as well as message formats and protocol bindings and trust. Here we aim to take a closer look at trust issues, because trust is the key to address scalability problems in the current identity federation systems. We focus this comparative analysis on the SSO use case because this feature is supported by all the studied identity management technologies.

As it was mentioned in section 2, SAML specifies a primary trust mechanism for a SSO operation. It consists of having a pre-existing trust relationship between the RP and the AP. The trust relationship establishment typically relies on PKI since it is recommended. Shibboleth adopts this same model so that federation implies the aggregation of large lists of providers that agree to use common rules and contracts. The drawbacks of this kind of trust model are well known: hard to deploy and maintain, and high dependence on central authorities.

In the case of OpenId, trust considerations are not addressed in the main specification and SSO can be performed between previously unknown parties without any configuration. However, a new OpenID specification called PAPE (Provider Authentication Policy Extension) [18], has been recently approved in order to enforce trust mechanisms. This extension provides means for a RP to request previously agreed upon authentication policies being applied by the OpenID Provider and for an OpenID Provider to inform an RP what policies were used. With PAPE, OpenID moves from a *trust-all-comers* philosophy to a situation in which the decision to trust can be based in the knowledge of the employed authentication mechanism. In other words, there is no trust model specified by

OpenID, RPs must decide for themselves which providers are trustworthy, being their responsibility to implement any policies related to the OpenID Provider's response. For these reasons OpenID is simple, lighter and more scalable.

The trust topologies considered by WS-Federation and LA resemble PKI trust models between Certification Authorities (CAs). In the case of WS-Federation, IdPs are equivalent to CAs. In the case of LA, the specification considers two possible contexts, business agreements and authentication, so IdPs and SPs are equivalent to CAs depending on the context.

These models are typically implemented by means of trust lists containing trustworthy authorities and, sometimes, maintaining also lists of untrustworthy entities. These lists are manually configured by an administrator. Thus, the establishment of trust relationships is managed with formal contracts specifying policies and restrictions surrounding this relationship.

In WS-Federation, an administrator or other trusted authority may designate that all tokens of a certain type are trusted (e.g. all Kerberos tokens from a specific realm or all X.509 tokens from a specific CA). The security token service maintains this as a trust axiom and can communicate this to trust engines to make their own trust decisions.

Liberty bases Identity Federation on the concept of "Circle of Trust" (CoT), which means that entities must establish business and trust agreements in order to enable future interactions. Thus, CoTs defined by LA specify different kinds of trust relationships that can exist between two entities depending on the context. If the context is authentication, we can have direct or indirect trust relationships. On the other hand, a business relationship can be: *pairwise*, when directly links the two entities; *brokered*, when an intermediary (*"broker"*) is required; or *community*, when no relationship of any kind exists. So Liberty entities have a TAL or Trust Anchor List with the trustworthy entities for authentication purposes, and also have a BAL or Business Agreement List, containing those parties which are related to the entity via a business agreement.

Authority lists just allow us to take boolean decisions, which means that if the list contains an entry for an authority or a trustworthy path to reach it, then the decision will be positive. On the other hand, if the authority is unknown and there is no path to it, the decision will be negative. This mechanisms limit interaction in open environments, where the presence of unknown users is common and there is no previous configuration before interaction.

In Table 1 we summarize the main trust features of each identity system. To conclude, all the analyzed technologies typically handle trust management by means of trust lists together with PKI. The only exception is OpenID, which does not require trust relationships to be established and just follows the *trust-and-accept all-comers* principle. So, it can be noted that none of the above identity management technologies include efficient trust models for dynamic environments, which implies an important challenge. Furthermore, the problem of establishing trust relationships between previously unknown entities willing to interact is not covered by none of the current frameworks or specifications.

Table 1. Summary of Trust models in Identity Federation

IdM Technology	Trust Model
OpenID	No trust model defined, *trust-all-comers* philosophy, no preconfiguration required.
SAML	PKI recommended. Typically implemented with trust lists.
Liberty Alliance	Trust architecture based on CoTs. Follows SAML recommendation (PKI). Two relationship contexts: authentication, business. Hierarchical, peer-to-peer, mesh and hybrid topologies considered. Typically implemented with trust lists.
Shibboleth	Follows SAML recommendation (PKI). Typically implemented with trust lists.
WS-Federation	Trust architecture based on WS-Trust. Peer-to-peer, mesh and hybrid topologies considered. Typically implemented with trust lists.

4   SAML Extension for Dynamic Federation

Trust is a fundamental issue to address scalability in identity management for open dynamic environments. In fact, the flexibility of every federation framework is tied to the underlying trust model, often poorly defined or even out of the specifications scope. Our goal is to design and incorporate dynamic trust models in order to facilitate the interaction between different actors involved in an identity management system.

A popular approach to addressing security challenges in these environments is the use of distributed trust mechanisms. By analyzing previously gathered information, such as certificates or reputation scores, a trust decision can be made to interact with other entities in the system with some assurance.

After reviewing the current frameworks for identity federation, we conclude that SAML is the most flexible to add extensions in order to achieve dynamic federation in a generic way. As described in section 3, all the approaches except OpenID need trust to be preconfigured. In OpenID, despite there is no need for previous configuration, the extension mechanism seems to be rudimentary and less modular. Furthermore, while all the solutions are mainly concerned with web scenarios and the SSO use case, SAML offers abstraction enough to be applied to a wider range of situations [20]. Also, SAML is the only standard nowadays and LA and Shibboleth are based in its specifications, so it is more logical to introduce modifications in SAML that could be later adopted by other technologies based on it. But SAML-based federations have challenges scaling IdP and metadata discovery, protocol binding choice, attribute and nameID usage, key management and trust establishment.

Therefore, we propose a SAML extension that allows entities to include external trust information but still maintains compatibility with the existing trust mechanisms that are mostly employed today. The main benefits of this extension are:

— Minimize dependence on central servers or previous configuration, making entities more autonomous and capable of taking trust decisions
— Model trust evolution over time, as it has a clear impact in risk management and trust decisions
— Take advantage of common knowledge, by means of requesting and collecting reputation information
— Enrich trust mechanisms (not only certificate-based trust but also reputation-based decision making)

To achieve the above goals we consider that SAML should be extended to allow the collection of reputation information, which means defining an XML representation of reputation data to be included in the Assertions or in the Metadata, and also describing an exchange protocol for requesting and sending reputation messages.

SAML is defined in a modular way, by including modifications in the abstract level we can assure its later application in more specific use cases. With this philosophy we propose an initial extension model, depicted in Fig. 2.

We present a generic solution, consisting of a SAML extension, as a first step towards dynamic federation. We focus on the point of trust negotiation before establishing a federation, which means that entity discovery is out of scope. Although the aim of this solution is to be generic enough to be applied to different profiles and use cases, we will focus on analyzing the SSO scenario to start building a prototype (see section 5).

Basically we introduce the idea of SAML entities maintaining a dynamic store instead of a static list with trust information. The new Dynamic Trust List (DTL) is automatically updated according to the establishment and evolution

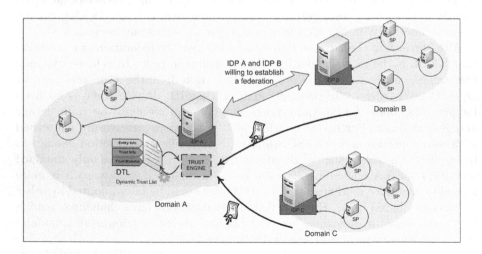

Fig. 2. SAML Extension for Dynamic Trust Establishment in Identity Federation

of trust relationships. This solution implies modifications to allow gathering of external trust information and also new functionality must be added to process these data and manage the DTL. Next, all the new concepts and implications are detailed.

4.1 Dynamic Trust List

In SAML implementations every entity is usually configured within a TAL before any interaction between parties takes place. This list contains the digital certificates associated to every other entity, which is considered trustworthy. Protocol messages whose digital signature cannot be validated against the TAL are rejected. Thus, trust decisions are just boolean. Trust does not evolve over time, because interaction experiences are not taken into account, community knowledge is not exploited, distrust and ignorance are treated in the same way, and the automatic establishment of trust relationships between unknown entities is impossible.

The preconfigured TAL model poses important obvious limitations in dynamic open environments. Instead of a static TAL, the system maintains an enhanced Dynamic Trust List with more complete information: entity data and its associated trust information (e.g. reputation scores, trust level, previous interaction results, etc.) and trust material (e.g. keys, credentials, etc.). The list will be dynamically updated under specific events such as receiving recommendations from other entities or when a successful interaction ends. In order to allow secure exchange of trust material, it is required to define key agreement protocols (e.g. by adding Diffie-Hellman) because these mechanisms are not included in none of the SAML specifications.

4.2 Trust Engine

In order to include the previously described dynamic features in the process of trust establishment, SAML entities must be also extended with a trust engine. This component is the responsible for processing external and internal trust information and performing DTL updating. Also, decisions to trust will be made by this logical block.

The internal trust information can be obtained from the DTL. Furthermore, other data as internal policies could be useful when applying custom trust levels. e.g. transient or attribute federation may have less requirements than permanent federation as it implies less personal user identity information disclosure.

On the other hand, external trust information can be obtained from other entities. To give an example, information can be gathered from entities belonging to the same domain of the target of federation, or even from entities of different domains. Such entities may have had a previous relationship with the target entity (as shown in Fig. 2). In this field, many distributed reputation solutions have been proposed [11] that can be suitable to implement in top of SAML to allow trust information exchange.

The trust engine can be enriched with more complex functionality, e.g. by adding a risk manager or a policy manager block. If we consider timing, analysis of cached trust material, update policies, etc., a better trust management can be achieved. To give an example, using policies to determine when to ask for reputation information offers the capability of implementing different dynamic trust models, in order to select the more appropiated for each situation.

5   Implementation Issues

In order to evaluate our proposal, we have deployed our own identity management infrastructure. As a first phase, we have chosen ZXID [24], a light open C library that implements the full SAML 2.0 stack, to set up a test scenario. So we have developed a SP under ZXID. For deploying the IdP, we are using Authentic [2], because ZXID does not include an IdP implementation. Authentic is a Quixote application. Quixote is a framework for developing web applications in Python. Authentic is a Liberty-enabled identity provider based on the lasso library [12] that also supports SAML 2.0 metadata. These libraries use OpenSSL as underlying cryptographic library and Apache2 as the web server.

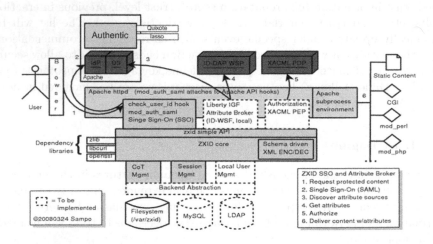

Fig. 3. SSO Test Scenario Architecture

The configuration of the mutual trust relationship between the SP and the IdP has required some slight changes in the SAML metadata generated by them. In the ZXID libraries, we had to remove the binding POST SIMPLE SIGN because it is a draft binding that has not been implemented by the lasso yet. Regarding the IdP metadata, we had to add the inclusion of a complete x509 certificate instead of the public key like the SAML specification defines.

After the successful integration between SP and IdP, we have tested as the SSO and Single Logout (SLO) of a user between these two providers is possible.

The user's identity information only contains a user and password, but this must be extended to include other kind of credentials. For this, the support of data managers by ZXID has to be improved, because so far ZXID only supports text files. The database or LDAP support is a future work, as reflected in the whole test scenario architecture (Fig. 3).

In the Fig. 3 we can also see that for making authorization decision SAML is based on XACML [16], which has not been implemented yet. We have our own XACML-compliant authorization system [1], which has been successfully deployed and extended for taking into account trust information in the access control policies. So we are going to integrate our SP with such system to grant or to deny permissions.

Now, we are firstly including dynamic trust lists (DTLs) and are also developing services in different administrative domains in order to test more complex scenarios, interoperability between federation domains, management of the user's identity information, etc. In this way, we could test our SAML extensions and evaluate their performance.

6   Conclusions and Future Work

We have reviewed the main current frameworks to achieve identity federation, identifying its main drawbacks to be deployed in dynamic open environments. Underlying trust models are too rigid to allow an agile way of establishing relationships between entities, specially when it comes to interaction with previously unknown parties.

Among all the current approaches, SAML offers the required flexibility, abstraction and modularity to be extended for its application in dynamic open environments. Thus, we present a SAML extension, which allows not only certificate based decisions but also reputation based decisions and permits the inclusion of reputation information in order to take richer trust decisions.

Now, we are implementing the proposed SAML extension, which require to define a XML syntax to express reputation information, to describe a mechanism of reputation information request/response, to enhance SAML entities with a trust engine capable of dealing with dynamic trust lists, and finally to perform evaluation experiments in a federation scenario.

Acknowledgement. The authors want to thank Rosa Sánchez for her work on the deployment of the SAML identity management infrastructure.

References

1. Almenárez, F., Marín, A., Campo, C., García, C.: TrustAC: Trust-Based Access Control for Pervasive Devices. In: Hutter, D., Ullmann, M. (eds.) SPC 2005. LNCS, vol. 3450, pp. 225–238. Springer, Heidelberg (2005)
2. Authentic: Liberty-compliant Identity Provider,
   http://authentic.labs.libre-entreprise.org/

3. Sharp, C., Shewchuk, J., Vedamuthu, A., Yalinalp, Ü., Orchard, D.: Web Services Policy 1.2 - Framework (WS-Policy). In: Bajaj, S., Box, D., Chappell, D., Curbera, F., Daniels, G., Hallam-Baker, P., Hondo, M., Kaler, C., Langworthy, D., Nadalin, A., Nagaratnam, N., Prafullchandra, H., Riegen, C., Roth, D., Schlimmer, J. (eds.) W3C Member Submission (April 2006)
4. Cantor, S. (ed.): SAML V2.0 Metadata Interoperability Profile, Working Draft 01 (2008)
5. Cooper, D., Santesson, S., Farrell, S., Boeyen, S., Housley, R., Polk, W.: Internet X.509 Public Key Infrastructure: Certificate and Certificate Revocation List (CRL) Profile. IETF Network Working Group, RFC 5280 (2008), http://www.ietf.org/rfc/rfc5280.txt
6. Harding, P., Johansson, L., Klingenstein, N.: Dynamic Security Assertion Markup Language: Simplifying Single Sign-On. IEEE Security & Privacy 6(2), 83–85 (2008)
7. Hardt, D., Bufu, J., Hoyt, J.: OpenID Attribute Exchange 1.0., http://www.openid.net
8. Hodges, J.: Technical Comparison: OpenID and SAML - Draft 06 (January 2008)
9. Internet2.: Shibboleth Architecture, http://shibboleth.internet2.edu
10. Internet2.: Distributed Dynamic SAML (October 2007), https://spaces.internet2.edu/display/dsaml/
11. Josang, A., Ismail, R., Boyd, C.: A survey of trust and reputation systems for online service provision. Decis. Support Syst. 43(2), 618–644 (2007)
12. Lasso, Liberty Alliance Single Sign-On, http://lasso.entrouvert.org/
13. LA.: Liberty ID-FF Protocols and Schema Specification, http://www.projectliberty.org
14. Maler, E., Reed, D.: Options and Issues in Federated Identity Management. IEEE Security & Privacy 6(2), 16–23 (2008)
15. Ragouzis, N., Hughes, J., Philpott, R., Maler, E., Madsen, P., Scavo, T. (eds.): Security Assertion Markup Language (SAML) V.2.0 Technical Overview. OASIS Comittee Draft 02 (March 2008)
16. OASIS.:eXtensible Access Control Markup Language, XACML (2003), http://www.oasis-open.org/apps/org/workgroup/xacml/
17. OpenID.: OpenID Authentication 2.0. (December 2007), http://www.openid.net
18. Recordon, D., Jones, M., Bufu, J., Daugherty, J., Sakimura, N.: OpenID Provider Authentication Policy Extension 1.0., http://www.openid.net
19. Stockholms Universitet, A profile for distributed SAML metadata management, Distributed Dynamic SAML version 7, Gruesome Gorilla (October 2007)
20. Tschofenig, H., Hodges, J., Peterson, J., Polk, J., Sicker, D.: SIP SAML Profile and Binding, draft-ietf-sip-saml-06.txt, IETF Internet-Draft
21. WS-Federation.: Web Services Federation Language version 1.1. (December 2006)
22. Nadalin, A., Kaler, C., Monzillo, R., Hallam-Baker, P. (eds.): Web Services Security: SOAP Message Security 1.1 (WS-Security 2004), OASIS Standard Specification (February 2006)
23. Nadalin, A., Goodner, M., Gudgin, M., Barbir, A., Granqvist, H.: WS-Trust 1.3, OASIS Standard (March 2007)
24. SymLabs.: ZXID: Open SAML implementation in C, http://www.zxid.org

DST-Based Detection of Non-cooperative Forwarding Behavior of MANET and WSN Nodes

Jerzy Konorski[1] and Rafal Orlikowski[2]

[1] Gdansk University of Technology, ul. Narutowicza 11/12, 80-233 Gdansk, Poland
jekon@eti.pg.gda.pl
[2] R&D Marine Technology Centre S.A, ul. Dickmana 62, 81-109 Gdynia, Poland
ro@ctm.gdynia.pl

Abstract. Selfish node behavior can diminish the reliability of a mobile ad hoc network (MANET) or a wireless sensor network (WSN). Efficient detection of such behavior is therefore essential. One approach is to construct a reputation scheme, which has network nodes determine and share reputation values associated with each node; these values can next be used as input to a routing algorithm to avoid end-to-end routes containing ill-reputed nodes. The main problem lies in handling possibly conflicting evidence of a particular node's behavior so as to enable rapid detection of all selfish nodes. To this end, we explore the Dempster-Shafer Theory of Evidence (DST) as part of a novel framework called DST-SDF and discuss some of its advantages and disadvantages. It differs from existing reputation schemes in that the well-known but faulty watchdog mechanism is dispensed with, and end-to-end acknowledgments are used instead. Sample simulation results illustrate the merits of DST-SDF under two proposed working modes related to the applied rule of evidence combination.

1 Introduction

Mobile ad hoc networks (MANETs) and ad hoc wireless sensor networks (WSNs) are collections of mobile nodes that exchange packets over a wireless transmission medium. There may be pairs of nodes out of each other's reception range, for which the only way of exchanging data is via in-range nodes acting as packet forwarders i.e., agreeing to relay packets on behalf of other nodes. However, packet forwarding costs extra energy and bandwidth, each being a scarce resource in wireless ad hoc devices. It has been conjectured for a few years [8,9,12,15,16] that under such circumstances, rational nodes with enough internal intelligence may try to save energy and bandwidth as much as possible, and the most obvious way of doing it is by refusing to forward packets. Such non-cooperative behavior is usually called *selfish*. Selfishness adds another unreliability factor besides those stemming from the very nature of wireless ad hoc communications. (Note that selfishness is to be distinguished from *malicious* behavior, as the latter brings its perpetrators no tangible benefit.)

Prevention, detection and/or mitigation of selfishness, as well as enforcement of cooperative behavior among MANET or WSN nodes have received considerable attention. Currently there exist a large number of solutions addressing the above goals. A promising class of them are *reputation-based schemes*, which rely on determination

J. Wozniak et al. (Eds.): WMNC 2009, IFIP AICT 308, pp. 185–196, 2009.

and sharing *reputation values* among all the network nodes or groups thereof. These values can next be used as input to a routing algorithm programmed to avoid end-to-end routes containing ill-reputed nodes; further provisions may have a node punish ill-reputed nodes by refusing to forward packets originated by them e.g., as in the *pathrater* mechanism [5] or in *Cooperative On-Demand Secure Route Protocol* [14].

Since MANET or WSN nodes can apply a wide diversity of packet forwarding strategies, detection of selfish nodes becomes a challenge. In this work we address the problem with the help of *Dempster-Shafer Theory of Evidence* (DST) [1, 2, 3, 4]. Our novel approach, called DST-SDF (*DST-based Selfishness Detection Framework*) differs from the existing ones in the following main respects:

- There is no need to overhear immediate neighbours nodes' transmissions to detect their cooperative or non-cooperative behavior – no additional tools to cover this functionality (such as the MAC-layer *watchdog* mechanism [5]) are needed in contrast with many known reputation-based systems [8, 9].
- Communication overhead is significantly reduced through an economy of scale – only data packets' source nodes are authorized to generate recommendations, each of which moreover pertains to a set of nodes, rather than a single one.
- Nodes' selfishness is evaluated based on evidence received both directly (as derived from successive packet acknowledgments or lack thereof) and indirectly via recommendation messages; while the idea sounds familiar, it is given a more systematic and consistent treatment using the subjective logic of DST.

The rest of the paper is organized as follows: Section 2 discusses related work and outlines some well-known methods of selfishness evaluation. Section 3 contains a brief introduction to DST and the methods of evidence combination under uncertain information, whereas Section 4 describes DST-SDF in more detail. Sample performance evaluation results obtained via simulation are reported in Section 5. Section 6 concludes and outlines future work.

2   Related Work

Enforcement of cooperative behavior in MANETs has been the subject of a number of works. Essentially, two types of schemes dealing with non-cooperative (selfish or, to a lesser extent, malicious) nodes are being proposed. The first type, based on micropayments in a virtual currency e.g., *Nuglets* [6] or *Sprite* [7], build in a direct way incentives for a node to reciprocate other nodes' forwarding services. Micropayments are conceptually attractive and flexible, as they assign quasi-monetary value to every single act of packet forwarding; however, they are usually too hard to implement in ad hoc networks, since they typically require tamper-proof hardware at each node or a trusted third party to ensure transaction security, and have difficulty handling inflationary/deflationary scenarios as well as the so-called topology handicap.

A more promising type of solutions are reputation-based schemes. The best known ones include *Cooperation of Nodes Fairness in Dynamic Ad-Hoc NeTworks* (CONFIDANT) [8], *Collaborative Reputation Mechanism (CORE)* [9], *Secure and Objective Reputation-based Incentive Scheme* (SORI) [10], *Observation-based Cooperation Enforcement in Ad Hoc Networks* (OCEAN) [11], *Reputation-based Mechanism for*

Isolating Selfish Nodes in Ad Hoc Networks [12], *Locally Aware Reputation System* (LARS) [13] and *Cooperative On-Demand Source Route Protocol* (COSR) [14]. They rely on generic concepts that can be briefly characterized as follows: each node gathers direct (first-hand) experience regarding the forwarding behavior of nodes it directly interacts with. Based on that information, it calculates local reputation values and possibly shares them (by dissemination of *recommendation messages*) with all other nodes in the network. The disseminated recommendation messages may account for behavior information extracted from previously received messages as well, thus incorporating indirect (second-hand) experience. Eventually, every node will have formed a reputation value regarding all the other nodes, whereupon it will be in a position to instruct the local routing algorithm to avoid non-cooperative nodes and/or to help ostracize such nodes by refusing to forward their packets.

Currently existing reputation schemes have two principal drawbacks, which our approach aims at overcoming: (1) Gathering first-hand experience involves mechanisms external to forwarding and routing, like the watchdog; it is employed in all the above mentioned reputation-based schemes except [12]. Each node is thus obliged to promiscuously overhear transmissions by its neighbor nodes to determine if they indeed have forwarded a received packet. Clearly, in a wireless collision domain with possible transmission power adjustment it is a faulty tool by nature [5]. Our DST-SDF approach eliminates it from the reputation scheme. (2) Mechanisms for non-cooperative (selfish) behavior detection typically are vulnerable to node collusion or DoS attacks, and hardly distinguish apparent non-cooperative behavior from real. Our approach does not remedy the problem completely, yet DST has a potential of marginalizing its effects if enough recommendation messages are exchanged.

Recently DST has received some attention as a mathematical background for reputation-based schemes, but the offered solutions [15, 16], are focused on aspects of information fusion and deceptive information distillation in their generality, and mostly address higher-layer (e.g., e-commerce) services. They are not specifically aimed at detecting non-cooperative forwarding, except COSR [14] that does address the problem. COSR however assumes that each network node is able to directly evaluate other nodes' behavior regarding forwarding and based on that information generates DST processed recommendations; the questions how to determine a specific node's behavior and in particular decide if it is selfish remain open.

3   Dempster-Shafer Theory of Evidence

DST, developed by A.P. Dempster and G. Shafer in the 1960s and 1970s [1, 2, 3, 4], offers an alternative to classical probability as a formal representation of uncertainty, and may be used to combine separate and independent pieces of evidence to quantify the belief in a given hypothesis. We decided to use DST as the underlying computational framework firstly because in the absence of direct mechanisms such as the watchdog, we deal with inherent uncertainty as to the behavior of other nodes; hence, detection of selfishness has to rely on incomplete evidence. Secondly, the incomplete evidence we are dealing with originates from multiple independent sources; as a consequence, there inevitably arise ambiguities and conflicting evidence (possibly, but not necessarily due to false recommendations), which DST handles in an intuitive way.

Hypotheses in DST are related to some universal set Θ and take the form of stating that a particular element x of Θ belongs to a set $X \subseteq \Theta$. Belief in a hypothesis derives from a DST primitive called *basic probability assignment* (bpa). It is a function mapping the powerset of Θ onto the interval [0, 1] i.e., m: $2^\Theta \to [0, 1]$, with the normalization constraint satisfied over the entire powerset. That is, with each $X \subseteq \Theta$ (i.e., $X \in 2^\Theta$) is associated a real number $m(X)$ between 0 and 1 inclusive that measures the amount of trust we put in the claim that (1) $x \in X$, and (2) no evidence supports a stronger hypothesis that $x \in X'$ for an $X' \subset X$. By convention, $m(\varnothing) = 0$ and

$$\sum_{X \in 2^\Theta} m(X) = 1. \tag{1}$$

Belief, or evidence value, associated with X is then defined as

$$ev(X) = \sum_{X' \in 2^\Theta | X' \subseteq X} m(X'). \tag{2}$$

As an example consider a network node that can be designated as selfish or nonselfish. Thus the universal set $\Theta = \{SELFISH, NONSELFISH\}$. Assuming there is enough information to claim that the node is selfish with probability 0.1 and nonselfish with probability 0.9, we can create the following bpa:

$$m(X) = \begin{cases} 0.1, & X = \{SELFISH\} \\ 0.9, & X = \{NONSELFISH\}. \end{cases} \tag{3}$$

Such an assignment can be regarded as a classical probability distribution over Θ. However, one might just as well assign a probability of 0.9 to not knowing at all whether the node is selfish or not. In that case we get

$$m(X) = \begin{cases} 0.1, & X = \{SELFISH\} \\ 0.9, & X = \{SELFISH, NONSELFISH\} \end{cases} \tag{4}$$

and the resulting distribution of evidence values, namely $ev(\{SELFISH\}) = 0.1$ and $ev(\{NONSELFISH\}) = 0$, is no longer a probability distribution over Θ.

A useful feature of DST is the formalism to express the bpa associated with a subset of Θ through the bpa's associated with other subsets of Θ. This enables combination of (possibly conflicting) evidence obtained from multiple sources into a new bpa. Several evidence combination rules have been proposed; hereafter we restrict attention to Dempster's and mixing combination rules [4].

3.1 Dempster's Combination Rule

Given two pieces of evidence in the form of bpa's m_1 and m_2 over 2^Θ, the resulting bpa for a set $X \subseteq \Theta$ is defined as

$$m(X) = (m_1 \oplus m_2)(X) = \frac{\displaystyle\sum_{Y, Z \in 2^\Theta | Y \cap Z = X} m_1(Y) m_2(Z)}{1 - \displaystyle\sum_{Y, Z \in 2^\Theta | Y \cap Z \neq \varnothing} m_1(Y) m_2(Z)}, \tag{5}$$

Coming back to our example, let m_1 be as in (4) and

$$m_2(X) = \begin{cases} 0, & X = \{SELFISH\} \\ 0.5, & X = \{NONSELFISH\} \\ 0.5, & X = \{SELFISH, NONSELFISH\}. \end{cases} \tag{6}$$

Based on (5) it is easy to calculate

$$(m_1 \oplus m_2)(\{SELFISH\}) = \frac{0+0+0.05}{1-(0.05+0)} \approx 0.0526,$$

$$(m_1 \oplus m_2)(\{NONSELFISH\}) = \frac{0.45+0+0}{1-(0.05+0)} \approx 0.4736, \tag{7}$$

$$(m_1 \oplus m_2)(\{SELFISH, NONSELFISH\}) = \frac{0.45}{1-(0.05+0)} \approx 0.4736$$

.

3.2 Mixing Combination Rule

The general formula for the mixing combination rule proposed in [4] is somewhat simplistic; for two given basic probability assignments m_1 and m_2 over 2^Θ :

$$(m_1 \oplus m_2)(X) = \frac{1}{2}(m_1(X) + m_2(X)). \tag{8}$$

Regarding our example with m_1 as in (3.4) and m_2 as in (3.6), the resulting bpa is:

$$(m_1 \oplus m_2)(X) = \begin{cases} 0.05, & X = \{SELFISH\} \\ 0.25, & X = \{NONSELFISH\} \\ 0.7, & X = \{SELFISH, NONSELFISH\}. \end{cases} \tag{9}$$

DST-SDF being intended to detect selfishness as it varies in time, one would like to emphasize on nodes' most recent behavior compared to the remote past, which is typically achieved by way of the EWMA (*Exponentially Weighted Moving Average*) algorithm. Therefore to emphasize a newer piece of evidence reflected by m_2 we modify current bpa m_1 using a learning constant $r \in [0,1]$ similarly as in [19]:

$$(m_1 \oplus m_2)(X) = r \cdot m_1(X) + (1-r) \cdot m_2(X). \tag{10}$$

4 DST-SDF

DST-SDF aims at detection of selfish nodes regarding packet forwarding. The concept of gathering direct (first-hand) experience dispenses with the watchdog and relies on end-to-end acknowledgments instead. One requirement it poses is that a source-to-destination route be known in advance to the source node, as satisfied e.g., by *Dynamic Source Routing* (DSR) [17]. Each time a source node S wishes to send a packet to a destination node D, a route $p_{S,D}$ from S to D of length $L_{S,D}$ is selected, consisting of a set $N_{S,D}$ of intermediate nodes. Having sent the packet, node S waits for an acknowledgement from node D. If it arrives within a predefined time, node S has reason to claim that no node on $p_{S,D}$ is selfish. Otherwise if there are no other indications of the route's faultiness, node S knows that there are selfish nodes on $p_{S,D}$. In either case, a recommendation message is sent out to inform the other nodes all over the network about the detected situation (selfish or cooperative behavior of the nodes on $p_{S,D}$, respectively).

Every network node is equipped with a dedicated *Evidence Manager Component* (EMC) executing a DST-based algorithm responsible for detection of selfish nodes based on the input information of two types: direct i.e., the node's own experience (arrival/lack of arrival of packets' acknowledgements), and indirect i.e., gathered from received recommendation messages. Inside the EMC, behavioral data for each node are converted and maintained in the form of bpa. Current evidence values for all the nodes are stored in an *Evidence Storage Component* (ESC). When a node becomes operational (joins the network) and before it receives input information (direct or indirect) for the first time, arbitrary initial bpa's are created. Throughout the node's lifetime within the network, they are updated according to subsequent input events (i.e., reception of direct or indirect behavioral information regarding other nodes). There are two modes in which EMC component can operate: DEC (*Dempster's Evidence Combination*) mode with Dempster's combination rule (5) in use, and MEC (*Mixing Evidence Combination*) mode with mixing combination rule (10) in use. EMC output data can be fed into the routing protocol's mechanisms to punish nodes designated as selfish similarly as in [8, 9].

4.1   Direct Information

At the outset, every network node sets initial bpa's for all the other nodes:

$$curr\_m_{Sj}(X) = \begin{cases} 0, & X = \{SELFISH\} \\ 0, & X = \{NONSELFISH\} \\ 1, & X = \{SELFISH, NONSELFISH\}, \end{cases} \tag{11}$$

where $curr\_m_{ij}$ denotes the current bpa at node i regarding node j's status. These arbitrary initial assignments simply tell node i that since it has no information about node j, it should treat node j as *SELFISH* or *NONSELFISH* with the same degree of uncertainty. As mentioned earlier, every time a source node S sends a packet to a destination node D and receives an acknowledgment in time, node S is certain that all nodes along the selected route $p_{S,D}$ have behaved cooperatively. Thus node S creates a new bpa for each node $j \in N_{S,D}$:

$$new\_m_{Sj}(X) = \begin{cases} 0, & X = \{SELFISH\} \\ 1, & X = \{NONSELFISH\} \end{cases} \tag{12}$$

and updates the current bpa's:

$$curr\_m_{Sj} := curr\_m_{Sj} \oplus new\_m_{Sj}, \tag{13}$$

where $\oplus$ denotes the evidence combination operator – Dempster's (5) or mixing (10) dependent on DST-SDF working mode.

If no acknowledgment for the packet arrives within the predefined time and there is no other indication (e.g. RERR – *Route Error Message*) that route $p_{S,D}$ is invalid or packet is lost, node S can only claim that there are selfish nodes in $N_{S,D}$. It does not know which nodes in $N_{S,D}$ are selfish, of course, nor does it know how many selfish nodes there are. Yet there is no doubt that relative to the particular packet in question, only one node in $N_{S,D}$ has behaved selfishly (other selfish nodes in $N_{S,D}$, if any, did not have a chance to manifest their selfishness for the packet did not reach them). Therefore, while one can imagine any kind of assumptions as to the number of selfish nodes in $N_{S,D}$, our approach relies on the simplest one: if no acknowledgement for a packet sent over $p_{S,D}$ has arrived in time, only one selfish node in $N_{S,D}$ has been experienced. Furthermore, the new bpa for the hypothesis that a particular node j in $N_{S,D}$ has just been experienced to be selfish is taken to be proportional to the current bpa for that node:

$$P_j = \frac{curr\_m_{Sj}(\{SELFISH\})}{\sum_{k \in N_{S,D}} curr\_m_{Sk}(\{SELFISH\})}.$$

(14)

(In view of these somewhat arbitrary and simplifying assumptions, it is appropriate to stress that our approach is expected to provide efficient detection of selfishness in the first place, generality and conceptual elegance being secondary considerations.)

Hence, should no packet acknowledgment arrive in time from node D, the following new bpa's will be created at node S regarding each node $j \in N_{S,D}$:

$$new\_m_{Sj}(X) = \begin{cases} P_j, & X = \{SELFISH\} \\ 1 - P_j, & X = \{SELFISH, NONSELFISH\}. \end{cases}$$

(15)

Node S next updates its current bpa's for each node $j \in N_{S,D}$ according to (4.3).

4.2 Indirect Information

Whenever a packet's source node receives within a predefined time an acknowledgment for a packet sent over $p_{S,D}$ or observes the predefined time expired, it spreads a suitable recommendation message all over the network. The message lists the set $N_{S,D}$ and contains an indication of the respective route's behavior status that can assume one of two values: *NONSELFISH* (if the acknowledgment has arrived) or *SELFISH* (otherwise). An important point to note is that unlike in traditional reputation-based systems, only packets' source nodes ever spread out recommendation messages. When a given node i receives from another node a recommendation message, it builds bpa's for all the nodes listed therein i.e., for $j \in N_{S,D}$, based on the route's behavior indication. Since there is uncertainty related to recommendations, they should not be treated in the same way as a node's direct experience. They ought to have smaller influence on the current bpa's *curr_m*. Our solution weighs incoming indirect information using a factor $u \in [0,1]$ that reflects how much trust a recipient of a recommendation message puts in it. In general, u can be different for each recommendation message (e.g., depending on the source node). If the route's behavior indication is *NONSELFISH* then

$$new\_m_{ij}(X) = \begin{cases} u, & X = \{NONSELFISH\} \\ 1-u, & X = \{SELFISH, NONSELFISH\}, \end{cases} \tag{16}$$

whereas if the route's behavior indication is *SELFISH* then

$$new\_m_{ij}(X) = \begin{cases} uP_j, & X = \{NONSELFISH\} \\ 1-uP_j, & X = \{SELFISH, NONSELFISH\}. \end{cases} \tag{17}$$

Node i next updates its current bpa *curr_m* for all nodes in $N_{S,D}$ according to (13). Finally, in the spirit of DST and since the hypotheses of interest are associated with singleton subsets of Θ, we have node i consider node j as:

- selfish, if $curr\_m_{ij}(\{SELFISH\}) \geq T$,
- nonselfish, if $curr\_m_{ij}(\{NONSELFISH\}) \geq T$, and
- undefined, if $curr\_m_{ij}(\{SELFISH\}) < T$ and $curr\_m_{ij}(\{NONSELFISH\}) < T$,

where $T \in (0.5, 1]$ is a selfishness threshold. It is very important to come up with an appropriate T value. Too low T contributes to false accusations, whereas too high T lengthens the time needed to detect selfish nodes and in the worst case can prevent DST-SDF from determining nodes' selfishness at all.

5 Simulation

In this section we briefly address via simulation the issues of convergence (how long it takes to detect all selfish nodes) and robustness to false recommendations. DST-SDF is implemented using the *j-sim* tool [18] in a simulation environment composed of IEEE 802.11-based nodes. The simulated scenario features 100 nodes arranged on a grid with each node pair's reception range confined to one hop, with source-to-destination routes containing L hops on average. We let a subset consisting of *SN* network nodes refuse to forward any packets. In the three tested scenarios SN amounted to 5%, 10%, and 15% of the network nodes. The other nodes were assumed to unconditionally forward all packets they were requested to. Both DEC and MEC modes featured $u = 0.9$. The threshold T was optimized experimentally and set to the lowest level guaranteeing that DST-SDF caused no node to be misdetected as selfish. Accordingly, $T = 0.95$ was set for DEC mode and $T = 0.6$ for MEC mode; additionally, $r = 0.4$ was set in MEC mode. The DST-SDF efficiency is presented in Fig. 1 (DEC mode) and Fig. 2 (MEC mode) for $L = 10$ and $L = 5$.

The simulations show that under Dempster's combination rule all the selfish nodes are detected much faster compared to the mixing combination rule. This is because Dempster's combination rule has a way of rejecting conflicting information about any particular node. In MEC mode conflicting information does enter the calculated evidence values, hence one needs more evidence to find out about the *NS* nodes – around 150 data packets in total are needed to be sent by all the network nodes in DEC mode compared to twice as many in MEC mode. Note that the convergence time is hardly sensitive to the average route length in either mode.

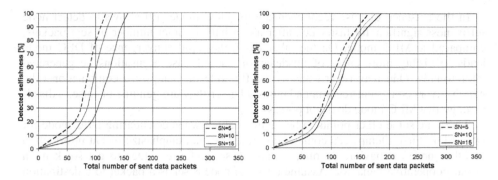

Fig. 1. Efficiency of selfishness detection in DEC mode; $L = 10$ (left), $L = 5$ (right)

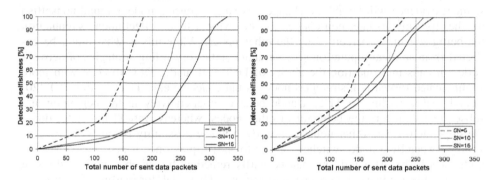

Fig. 2. Efficiency of selfishness detection in MEC mode; $L = 10$ (left), $L = 5$ (right)

To demonstrate how robust DST-SDF is to false recommendations we let a certain proportion of nodes, ranging from 0 to 30%, deliberately act in reverse: all source nodes within this group spread a recommendation message with *SELFISH* route's behavior indication whenever a packet acknowledgment arrives in time and *NON-SELFISH* otherwise. Simulation results for $L = 10$ (Fig. 3) show that DST-SDF is moderately sensitive to *SN*.

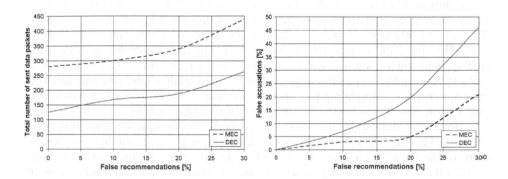

Fig. 3. Influence of false recommendations on detection time (left), false accusations (right)

In the presence of false recommendations, DEC mode still allows to detect node selfishness significantly faster i.e., after about half the total number of sent packets, compared to MEC mode. On the other hand, the rapid convergence of DEC mode leads to visibly more cases of misdetection should false accusations be generated intentionally. One may regard this as the flip side of Dempster's combination rule, which favors consistent recommendations whether they are genuine or false, a disadvantage that MEC mode does not possess.

There are some other points to be cautious about when considering DEC mode for DST-SDF. First, Dempster's combination rule does not emphasize recent information of other nodes' behavior. Second, it may cause evidence values to get stuck upon initial cooperative behavior. Assume a source node S sends a packet to a destination node D over route $p_{S,D}$ and receives in time an acknowledgment. Combining initial bpa (11) with the new one (12), node S arrives at $curr\_m_{Sj}(\{SELFISH\}) = 0$. No subsequent evidence will be able to change this value – once a node is found nonselfish for the first time, its later non-cooperative behavior will not reflect upon its reputation for being selfish, as perceived by the other nodes. While this is what might be expected from an ideal reputation scheme under static forwarding strategies of network nodes, it calls into question the ability of DST-SDF under DEC to cope with more sophisticated forwarding strategies nodes could conceive e.g., TFT or Anti-TFT [20] (recall that according to TFT, a node behaves cooperatively in the beginning and subsequently mirrors the behavior of other nodes, whereas Anti-TFT dictates doing the opposite of what other nodes do; an even worse possibility is the "grim trigger" strategy whereby a node switches forever to selfish play once it perceives any other node as selfish). Although this can be partly remedied by suitable modifications of (12) and (14), we leave the present description of DST-SDF for reasons of clarity and space, leaving an enhanced version to a later paper. All things considered, it seems that a mixing combination rule using EWMA is a better solution for DST-SDF selfishness detection framework.

6   Conclusion and Future Work

This paper investigates selected aspects of detecting selfish forwarding behavior of MANET or WSN nodes. Against the numerous existing solutions to detection (and later punishment) of selfish nodes, the novelty of our DST-SDF framework consists in the application of DST combined with end-to-end acknowledgment-based gathering of first-hand behavior information. Preliminary simulations show that DST-SDF in both MEC and DEC modes allows to detect all selfish nodes in the network fairly quickly, after each node has sent as few as 2 to 5 packets on average. Yet a clear tradeoff between the speed of detection and robustness against false recommendations has been observed, stimulating further work on tailoring Dempster's combination rule and/or formation of new evidence to the needs of reputation schemes.

A number of simulation model extensions and DST-SDF optimizations have been undertaken but are not reported here for lack of space e.g., node mobility, cumulative acknowledgments to reduce the recommendation message overhead, or non-cooperative behavior regarding recommendation message forwarding. There are more serious impediments that have to be overcome. In particular, more work needs to be done on introducing appropriate weighting of recommendations, proper configuration of DST-SDF (e.g., of the T and u parameters) to ensure higher selfishness detection ratings in different network structures. A challenging issue is to accommodate node anonymity (lack of permanent identities), while retaining the main advantages of DST-SDF. In this context, protocols like *Anonymous Packet Forwarding* combined with a *Congestion Control Mechanism* [19] need to be re-considered.

Acknowledgment

Effort sponsored by the Air Force Office of Scientific Research, Air Force Material Command, USAF, under Grant FA8655-08-1-3018, also supported in part by the MNiSW Grant PBZ-MNiSW-02/II/2007.

References

1. Shafer, G.: A Mathematical Theory of Evidence. Princeton University Press, Princeton (1976)
2. Zadeh, L.A.: A Simple View of the Dempster-Shafer Theory of Evidence and its Implication for the Rule of Combination. The AI Magazine 7, 85–90 (1986)
3. Yager, R.R., Kacprzyk, J., Fedrizzi, M.: Advances in the Dempster-Shafer Theory of Evidence, New York. John Wiley & Sons, Inc., Chichester (1994)
4. Sentz, K., Ferson, S.: Combination of Evidence in Dempster-Shafer Theory. SAND2002-0835 Technical Report. Sandia National Laboratories, New Mexico (2002)
5. Martini, S., Giuli, T.J., Lai, K., Baker, M.: Mitigating routing misbehavior in mobile ad hoc networks. In: Proceedings of the 6th annual international conference on Mobile computing and networking, pp. 255–265 (2000)
6. Buttyan, L., Hubaux, J.P.: Nuglets – A Virtual Currency to Stimulate Cooperation in Self Organized Mobile Ad Hoc Networks, Technical Report DSC/2001, EPFL, Lausanne (2001)
7. Zhong, S., Chen, J., Yang, R.: Sprite: A Simple, Cheatproof Credit-Based System for Mobile Ad hoc Networks, In: Proceedings IEEE Infocom, San Francisco CA (2003)
8. Buchegger, S., Le Boundec, J.Y.: Performance Analysis of the CONFIDANT Protocol: Cooperation Of Nodes – Fairness In Distributed Ad-hoc NeTworks. In: Proceedings of IEEE/ACM MobiHOC, Lausanne, Switzerland, pp. 226–236 (2002)
9. Michiardi, P., Molva, R.: CORE: A Collaborative Reputation Mechanism to Enforce Node Cooperation in Mobile Ad Hoc Networks. In: Proceedings IFIP TC6/TC11 Sixth Joint Working Conference on Communications and Multimedia Security: Advanced Communications and Multimedia Security, Portoroz, Slovenia, pp. 107–121 (2002)
10. He, Q., Wu, D., Khosla, P.: SORI: A Secure and Objective Reputation-Based Incentive Scheme for Ad Hoc Networks. In: Proceedings IEEE Wireless Communication and Networking Conference, Pittsburgh, PA, USA, pp. 825–830 (2004)
11. Bansal, S., Baker, M.: Observation-Based Cooperation Enforcement in Ad Hoc Networks, Technical Report Arxiv preprint cs. NI/0307012 (2003)
12. Refaei, T.M., Srivastava, V., Dasilva, L., Eltoweissy, M.: A Reputation-based Mechanism for Isolating Selfish Nodes in Ad Hoc Networks. In: Proceedings of the Second Annual International Conference on Mobile and Ubiquitous Systems: Networking and Services, San Diego, CA, USA, pp. 3–11 (2005)

13. Hu, J., Burmeister, M.: LARS: A Locally Aware Reputation System for Mobile Ad Hoc Networks. In: Proceedings of the 44th annual Southeast Regional Conference, Melbourne, Florida, pp. 119–123 (2006)
14. Fie, W., Yijun, M., Benxiong, H.: COSR: Cooperative On-Demand Source Route Protocol in MANET. In: Proceedings of the International Symposium on Communication and Information Technologies, Bangkok, Thailand, pp. 890–893 (2006)
15. Bin, Y., Munindar, P.S.: Detecting Deception in Reputation Management. In: Proceedings of the second international joint conference on Autonomous agents and multiagent systems, Melbourne, pp. 73–80 (2003)
16. Dong, Y., Deborah, F.: Alert Confidence Fusion in Intrusion Detection Systems with Extended Dempster-Shafer Theory. In: Proceedings of the 43rd ACM Southeast Conference, Kennesaw, GA, USA (2005)
17. Johnson, D.B., Maltz, D.A.: Hu Yih-Chun: The Dynamic Source Routing Protocol for Mobile Ad Hoc Networks, DSR (2004),
   `http://www.ietf.org/internet-drafts/`
   `draft-ietf-manet-dsr-10.txt`
18. `http://www.j-sim.org`
19. Konorski, J., Orlikowski, R.: Distributed Reputation System for Multihop Mobile Ad Hoc Networks. In: Proceedings of the 5th Polish-German Teletraffic Symposium, PGTS 2008, Berlin, Germany, pp. 161–167 (2008)
20. Felegyhazi, M., Hubaux, J.P., Buttyan, L.: Nash equilibria of packet forwarding strategies in wireless ad hoc networks. IEEE Transactions on Mobile Computing 5, 463–476 (2006)

Competitive Spectrum Sharing in Wireless Networks: A Dynamic Non-cooperative Game Approach

Omar Raoof, Zaineb Al-Banna, and H.S. Al-Raweshidy

Brunel University, Wireless Networks and Communications Centre (WNCC),
London, UB8 3PH, UK
{Omar.Raoof,Zaineb.Al-Banna}@brunel.ac.uk

Abstract. "Game Theory" is a promising mathematical tool to improve the utilization of radio frequency spectrum in wireless networks. In this paper, we consider the problem of spectrum sharing between a primary user and a group of secondary users. We formulate our solution in such a way that one of the secondary users will be a secondary primary user by sharing the spectrum offered from the primary user and offer his share to be shared by the rest of the secondary users in the network. A theoretical non-cooperative game model is introduced to study node behavior in wireless networks based on their reputation. The only way for a node to guarantee a share in the spectrum is by enhancing its reputation, which is done by serving other nodes in the network. Game theory can be used by individual selfish nodes to determine their optimal strategy for participation level in the network. Furthermore, game theory produces information about the overall nature of nodes' interaction and system efficiency, showing how system efficiency can be improved.

Keywords: Spectrum sharing, game theory, spectrum pooling, non-cooperative game.

1 Introduction

Today's wireless communication occupies the bulk of the research in the field of communications because of the increasing demand of wireless services. It has been predicted that in the next five years, about 55% of all users will access the Internet wirelessly [1] which leads to the need for more efficient use of frequency, where the scarce resources require wisdom of distribution and utilization. Cognitive technique is considered a candidate for the optimum use of the frequency spectrum.

Spectrum users can be classified to be Primary users (PU) and Secondary users (SU) (i.e. licensed and unlicensed users respectively). However the licensed frequency is not used completely by the primary users there are holes left in the licensed band. In a cognitive radio system, secondary users seek for these holes and opportunistically use them. The benefit then returns to both primary user which can get revenue from selling its own spectrum, and secondary users by subleasing available channels. In [2] the author mentioned that Neel and Reed have studied how to apply game theory to a radio as a member of a cognitive network. With game theory they were able to analyze protocols to determine whether cognitive radio behaviour

J. Wozniak et al. (Eds.): WMNC 2009, IFIP AICT 308, pp. 197–207, 2009.

could result in a stable network behaviour, and they conclude that it can. Generally, the PU's are the license holders and the SU are allowed to use frequency band when the PU are inactive however they are required to vacate when the PU are active again. The arrival of PU causes the SU to lose that band and therefore link maintenance become a key issue. Spectrum pooling (shown in figure 1) is one of the solutions to this problem, where the entire band is divided into large number of sub-channels in such a way that they lie in a different PU's bands. This solution minimizes the chance of losing the entire sub-channel upon the arrival of the PU, as no PU can cause the complete breakdown of the SU link.

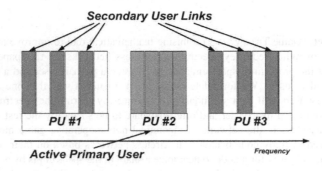

Fig. 1. Spectrum Pooling

In this paper we used game theory to model the utilization of the spaces and holes which were found in the licensed band in a competitive manner among multi secondary users, as shown in figure 1. It is assumed that the search and detection of vacancies within the licensed spectrum is already carried out by implementing the methods of spectrum sensing. Then secondary users will compete to share the spectrum frequency with the primary user according to a pricing function, which is computed by the primary user's administrator. It also depends on the dimensions similar to those which are considered in assessment of the frequency license as [2]:

a) Bandwidth required (in MHz).
b) Demand within the service region.
c) Duration of the contract.
d) Opportunity for growth of services within the service region.
e) Cost of installing and providing service.

The cost then has been assumed as units with five categories. Each SU takes one of these five categories and will compete with the other SU according to number of units it obtained. The first priority will be granted to the SU which has maximum number of units and assigned the first category and second priority will be granted to the SU which collected maximum units of all the rest and so on.

The rest of the paper is organized as follows; the following part of this paper presents some related work to this area, giving brief details of what have been achieved as well. Following from this, section '3' explains how to use game theory as a tool to predict and strategies node behaviour in wireless communications. It presents a game theoretical model that controls how SUs' access the shared spectrum. Nash equilibrium strategy is defined in section '4', and properties of this strategy are discussed in section '5'. Finally, conclusion and proposed future work are presented in the last section.

2 Related Works

An introduction to cognitive radio was provided in [3] where the fundamental cognitive tasks as well as the emergent behavior of cognitive radio were discussed. The problem of spectrum pricing in a cognitive radio network where multiple PU's providers compete with each other to offer spectrum access to SU's is presented in [4] and [5]. By using equilibrium pricing scheme, each of the primary service providers aims to maximize its profit under quality of service (QoS) constraint for PU's. The authors adopted a utility function to obtain the demand function. The stability of the proposed dynamic game algorithms in terms of convergence to the Nash equilibrium is studied. However, Nash equilibrium is not efficient in the scene that the total profit of the primary service providers is not maximized. Finally, PU's may deviate from the optimal solution reached by the Nash equilibrium stage; a punishment mechanism may be applied to the deviating primary service providers.

In [6] the problem of the spectrum sharing between the PU and SU were considered. The authors formulate the named problem as an oligopoly market competition and use a non-cooperative game to obtain the spectrum allocation for SU's. Numerically the authors investigated the stability conditions of the dynamic behavior for the spectrum sharing scheme.

3 Game Theory as a Tool to Predict and Strategize Spectrum Usage

In this section a theoretical game model is presented for the case where all the SU's can completely observe the strategies and the payoff of other SU's. Mainly, the SU's are fighting with each other in order to get to share the spectrum with the PU (*i.e. the players in the game are the secondary users*); here we assume that one of the SU's will try to be a secondary PU (*i.e. the SU will become responsible of the offered spectrum by the PU and will decide whether to share it with the other SU's by pooling the offered spectrum again or not*). In this case the SU will use the spectrum that the PU is offering and shares it with the other SU's simultaneously. This mechanism will save time and assures that all the SU will receive a fair share of the spectrum. A node with low priority traffic will not be interested in paying more money to share the spectrum with the PU and compete with the other SU's rather than sharing it with the SU, and the opposite works for the node with high priority traffic.

Game theory mostly attractive to analyze the performance of all kind of problems related to wireless networks. The reason is that it has the ability to model individual, independent decision makers whose actions potentially affect all other decision makers. Furthermore, game theory is a field of applied mathematics that describes and analyzes interactive decision situations. It consists of a set of analytical tools that predict the outcome of complex interactions among rational entities, where rationality demands a strict adherence to a strategy based on perceived or measured results [7].

Game theory studies strategic interaction in competitive, non-cooperative, and co-operative environments. Half a century old, it has already revolutionized economics, and is spreading rapidly to a wide variety of fields. In the early to mid 1990's, game theory was applied to networking problems including flow control, congestion control, routing and pricing of Internet services [7]. Non-cooperative and Cooperative games were introduced by John Nash (in papers between 1950- 1953), his most contribution was his existence proof of an equilibrium state in non-cooperative games, the Nash equilibrium [7]. More recently, there has been growing interest in adopting game-theoretic methods to model today's leading communications and networking issues, including power control and resource sharing in wireless and peer-to-peer networks. Moreover, leading computer scientists are often invited to speak at major game theory conferences, such as the World Congress on Game Theory 2000, 2004 and 2008 [8]. So far, evidence of previous researches shows that game theory may be an appropriate tool to solve some problems in communications systems; this section presents some of the most important concepts of game theory and some particular concepts which is very important in our proposal. Normally, any game G has three components: a set of players, a set of possible actions for each player, and a set of utility functions mapping action profiles into the real numbers. In this paper, the set of players are denoted as I, where I, is finite with, $i = \{1, 2, 3, \ldots \ldots, I\}$ }. For each player $i \in$ I is denoted by A_i the set of possible actions that player i can take, and A, which is denoted as the space of all action profiles is equal to:

$$A = A1 \times A2 \times A3 \times \ldots \times AI . \tag{1}$$

Finally, for each $i \in$ I, we have $U_i : A \to R$, which denotes i's player utility function. Another notation to be defined before carrying on; suppose that $a \in A$ is a strategy profile and $i \in I$ is a player; and then $ai \in Ai$ denote player $i's$ action in a and $a - i$ denote the actions of the other $I - 1$ plyers.

The most important equilibrium concept in game theory is the concept of Nash Equilibrium [7]. A *Nash equilibrium* is an action profile at which no user may gain by unilaterally deviating. So a *Nash equilibrium* is a stable operating point because no user has any incentive to change strategy. More formally, a Nash equilibrium is a strategy profile a such that for all $ai \in Ai$,

$$U(ai, a - i) \geq U(\tilde{a}i, a - i) . \tag{2}$$

The $\tilde{a}$, denote another action for the player $i's$.

Pareto efficiency is another important concept for our application of game theory. An action profile $a \in A$ is said to be Pareto if there is no action profile $\tilde{a} \in A$ such that for all i,

$$U(ai) \geq U(\tilde{a}i) . \tag{3}$$

In another word, an action profile is said to be Pareto efficient if and only if it is impossible to improve the utility of any player without harming another player.

In this paper, we assume that the network lifetime is infinitely long and divided into individual time periods, represented by t for each $t = 0, 1, 2, \ldots \ldots, \infty$. At each time period t a node sends a request for share in the spectrum from the PU. Other activities that each node follows are to share its requirement of the spectrum with the other nodes. By this the node will decide whether it's worth it to share the spectrum with the PU and become a secondary PU, or give a chance to the other nodes to do so.

In this paper we will represent the infinitely repeated version of game G (*i.e. this is the case when G is going to be played over and over again in successive time periods*) by G^∞. In this paper we are assuming that the PU is offering a single frequency band to be shared by other SU's. However, if the PU is planning to offer more bands then the proposed mechanism must be repeated for the other bands between the secondary users.

We will define the node reputation as R which will depends on the node performance during any time period t as well as in prior time periods. Reputation of player i in some time period t is denoted by R_t^i. Formally, we define node reputation as follows:

$$R_t^i = R_{t-1}^i (1-\alpha) + w \times \alpha \qquad 0 \leq \alpha \leq 1, t \geq 2 . \tag{4}$$

Where the following "α" is the history of the node, it depends on the node reputation in the previous time periods according to node movements. The "w" is equal to "1" when player i at time t is interested in sharing the offered spectrum and "0". Therefore, $0 \leq R_t^i \leq 1$, i.e. the reputation of each player is a value form "0" and "1" (including) ($R_t^i \in [0,1]$). Moreover, the reputation value of all players is equal to "0" when $t = 0$. A high value of α means the more importance is assigned to a player's need to share the spectrum with the PU (higher priority) in the current time period than its previous need record, and vice versa. Thus, when α is high, a node with even low reputation value in the current time period t, can significantly improve its reputation when it realises that it needs a better share of the spectrum.

4   Nash Equilibrium of the Repeated Game G^∞

As was defined the Nash equilibrium case earlier, the evaluation of the Nash equilibrium of the repeated game G^∞ will be engaged. By finding the Nash equilibrium of G^∞ it leads to the deduction of the Nash equilibria of G. The proposed incentive mechanism is based on a player's reputation R links, the benefit of which is that a player draws from the system to its contribution the benefit is a monotonically increasing function of a player's contribution. Thus, this is a non-cooperative game among the players, where each player with high priority traffic wants to maximize its utility. The classical concept of Nash equilibrium points a way out of the endless cycle of speculation and counter-speculation as to what strategies the players should use. The intent is to deduce a symmetric Nash equilibrium because all the players belong to the same population/network (i.e., assume the same role) and it is therefore easier (i.e., require no coordination among players) to achieve such an equilibrium. If the players in a game either do not differ significantly or are not aware of any differences among themselves (i.e., if they are drawn from a single homogeneous population) then it is difficult for them to coordinate, and a symmetric equilibrium, in which every player uses the same strategy, is more compelling.

The argument of a single homogeneous population implies that all the peers in a wireless mesh network have equivalent responsibilities and capabilities as everybody else. We assume that if the player chooses the action {want to share} to the access point, this will assign him a probability of p, and if the player chooses the action {does not want to share}, this will assign one a probability of $1 - p$.

It must be mentioned that in the action profile, all the players choose the action {does not want to share}; this is a time and money saving Nash equilibrium case. As this will mean that the nodes are not interested in sharing the spectrum for the entire communication time at all. That is to say nodes have low priority traffic and accessing the spectrum will be by chance, nodes will not compete to send their data and will not offer more money to the PU to get the spectrum. If any other player i decided to switch to the action {want to share} instead, its payoff will be $-C$ which is less than a payoff of "0" that the node gets when decide not to share the spectrum. If all the players choose the action {want to share}, this is an undesirable Nash equilibrium case. This is easy to see because all the nodes will have to compete against each other again, this will waste time and the winner will be the PU, as one of the SU's should offer more to share the offered spectrum.

The expected payoff of any player in time period t when it selects the action {want to share} is:

$$p\left(-C + R_t^{share} \times U\right). \tag{5}$$

This payoff is denoted as $Payoff_{share}$. Similarly, the payoff for any player selects the action {does not want to share} will be:

$$(1 - p)(R_t^{don'tshare} \times U). \tag{6}$$

This will be denoted as $payoff_{don't\,share}$. It can be easy to show that the term $R_t^{share} \times U$ captures the notation that the probability of SU becoming a secondary PU by sharing the offered spectrum is directly proportional to node's reputation.

R_t^{share} is player   reputation when it wants to share the offered spectrum at time t (i.e. $w = 1$ in equation 4), and $R_t^{don'tshare}$ is player i reputation when it decides to take the action {does not want to share} at the same time period t (i.e. $w = 0$ in equation 4), from equation 4, we can get:

$$R_t^{share} = R_{t-1}(1-\alpha) + \alpha$$

and

$$R_t^{don'tshare} = R_{t-1}(1-\alpha). \tag{7}$$

Generally, each player's expected payoff in equilibrium is its expected payoff to any of its actions that it uses with positive probability. The above useful characterization of mixed-strategy Nash equilibrium yields to;

$$payoff_{share} = payoff_{don'tshare}. \tag{8}$$

Using equations 5, 6, and 7;

$$p\left(-C + (R_{t-1}(1-\alpha) + \alpha) \times U\right) = (1 - p)(R_{t-1}(1-\alpha) \times U). \tag{9}$$

Solving equation 9 to get the final value of p;

$$p = \frac{R_{t-1} \times U \times (1-\alpha)}{-C + 2R_{t-1} \times U \times (1-\alpha) + U \times \alpha}. \tag{10}$$

It must be mentioned that the value p obtained above is not a constant, but varies in each time interval depending upon a node's reputation at the end of the previous time interval $t - 1$.

Finally, the mixed strategy pair $(p, 1 - p)$ for actions {want to share, does not want to share} respectively, is a mixed strategy Nash equilibrium for the players (i.e. nodes in the network). Assuming no collusion among nodes, if all the other nodes follow the above strategy, then the best strategy for any node is to also follow the above strategy. Actually, this is a symmetric mixed strategy Nash equilibrium for any G, as well as G^∞. In fact, it is a more stable equilibrium than the one in which no node is interested in sharing the offered spectrum. This is due to two reasons. First, when none of the nodes is interested in sharing the spectrum, the network is not useful to any user. Second, in practice nodes, which derive finite utility from altruism, would always send some messages irrespective of how much they obtain in return. Therefore, it is unlikely to have a scenario in which no node is looking to contact the PU to share the spectrum.

5 Properties of the Proposed Nash Equilibrium

In this section, we will present some of the interesting properties of the Nash equilibrium derived in the section above (equation 10).

5.1 Simplicity of Calculating the Nash Equilibrium

In section '4' we have calculated the probability of achieving the equilibrium point between the nodes based on which node will decide to share the spectrum with the PU and become a secondary PU. In each play of the game (or time period t), players (wireless nodes) based on their reputation at the end of the prior time period decide whether they should ask to share the offered spectrum or not. This probability as one can see does not remain constant from one period to another, and depends on a player's reputation at the end of the last time period. Players can calculate their reputation using equation 4, since they know precisely their actions at each play of the game. Thus, determining the Nash equilibrium strategy is fairly straightforward for any player. However, it must be noted that there is an inherent assumption that nodes get serviced based on their current reputation.

Figure 2, shows how players' reputations change in every time interval depending on their Nash strategy. At the beginning of the communication time, both, player 1 and 2 are competing with each other to guarantee access to the offered spectrum. However, player 1 uses the spectrum but at the same time managed to help player 2 (i.e. player 1 will be the secondary PU and will manage the access of players 2 and 3 to the offered spectrum; player2 will have a better chance as it has a better reputation). Player 3 shows his interest in the offered spectrum after the third time interval, and managed to use the spectrum once both player 1 and 2 finished using it or they are not interested anymore in sharing it. The figure shows the players (nodes) reputation values $0 \leq R_t^i \leq 1$ over ten time intervals.

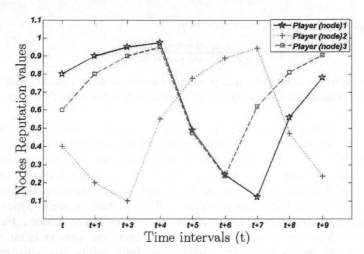

Fig. 2. Change in player's reputation controlled by their Nash equilibrium strategies

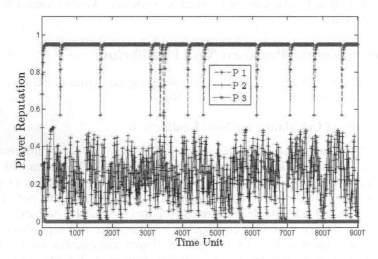

Fig. 3. Changing player reputation over a longer time period

On the other hand, figure 3 above shows the same result but over a longer time period, around nine hundred time intervals. Similarly, three nodes are competing with each other, with player one with highest reputation and player three with the lowest. Player 1 will act as the secondary PU over the other two users (i.e. player 2 and 3). In this figure we used a random matrix generator to show different reputation when player 1 is interested to share the spectrum for 80% of the time, player 2 for 50% of the time and player 3 for 8% of the time only.

5.2   Addressing the Spectrum to the Right Node Faster

The simple game theoretic model presented in this paper, wherein node reputation is used as a basis for deciding who will share the offered spectrum, predicts that it is in every peer's best interest to serve others. This includes the nodes that are not interested to share the spectrum at the current time period. Our simulations support this behavior as we found that the total service received by a node is balanced by the total service that it has to offer to others, as shown in Figure 2.

5.3 Addressing the Problem of Competitive Sharing

An important property of the equilibrium emerges from equation 10 that predicts the probability with which one node will be a secondary PU and it should serve others. If we set the value of C in away such that, $C <<< U$ (i.e. C can be ignored from equation 10), then equation 10 would be;

$$ p = \frac{R_{t-1}(1-\alpha)}{2 R_{t-1}(1-\alpha)+\alpha}. \tag{11} $$

That would lead us to the conclusion that $p < 0.5$, hence showing that if the cost of providing a service to other nodes (i.e. sharing the shared spectrum), Nash equilibrium of the proposed game predicts that players should help each other less than 50 percent of the time when the spectrum is offered by the PU. This, although it appears to be very restrictive, is a consequence of the fact that all nodes are selfish and are better off trying to share the spectrum than serving others. Intuitively, if a node knows that everyone else in the network behaves selfishly, i.e., provide as little service as possible, then the best strategy for the named node cannot be to serve others most of the time (i.e., with probability greater than 0.5).

5.4 Fairness, and Equal Sharing of Cost and Spectrum

We concluded from the previous section that serving with a priority of less than 50 percent (i.e. when $C <<< U$) is an optimal point, the observer can notice that the overall system efficiency is severely reduced. This is because most of the nodes in the network act selfishly and at least half of the service requests from other nodes will are not fulfilled. On the other hand, this equilibrium strategy provides fairness in the sense that the cost of system inefficiency is not burn by a single node (i.e. has one positive side), but it is shared among all nodes. This is because each node's request is likely to be turned down by the serving node (i.e. selfish secondary PU). In this paper, we assume that if a node's request at one node is turned down, the node tries at some other candidate node capable of serving the request. On average, the probability that a node's request is successfully served in a time period is proportional to its current reputation.

5.5 Decreasing α for a Better Share of the Spectrum

Figure 4 shows the effects of α on the reputation probability of the nodes in the case where the node is not interested in sharing the spectrum. On the other hand, the node in figure 5 is looking to keep its share of the spectrum (derived from equation 7).

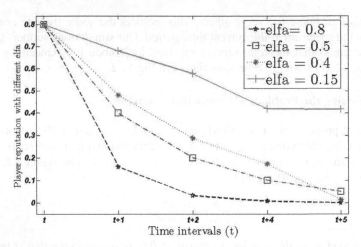

Fig. 4. Players reputation with respect to ∝, and the node is not interested in sharing the offered spectrum

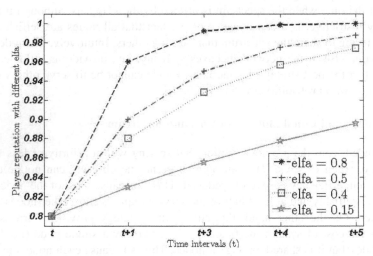

Fig. 5. Players reputation with respect to ∝, and the node is definitely interested in sharing the offered spectrum from the PU

As can be seen from figure 4 and 5, a lower value of α shifts the reputation probability curve upwards. However, that all depends on whether the node is interested in using the offered spectrum or not. If the node is looking to give its share of the spectrum to the other nodes, a low value of ∝ will gradually help the node to lose its share, however a high value of ∝ will guarantee a faster release of the spectrum. This is true for figure 5 as well, which is to be expected since ∝ determines how much importance is given to a node's current performance as compared to its past service record. A low value of ∝ (i.e., giving more importance to nodes past actions up to the current time period t) means that nodes need to continually provide service to be able to maintain high reputation and access spectrum offered from the PU. If however ∝ is high, nodes can easily increase their reputation in any period in which they provide service to other nodes. This is irrespective of how cooperative they have been in the past with regards to providing service to others. Therefore a simple way to improve the system efficiency is to set $\propto$ as low as possible.

6 Conclusion and Future Work

In this paper, we have proposed a simple non-cooperative game mechanism based on a node's reputation to overcome the competitive spectrum sharing problem prevalent in wireless networks. The model presented here addresses the spectrum sharing problem as it predicts that even for selfish users serving others is the best strategy. Game theory is used to predict the optimum (Nash equilibrium) strategies of selfish nodes such that their profits are maximized. Game theory is also used to provide valuable insight into the behavior of individual nodes, as well as the performance of the overall system. Interestingly, game theory provided us proof for some of the intuitive results, such as the strategy of serving less than 50 percent of the times when the cost of providing service is negligible. The proposed game theoretic solution of the spectrum sharing problem has several significant advantages; fairness, simple implementation, and ease of calculating optimum strategies. For simplicity we assumed that the cost and utility attached with serving and obtaining services is the same for all service types and for all users. In future, developing more elaborate models that take into account the heterogeneity of service types and users would be desired. Also, we would investigate the applicability of game theory in cognitive radio systems when heterogeneity is the overriding factor in the designing of protocols for system operations.

References

1. Nagaraj, S.V.: Entropy based spectrum sensing in cognitive radio. San Diego State University, 5500 Campanile Dr., San Diego (2008)
2. Arslan, H. (ed.): Cognitive Radio, Software Defined Radio, and Adaptive Wireless Systems, University of South Florida, Tampa. Springer, Heidelberg (2007)
3. Haykin, S.: Cognitive radio: Brain-empowered wireless communications. IEEE J. Select. Areas Comm. 23(2), 201–220 (2005)
4. Niyato, D., Hossain, E.: Competitive Spectrum Sharing in Cognitive Radio Networks: A Dynamic Game Approach. IEEE Transactions on Wireless Communications 7(7), 2651–2660 (2008)
5. Niyato, D., Hossain, E.: Equilibrium and Disequilibrium Pricing for Spectrum Trading in Cognitive Radio: A Control-Theoretic Approach. In: IEEE GLOBECOM (2007)
6. Niyato, D., Hossain, E.: Competitive Pricing for Spectrum Sharing in Cognitive Radio Networks: Dynamic Game, Inefficiency of Nash Equilibrium, and Collusion. IEEE Journal on Selected Areas in Communications 26(1), 192–202 (2008)
7. MacKenzie, A.B., DaSilva, L.A.: Game Theory for Wireless Engineering, 1st edn. (2006)
8. Games 2008, Third World Congress of the Game Theory Society, http://www.kellogg.northwestern.edu/meds/games2008/

A Novel AoA Positioning Solution for Wireless Ad Hoc Networks Based on Six-Port Technology

Peter Brida[1], Norbert Majer[1], Jan Duha[1], and Peter Cepel[2]

[1] Dept. of Telecommunications and Multimedia, University of Zilina, Univerzitna 1,
010 26 Zilina, Slovakia
[2] Siemens Program and System Engineering s.r.o., Telco & Media Mobile, Hurbanova 21,
010 01 Zilina, Slovakia
{Peter.Brida,Norbert.Majer,Jan.Duha}@fel.uniza.sk,
Peter.Cepel@siemens.com

Abstract. Mobile device positioning in ad hoc networks is a crucial issue due to the network properties and hardware limitations of particular devices (nodes). In this paper, a new conception of mobile device based on software defined radio architecture for mobile positioning in ad hoc networks is proposed. Six-port technology is implemented into our software defined radio receiver. Angle of arrival localization method as self positioning approach is implemented. The aim of the paper is to present an interaction of specific positioning method with proposed receiver conception. Moreover, we analyze various situation scenarios and their impact on the positioning accuracy, e.g., mutual position of the nodes, radio channel characteristics.

Keywords: Angle of arrival method, mobile positioning, positioning error, wire-less ad hoc networks, software defined radio receiver, six-port technology.

1 Introduction

Positioning in wireless ad hoc networks plays a major role in the development of geographic aware routing and multicasting protocols that result in new more efficient ways for routing data in multihop networks. The positioning in these net-works is specific. It is given by network properties and hardware limitations of network devices. Most of proposed positioning methods for ad hoc networks are based on actual capabilities of devices; therefore the positioning accuracy is not excellent [1], [2], [3] and [4]. The capabilities of individual devices are very limited at the present time. In the near future, the practical use of receivers based on Software Defined Radio (SDR) will increase abilities of given devices and also facilities of offered applications. Implementation of SDR is necessary in ad hoc net-works, which could work at the different frequency bands and they are based on various communication platforms. This application could contribute for improvement of particular functions, e.g. positioning of devices in the networks.

In this paper, we propose the application of SDR based on six-port technology determined for wireless ad hoc networks positioning. Our proposal is based on the Angle of Arrival (AoA) positioning method. The combination of SDR and Six-Port

J. Wozniak et al. (Eds.): WMNC 2009, IFIP AICT 308, pp. 208–219, 2009.
© IFIP International Federation for Information Processing 2009

Technology (SPT) provides a great flexibility in system configuration, a significant reduction in system development cost, and also a high potential for soft-ware reuse. We decided to use this receiver solution because we think that SDR means future in the wireless communication. Selection of positioning method results from good ability to measure phase difference i.e. to determine angle of arrival by means of SPT. On the other hand, this method is only exemplar solution for positioning utilization. Implementation of SDR in ad hoc networks is not new idea, but conception of SDR, AoA and ad hoc network is solution of the future. Many researchers have paid an attention on the issue of angle of arrival estimation by six-port technology [13], [14] and [15]. But any of the works investigated the impact of radio channel on a node mobile positioning. It is aim of this paper, the investigation of radio channel impact on mobile positioning.

This paper depicts untraditional realization of AoA method because the pro-posed method is implemented as a self positioning solution. In the case that localized node is realized as SDR. The AoA method is implemented as remote-positioning solution in almost previous works [5], [6], [7] and [8].

The rest of the paper is structured as follows. The following section introduces positioning by angle of arrival method. Then SDR conception and SPT is presented. Implementation of the AoA method into SDR is presented. Simulation results are presented and discussed. Conclusion concludes the paper.

2 Angle of Arrival Positioning Method

Node with the information about position coordinates is called "Reference Node (RN)" and node without this information is called "Blindfolded Node (BN)". AoA method represents measuring the angle of arrival of a signal propagating from RN to a BN or vice versa. Single measurement produces a straight line between two mentioned nodes.

Position estimation is determined as the point of intersection of minimal two straight lines drawn from the RNs (different locations).

The information about AoA can be obtained by means of following approaches. The first fundamental approach is based on measuring of the incoming signal phase differences. Another approach is based on a beam-forming.

In our case, we decided to use the phase difference measuring approach. Mod-eled blindfolded node utilizes six-port technology because of its positive proper-ties for simple phase differences measuring and on the other hand the six-port technology can be also necessary part of SDR receiver. The principle of phase measuring by means of six-port technology will be described later.

3 Software Defined Radio

Software defined radio has been identified as one of the potential methods to enhance the flexibility of wireless communication systems. In the past, the operating speed limitation of analog digital converter and processing ability limitation of reconfigurable chips for signal processing have slowed down the development of SDR for useful commercial application. The development of re-configurable de-vices such as digital signal processors and field programmable gate arrays caused that SDR has now be-come practical for use in a lot of system solutions.

The essence of an SDR is the ability, without introducing new hardware, to change operating characteristics such as operating frequency range, modulation type, band-width, maximum radiated or conducted output power and network protocols by chang-ing the software programs executing in processing resources [8]. Operating parameters in SDR are determined by software [9]. This enables a single wireless device to be reprogrammed to use different modulation, access proto-cols etc. An ability of pro-gramming could also enhance interoperability between different radio services.

3.1 Six-Port Technology (SPT)

An SPT enters a progress lately, promising its mass application in the SDR and mo-bile multimedia networks. SPT can be applied as a broadband input equipment of receivers, performing direct baseband conversion without demodulation [11].

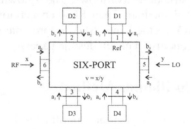

Fig. 1. Principle of the SPR [10]

The linear six-port is a basis of the Six-Port Reflectometer (SPR) (see Fig. 1). Port No. 5 is connected to the Local Oscillator (LO), port No. 6 is connected to the signal source (RF signal from an antenna) and the remaining four ports are connected to power detectors (D_1 to D_4).

Power detector (diode) D1 is used only for the power changes detection at the re-ceived travelling wave. Diodes $D_2 \div D_4$ detect the state of standing wave, which is created due to interference of two waves ($x(t)$ - from the port No. 6 and $y(t)$ - from the local oscillator). The detailed principle of SPR can be seen in [10].

Finally, outgoing signal from SPR is defined as a complex ratio of input RF signal phasor $x(t)$ and LO signal phasor $y(t)$

$$v(t) = x(t) / y(t) = (X_0 / Y_0)e^{j\psi_0} M(t). \qquad (1)$$

The RF signal received by the antenna elements is amplified by low noise amplifier and mixed with a LO to convert the incoming RF signal to a suitable intermediate frequency or directly to baseband. To remove other intermodulation products, this signal is low pass filtered and sampled by an analog to digital converter. The base-band signal is extracted and processed [8].

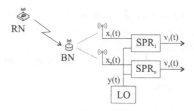

Fig. 2. Antenna array with SPRs

In our case, the signal is especially processed for AoA estimation. In our proposal is used SDR receiver model without baseband conversion, without any training sequence, and without demodulation system. Important part of principle regarding SDR receiver based on SPRs is shown in the Fig. 2.

Each receiving antenna is connected to the input of each SPR (port No. 6). The signal received from antenna (x_i) and the signal from the local oscillator (y) are supplied to SPR_i and the signal received from antenna (x_j) and the signal from the local oscillator (y) are supplied to SPR_j. Receiving antennas form a uniform linear array with the defined distance between individual elements of antenna array. The final signal (after SPRs processing) is given by combination of particular signals from each SPR. The particular equations result from equation (1)

$$v_n(t) = x_n(t) / y(t) = (X_{0n} / Y_0) e^{j\psi_0} M(t) = \alpha_n M(t), \tag{2}$$

where n is the number of SPR, $M(t) = I(t) + jQ(t)$ is a modulation signal in the baseband and α is complex constant resulting from state in the six-port. Positioning information *AOA* is obtained from equations (2).

SPR output signal is directly proportional to modulation signal and represent processing result. The receiver is tuned by local oscillator, i.e. the frequency of the received signal is equivalent to the frequency of the local oscillator [10].

3.2 Angle of Arrival Estimation by Antenna Array with Six-Port Technology

The principle of angle of arrival estimation consists of phase difference determination of the signal impinging to elements of antenna array.

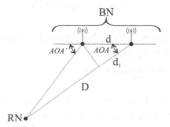

Fig. 3. Illustration of AOA

The distance of antenna elements d is known (see Fig. 3). The received signal $x(t)$ impinges on antenna elements in a different time, i.e. modulation signal from SPR_i will be delayed, or forwarded against the modulation signal from SPR_j. A phase shift (φ_{ij} in degrees) of carriers will be calculated by following equation

$$\varphi_{ij} = phasor(v_i) - phasor(v_j),\qquad(3)$$

where v_1, v_2 – are complex variables obtained from eq. (2).

Since the SPR can assign the shift phase of the carrier frequency in a real time, we need to calculate equation (3). The *AOA* of signal traveling from RN to BN can be calculated

$$AOA = \cos^{-1}(d_1/d) = \cos^{-1}\left(\varphi_{ij}c/2\pi f_n d\right)\qquad(4)$$

where f_n is the carrier frequency in Hz, c is speed of light and d represents distance of antenna elements.

The equation (4) is valid under the condition that the distance D between observed nodes is much longer comparing with the distance d.

After this step, estimated angle of RN and BN is known with respect to reference direction. The estimated angle produces straight line. The next estimation provides a second straight line. The position estimation of the BN is determined as the point of intersection of minimal two straight lines drawn from the different reference nodes.

4   System Model

WANs consist of peer-to-peer links between nodes (devices). Pairwise measurements can be done from any of these links, but only a small number of nodes have coordinate knowledge. Thus measurements are made between pairs of nodes. The coordinates of one node are known.

Precise *AOA* measurement by means of SDR defines fundamental limitations on the performance of proposed positioning systems. This ability depends on a correct calibration of the SPR. We will assume that the SPR does not add any noise to the positioning process in our model.

4.1   Channel Model

We consider three different channels in our simulation:

- Additive White Gaussian Noise (AWGN) channel - absolute LoS propagation,
- Ricean channel - dominant LoS propagation,
- Rayleigh channel - NLoS, multipath propagation.

These three channels were chosen purposely because of their different properties. The channels have absolutely different signal propagation conditions. Therefore it can be said that performance of the method was validated in the representative samples of environment. Basic assumptions of the model have to be defined: perfect spherical radio propagation and identical transmission power for all nodes.

In case of AWGN channel various values of Signal to Noise Ratio (SNR) in [dB] has been used during the simulation. In cases of Ricean and Rayleigh channel also the signal fading was generated in addition. The fading was generated by Ricean distribution with different value of K-factor for particular channel. In terms of the Ricean K-factor, the envelope probability density function (pdf) of the amplitude can be expressed

$$p(r) = \frac{r}{\sigma^2} e^{-K} e^{-\frac{r^2}{2\sigma^2}} I_0(2rK),$$ (5)

where $I_0(.)$ is the modified Bessel function of order zero, σ^2 is variance. The Ricean K-factor is the relation between the power of the LoS component and the power of the Rayleigh NLoS. When $K \rightarrow 0$ the envelope pdf approaches the Rayleigh pdf, it is case of our simulated Rayleigh channel. Ricean channel was simulated with $K = 5$.

4.2 Simulation Environment

Proposed simulation model takes into consideration a network of two RNs and one BN. The relative location problem corresponds to the estimation of BN coordinates. For simplification we consider the location in 2D plane. Let $[x_i; y_i]^T$ $i = 1,2$ are coordinates of RNs and $[x_j; y_j]^T$ $j = 1$ are coordinates of BNs. Pairwise measurements $\{AOA_{i,j}\}$ are done, where $AOA_{i,j}$ is an angle of arrival measurement between nodes i and j. Other conditions of the system model are following:

- signals from RNs are measured at a BN,
- received signal from each node is independent to each other,
- all RNs in the system are deployed with omni directional antennas,
- the BN is realized as a SDR receiver and calculate its position,
- position of RNs is based on relevant experiment in the specified area.

We were running 1000 independent trials. The first step of every trial was generating the positions of BN and RNs. In the next step were calculated $AOAs$ between each RNs and BN on the base of equation (4). Then were calculated mutual intersections of mentioned lines.

The accuracy of position estimation is evaluated by means of $RMSE$ (Root Mean Square Error)

$$RMSE = \sqrt{(x_r - x_{est})^2 + (y_r - y_{est})^2} \ [m]$$ (6)

where $[x_r; y_r]$ are coordinates of real (precise) position and $[x_{est}; y_{est}]$ are coordinates of estimated position.

5 Simulation Results

In this section, we discuss simulation results. The main goal of simulations is to evaluate the performance of AoA method implemented in SDR receiver. The simulations are realized for different nodes deployment scenarios. This condition is very important from optimization of RNs selection for positioning point of view. The following criterions were investigated:

- impact of SNR on the *AOA* estimation,
- impact of distance between nodes on the *AOA* estimation,
- impact of BN and RNs positions on the location accuracy,
- impact of radio channel on the location accuracy.

In the first experiment was investigated an influence of propagation conditions (channel quality) on the accuracy of *AOA*, i.e. influence of SNR on the *AOA* estimation. This dependency is very important for the next processing. In this model, the channel parameters means only one direct error source in the positioning process. Therefore it is necessary to observe the *AOA* positioning error vs. SNR dependency. This experiment was realized for two nodes (one RN and one BN) and *AOA* was estimated in BN. Distances between nodes were 100 and 250 m.

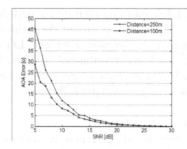

Fig. 4. *AOA* positioning error [°] versus SNR [dB]

Fig. 4 depicts error of *AOA* estimation versus SNR. On the basis of simulation results, it can be concluded that the error of *AOA* is an exponential function of SNR. Ascending value of SNR means decreased *AOA* estimation error.

In the following experiments will be used only AWGN channel, because the basic dependencies between nodes position, its distances and location accuracy is more clear for understanding with additional fading. The following values of SNR: 6, 9, 15, and 24 dB were used for simulation. Moreover, we can say that the increasing distance between nodes means the *AOA* error increase. This fact is significant by reason of selection of nodes for positioning. Nearer RNs to BN will be preferred for positioning.

The goal of the following simulation is observe relation between mutual position of nodes and positioning accuracy. Hence, we decided to use minimum RNs for BN positioning, i.e. only two RNs. The number of RNs is sufficient for verification of defined goal.

Two different scenarios were used (Fig. 5). The deployment of RNs was the same for both scenarios, the coordinates of RNs were [0; 125] and [0; 375]. The scenarios are different from trajectory of BN point of view. In scenario a), the BN positions were generated in the following way: the *x* coordinate was varying from 0 to 500 m and *y* coordinate was permanently 250 m, i.e. BN was moved along a horizontal line (see left part of Fig. 5).

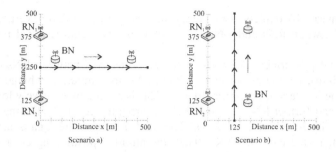

Fig. 5. Deployment of reference nodes and trajectory of blindfolded node

In scenario b), the BN positions were generated similarly compare to previous scenario, but the *y* coordinate was changed and *x* coordinate was constant during simulation. It was done in the following way: the *x* coordinate was permanently 250 m and *y* coordinate was changing from 0 to 500 m, i.e. BN was moved along a vertical line (see right part of Fig. 5).

Fig. 6 shows the dependency of positioning error on horizontal distance (*distance x*) and vertical distance (*distance y*) between reference and blindfolded nodes. It is scenario a).

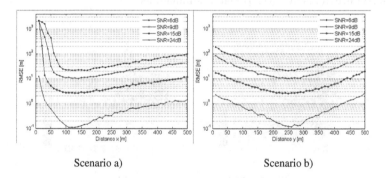

Scenario a)   Scenario b)

Fig. 6. RMSE [m] versus distance between nodes

The simulation was done for various SNR values. In the Fig. 6-Scenario a) can be seen that the change of distance and SNR plays an important role in positioning accuracy or error (RMSE). The most accurate results were obtained with the biggest SNR (24 dB). We consider that the real environment approximately correspond to the 9 dB SNR. In this case, the minimal positioning error is about 10 m.

The minimum RMSE has been achieved approximately at the coordinate [125; 250] for all values of SNR, i.e. *distance x* represents one half of the distance between RNs. Positioning error was the biggest in the case that *x* coordinate was small. It means the case when the BN is located in the middle of RNs. These differences are caused by mutual configuration of RNs and BN. Hence, the suitable geometric configuration of RNs and BN is very important for the positioning error elimination.

Fig. 6-Scenario b) depicts dependency of positioning error on vertical distance (*distance y*) between RNs and BN. The simulation was again realized for the same SNR values.

The minimal positioning error was obtained in the area between RNs (coordinate *y* is about 250 m), i.e. the distances from both RNs are similar. The biggest error was achieved at the borders of the observed area. This is caused by relatively long distance between BN and further RN. A further RN has brought bigger positioning error comparing with closer RN (see Fig. 4). The best results were again achieved for SNR = 24 dB. In the case of SNR = 9 dB, the minimal positioning error is approximately 10 m.

Finally, Fig. 7 shows a complex view on the observed area and depicts the RMSE dependence on a BN position. These simulations were done for SNR = 9 dB and two distances between RNs, 100 m and 250 m, respectively. First, it is necessary to note that the RMSE value bigger than 100 m was depicted with white color, because positioning error is very high in that case. On the other hand, black color defines the most accurate results. The achieved results for both cases are similar. These results demonstrate the fact that the increasing of distance between RNs means increasing of the positioning error. The more accurate results were in the case of smaller distance between RNs. The highest RMSE has been observed at the border of the area (especially close to RNs). The smallest error has been achieved at about one half of RNs distance. These facts confirm results from previous simulations.

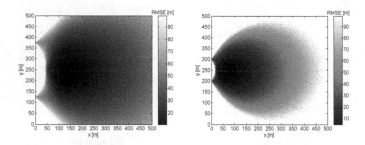

Fig. 7. RMSE [m] versus BN position

On the basis of obtained results we can conclude that the distance between RNs affects positioning accuracy and range of positioning. The range of positioning means a range with the satisfactory positioning accuracy. These facts are important for development of an algorithm dedicated to the RNs selection. An additional measurement or central information about positions of all RNs should be added to the mentioned algorithm.

In the next simulations, we would like to investigate impact of various channels on location accuracy, i.e. impact of fading on accuracy. Therefore, there was performed only one trial. The average value of more trials could eliminate the immediate fading, we need to know immediate situation in this case. The deployment of RNs and trajectory of BN was same as is shown in Fig. 5-Section a). The BN position estimation was done each 1 meter.

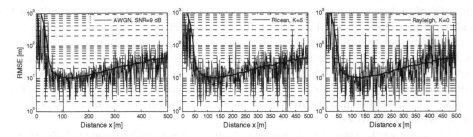

Fig. 8. RMSE [m] versus BN position for different channels

Fig. 8 depicts dependency of positioning error on horizontal distance between RNs and BN. The simulation was again realized for AWGN channel (SNR = 9 dB), Ricean channel (K = 5) and Rayleigh channel (K = 0), see eq. (5). In the Fig. 8 dependency from Fig. 6-Scenario a) is shown for comparison of the immediate and reference value (AWGN, SNR = 9 dB). From the figure can be seen immediate change of amplitude envelope and their impact on location accuracy. The changes are the smallest in the AWGN channel. In case of the Ricean channel the propagation properties are worse compare to AWGN. It results to bigger influence on the location accuracy. According assumption, the Rayleigh channel causes the most unstable situation. The biggest problem occurs in the presence of deep fading. It results in sudden change of trajectory, but this error can be eliminated by prediction of the node motion. This phenomenon is evident especially in case of Rayleigh channel.

6 Conclusions

The goal of this paper is a suggestion of a modern positioning solution for wireless ad hoc networks. Proposed solution combines technology of future - software defined radio receiver and well-known angle of arrival positioning method. Software defined radio receiver is based on the six-port technology in this proposal. Angle of arrival of incoming signal is obtained by means of SPT antenna array. Proposed positioning solution is implemented as self-positioning, i.e. measurements and calculation of position is done in a mobile node implemented as SDR. Practical implementation of the SPT based receiver is possible and it will be feasible in the near future.

We analyzed an influence of radio channel properties (specified by varying value of SNR and type of channel), spatial deployment of reference and blindfolded nodes and impact of these parameters on the positioning accuracy.

According to the reached results we can conclude that the performance of proposed method depends on spatial arrangement of the nodes. It seems necessary to make a proper selection of the reference nodes used for positioning. This selection should result from immediate parameters as a distance between reference nodes and particular distance between reference nodes and blindfolded node. The channel parameters are also very important for positioning accuracy. There is an assumption that the mobile positioning is more precise in outdoor environment in comparison with indoor conditions. It results from a nature of radio signal propagation in this frequency band.

In case of multipath propagation there is necessary to compensate impact of signal fading. For example, it could by done with Kalman filter.

Available positioning accuracy is high due to very precise calibration of six-port reflectometer. An advantage of software defined radio receiver is an ability to compensate errors from six-port calibration. These errors of scattering parameters are compensated by means of software processing. Proposed solution maybe seems unrealistic, but we believe that it can provide an effective mobile positioning in heterogeneous wireless networks.

Our future work consists of a proposal dealing with complex optimized algorithm for selection of reference nodes for localization. A new algorithm should be based on results presented in this paper.

Acknowledgments. The work on this paper was supported by the grant VEGA 1/4065/07 of Scientific Grant Agency of the Slovak Republic.

References

1. Niculescu, D., Nath, B.: Ad hoc Positioning System (APS). In: Proceedings of IEEE GLOBECOM 2001, vol. 5, pp. 2926–2931 (2001)
2. Bachrach, J., Taylor, C.: Localization in Sensor Networks. In: Stojmenović, I. (ed.) Handbook of sensor networks: algorithms and architectures, pp. 277–310. John Wiley & Sons, Inc., Chichester (2005)
3. Jayashree, L.S., Arumugam, S., Anusha, M., Hariny, A.B.: On the Accuracy of Centroid based Multilateration Procedure for Location Discovery in Wireless Sensor Networks. In: Proc. Wireless and Optical Communications Networks (2006)
4. Ilyas, M., Mahgoub, I.: Handbook of Sensor Networks: Compact Wireless and Wired Sensing Systems, 672 pages. CRC Press, Boca Raton (2004)
5. Voddiek, M., Wiebking, L., Gulden, P., Wieghardt, J., Hoffmann, C., Heide, P.: Wireless local positioning. IEEE Microwave Magazine 4, 77–86 (2004)
6. Drane, C.: Positioning GSM telephones. IEEE Communications Magazine 36(4), 46–59 (1998)
7. Caffery Jr., J.J.: Wireless Location in CDMA Cellular Radio Systems, 1st edn. Kluwer Academic Publishers, University of Cincinnati (2000)
8. Rappaport, T.S., Reed, J.H., Woerner, B.D.: Position Location Using Wireless Communications on Highways of the Future. IEEE Communications Magazine 34(10), 33–41 (1996)
9. Xu, X., Wu, K., Bosisio, R.G.: Software Defined Radio Receiver Based on Six-Port Technology. In: Microwave Symposium Digest, 2003 IEEE MTT-S International, vol. 2, pp. 1059–1062 (2003)
10. Bilík, V.: Six-port Measurement Technique: Principles, Impact, Applications, http://www.s-team.sk/download/SixPortTechnique.pdf (2008-03-20)
11. Kernévès, D., Huyart, B., Begaud, X., Bergeault, E., Jallet, L.: Direction Finding with SIX-PORT Reflectometer Array. In: 8th COST 260 Management Committee and Working Groups Meeting in Rennes, France, (2000)
12. Janaswamy, R.: Radiowave Propagation and Smart Antennas for Wireless Communications, p. 297. Kluwer, Dordrecht (2001)

13. Tatu, S.O., Wu, K., Denidni, T.A.: Direction-of-arrival estimation method based on six-port technology. IEE Proc.-Microw. Antennas Propag. 153(3), 263–269 (2006)
14. Yakabe, T., Fengchao Xiao Iwamoto, K., Ghannouchi, F.M., Fujii, K., Yabe, H.: Six-Port Based Wave-Correlator with Application to Beam Direction Finding. IEEE Transactions on instrumentation and Measurement 50(2), 377–380 (2001)
15. Huyart, B., Laurin, J.J., Bosisio, R.G., Roscoe, D.: A direction-finding antenna system using an integrated six-port circuit. IEEE Transactions on Antennas and Propagation 43(12), 1508–1512 (1995)

On the Impact of Node Placement and Profile Point Selection on Indoor Localization

Israat Tanzeena Haque, Ioanis Nikolaidis, and Pawel Gburzynski

Computing Science Dept.,
University of Alberta,
Edmonton AB T6G 2E8, Canada
{israat,yannis,pawel}@cs.ualberta.ca

Abstract. We present an indoor localization technique based on RF profiling using the received signal strength (RSS) measurements from a set of pre–selected reference points. We do not attach any interpretative significance to the measurements other than use them to calculate their difference from the measurements of the reference points. We study the performance of our technique in an environment with multiple adjacent rooms and find that it gives better results compared to the application of the k-Nearest Neighbor algorithm that has been used in the literature for the same task. We also study the proposed scheme and two other well–known localization schemes, with respect to the sensitivity of the localization on the number and layout of the reference points, as well as on the number and layout of the deployed fixed points (pegs) from where the measurements are collected. We find that one can achieve good localization performance with either fewer reference points or with fewer pegs as long as their layout is chosen carefully.

1 Introduction

Location-based services are a family of, predominantly wireless, network services that, in order to operate effectively, have to rely on some form of location information [4]. While the problem of determining location information has been well supported by the Global Positioning System (GPS) [7], location tracking inside buildings poses specific challenges. An alternative is to determine the location of a node without relying on GPS, but instead based on some ground-deployed permanent (or semi-permanent) wireless infrastructure, i.e., via indoor *radiolocation* techniques. The vast majority of the literature in this area considers WiFi wireless access points as a convenient (semi-) permanent infrastructure. The co-ordinates of the (semi-) permanent infrastructure nodes are assumed to be known and act as the reference by which the location of other nodes will be determined. Access points are already deployed to extend Local Area Networks (LANs) to wireless users, and localization is a secondary task carried out via the same infrastructure. A relevant, usually unstated, assumption is that the placement of access points is restricted and/or performed for completely different reasons than localization, i.e., to ensure adequate coverage for roaming users. Yet another assumption is that the number of access points cannot be arbitrary and is usually

J. Wozniak et al. (Eds.): WMNC 2009, IFIP AICT 308, pp. 220–231, 2009.
© IFIP International Federation for Information Processing 2009

restricted by cost and connectivity considerations. We would rather not have to deal with restrictions on the number and placement of the access points used for radiolocation.

One heavily explored research direction is that of determining which access point–based localization algorithm is best and under what circumstances, e.g. [8]. Instead, in this paper we examine the impact on localization from decisions about the number and placement of the (semi-) permanent nodes (called *pegs* in the sequel). To this end, we use a localization scheme, dubbed LEMON[1] which we previously proposed in [3] and which was found to provide good localization accuracy. In essence, our research is motivated by cost considerations. Namely, a large number of simple low-cost wireless transceivers, like those usually assumed to be part of a wireless sensor node, can be purchased for the cost of a single access point. For example, for the cost of a single access point implementing the most advanced version of the IEEE 802.11 standard (currently approximately USD $200.00) one can purchase a dozen or so boards equipped with the a low power RF transceiver and microcontroller (like, respectively, the Texas Instruments CC1100 and MSP430). With low power battery–based operation and fixed placement of the pegs, e.g., in furniture and walls, we could conceivably have a finer granularity of localization information than what is provided by a handful of WiFi access points that can be acquired with the same expense. The devices to be localized, which we subsequently call *tags*, are essentially disposable, which compares favorably with the cost of a (raw) GPS module (which cannot be used indoor, anyway).

While the cost–benefit relation appears to be self–evident, a shortcoming of populating a space with low cost wireless transceivers is that their readings are not reliable, or even consistent. The low production cost, usually implies poor accuracy when it comes to taking the measurements from such inexpensive nodes "seriously." Nevertheless, the hope is that, collectively, many poor quality measurements will be better than few, somewhat more accurate, ones. Additionally, the attainable bit rates of the low cost transceivers are usually low. For these reasons we forego localization techniques that are known to be very sensitive to the noise measurement (like lateration and angulation) or that demand extensive data transfers between pegs and localized tags. In fact in environments that exhibit strong influence by multipath propagation effects, even the accuracy of angulation and lateration techniques is questionable. We therefore consider only signal strength pattern matching techniques, i.e., techniques that do not attempt to link the signal strength measurements to any model–based estimate for the actual distance between pegs and localized tag. Specifically, we restrict our attention to location fingerprinting/profiling techniques based on the readings collected at a number of locations, called the *reference points*.

We note that the simplest form of information that can be reasonably extracted from a measurement is the received signal strength (RSS) [1]. RSS measurements do not require specialized additional hardware, as they are routinely available already from most RF transceivers. The commonly acknowledged

---

[1] Location Estimation by Mining Oversampled Neighborhoods.

disadvantage of RSS is its poor and unpredictable correlation with distance resulting from the multi-path effect, which is typically quite serious inside a building. On the other hand, RSS-based measurements are available from almost any transceiver design used today, regardless of which physical layer protocol it implements. In location fingerprinting, the perceived attributes of the tag's signal are compared against a pre-collected set of samples from known (profiled) reference points [1]. By resorting to profiling, one can hope to compensate for the intrinsic characteristics of the environment which render direct transformations of the perceived attributes of RF signals into distances or angles highly unreliable. This hope underlies our work. It appears that we would need to measure a large number of reference points to counter the imprecise nature of the measurements. This brings up the second issue discussed in this paper, which is determining the smallest number of reference points that are sufficient for a given accuracy of localization. Each reference point measurement could be considered an overhead, since a user (profiler) would have to perform it and associate it with actual coordinates. Whenever the environment changes drastically, e.g., when furniture are rearranged or changed, re-measurement would be necessary. Therefore, the smaller the number of required reference points, the better.

In the next section we present LEMON in detail and summarize the results from [3]. In Section 3 we detail the experiments that were carried out for the present study and the localization errors observed. Finally, Section 4 summarizes our findings and outlines future research directions.

2   LEMON

LEMON can be viewed as a combination of a range-free approach with "traditional" profiling: the scheme is driven by a (somewhat fuzzy) concept of neighborhood, while its objective is to produce an "educated" estimation of the actual location from the coordinates of selected reference point readings. Technically, in LEMON, the location of the static nodes (pegs) need not be known. A tag (the node to be localized) can be essentially a node of the same type as a peg. During the profiling stage, the network collects and stores in a database (maintained on a central server) *samples* acquired from tag devices placed at pre-selected points (the *reference points*) within the monitored area.

A single sample stored in the database can be viewed as a triplet $< C, \Omega, \tau >$, where C stands for the known coordinates of the sampled point, Ω is the so-called association set, and τ, called the sample's *class*, identifies the (settable) RF parameters of the transmitter (typically transmission power, bit rate, and channel number). τ's role is to discern samples collected under different "options" of the tag's transmitter, such that they will only be matched to (future) readings acquired under the same options. For now, we only consider a planar version of the problem, i.e., $C = (x, y)$. The association set Ω consists of pairs $< p, r >$, where p identifies a peg, and r is the RSS value measured by that peg. A tracked tag periodically emits RF packets that include a sequence number used to uniquely identify them. A peg receiving such a packet will forward to the central server a report consisting of its own identifier, the tag identifier, the packet number and class. Having received a number of such reports referring to the same packet, the server will built an association set Φ representing the combined momentary measurements of the tag's RSS by all the pegs that can hear the tag.

The initial step of the localization algorithm is to select from the database a subset of samples representing the best match to the association list Φ. First of all, only the samples of the same class as the received reports are subject to selection. To further narrow down the search, the server finds in Φ the pair $< p_m, r_m >$, such that r_m is the highest among all pairs. Then, it only considers those samples form the database whose association lists include p_m as one of the pegs. This pre-selection boils down to the postulate that the peg p_m appearing to be very close to the tracked tag be a member of all samples that will be used for estimating the tag's location.

Let $\Omega = \{\omega_1, \ldots, \omega_k\}$ and $\Psi = \{\psi_1, \ldots, \psi_m\}$ be two associations sets. By the distance between these sets, we understand

$$D(\Omega, \Psi) = \sqrt{\sum_{j=1}^{N}(R_\Omega(j) - R_\Psi(j))^2} \qquad (1)$$

where N is the total number of pegs in the network and $R_\Omega(j)$ is defined as r_j, if the pair $< p_j, r_j >$ occurs in Ω, and 0 otherwise.

In the second step, the server evaluates the distance of each pre-selected sample (its association list) from the current association list Φ. Then, it selects K samples with the smallest distance.

In the last step, the coordinates of the selected samples are averaged to produce the estimated coordinates of the tag. The averaging formula biases the samples in such a way that the ones with a smaller distance from Φ contribute proportionally more. Let D_{max} be the maximum distance from Φ among the best K selected samples and $S_D = \sum_i^K D_i$ be the sum of all those distances. The tracked coordinates are estimated as:

$$x_{est} = \frac{\sum_{i=1}^{K} x_i \times (D_{max} - D_i)}{K \times D_{max} - S_d} \quad \text{and} \quad y_{est} = \frac{\sum_{i=1}^{K} y_i \times (D_{max} - D_i)}{K \times D_{max} - S_d} \qquad (2)$$

where (x_i, y_i) are the coordinates associated with sample i.

Note that the above approach does not link RSS to any distance metric but treats it as a purely numerical attribute of a sample whose value should be close to the observed value. The averaging formula factors in the magnitude of discrepancy between RSS values (in terms of distance between points in Euclidean space), but this is a purely numerical interpolation and not an application of any RF propagation model. LEMON does not impose any restriction on the number of pegs or reference points. It relies on the matching rules to locate those samples that best apply to a particular "instance" of localization. If the number of samples is large, the role of the last-step interpolation becomes secondary: we do not purport to expect that the specific RSS values encode useful information about the distance. In particular, it may make sense to oversample the area, e.g., collecting multiple samples from the same point. For example, those multiple samples may correspond to the different orientations of the tag, as in [1]. In the rest of this paper we also study how localization is affected when multiple orientations (at the same point) are identified as such vs. what happens when they are ignored.

3   Experiments

A prototype LEMON system was implemented and tested at our university campus. Early results have been reported in [3]. The devices for both tags and pegs are the EMSPCC11 from Olsonet Communications,[2] which is a low-cost low-power mote for wireless sensor networking, programmable in PicOS [2]. The node employs the CC1100 RF module from Texas Instruments operating within the 916 MHz band. The RF module of EMSPCC11 offers several settings. The transmission power can vary from -15 dBm to 10 dBm (in 8 discrete steps), the bit rate options are 5 kbps, 10 kbps, 38 kpbs, and 200 kbps, and there are 256 different channels (numbered 0 to 255) with 200 kHz spacing. All combinations are possible and, in principle, sensible. The experiments reported on in the rest of this paper were carried out at the lowest power setting with 5 kps transmission rate using channel 0.

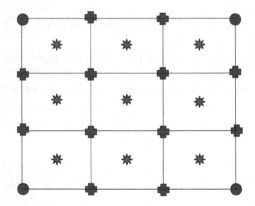

Fig. 1. Layout of pegs (circles), reference points (crosses), and localized points (stars)

In our earlier work [3] we demonstrated how localization in three distinctly different rooms resulted in a localization error less than one meter. Even though LEMON's accuracy has not been yet maximized, it already challenges the accuracy of location estimation reported in [5] which was about 2 m using an even finer grid than ours. Also, our analysis of data revealed that most of the problems with localization resulted from assigning too much relevance to low RSS

---

[2] See http://www.olsonet.com/Documents/emspcc11.pdf

values, i.e., corresponding to weak reception, which would exhibit large statistical fluctuations. The numerical RSS readings presented by the RF module of EMSPCC11 are positive numbers, roughly between 80 and 150, representing a shifted dB signal level of the received packet. Assuming that those readings are always between *MIN* and *MAX*, we applied the following scaling formula:

$$R_s = \left(\frac{r - MIN}{MAX - MIN} \right)^\alpha \tag{3}$$

where $\alpha > 1$. Note that for $\alpha = 1$ we effectively obtain the original (not rescaled) case, as a linear transformation of all RSS readings, does not change the outcome of our algorithm. The best results have been observed for $\alpha = 3$ which allowed us to estimate the tag location with an error of less than 1 m in 90% of the cases.

In this paper we restrict our attention to the problem of localization in multiple (adjacent) rooms focusing on the layout of pegs and the number and location of reference points, subject to certain regular patterns.

3.1   Multiple Room Localization

We conducted an experiment using three adjacent rooms. The dimension of each room is about 3 m × 5 m. Each room includes four wooden tables and chairs, a metal file cabinet and two desktop computers. We placed four pegs at the corners of each room and generated a grid with the edge size around 1 m. In particular, in each room twelve readings were taken at the reference points (indicated by crosses in Figure 1) and nine locations were estimated: by placing tags at the nine points positioned at the center of each grid cell (indicated by stars in Figure 1).

It was noted previously [1] that orientation impacts localization. To verify that observation, we collected tag readings for four different directions (North, South, East, and West) at each profiled or localized point. At the localization stage, to take the orientation aspect into account, we would only use for estimation the profiled data collected from tags oriented in the same way as the

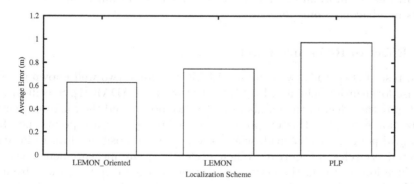

Fig. 2. LEMON (with & without orientation) vs. PLP's k-NN

one being localized. Even though the line of sight (LOS) was not interrupted regardless of the orientation of the tag (suggesting that the orientation should have no impact on location estimation), the observed error was larger when the orientation aspect was ignored (i.e., all profiled data were used regardless of the orientation of the localized tag). The observed difference was about 12 cm on the average (Figure 2), in favor of the localization based solely on the reference points collected with the same orientation. While the difference is not huge, it crisply illustrates how an even minor change of tag placement (namely rotation on the same spot) influences localization. Hence, baring information about orientation, we should not expect localization accuracy to be any tighter than the difference found between oriented vs. non-oriented localization.

A second element of this study is a comparison against a localization scheme based on the k–Nearest Neighbor (k–NN) algorithm. Specifically, the PowerLine Positioning (PLP) proposed in [6], employs k–NN for multi–room localization. It is one of the few papers in the bibliography where localization is designed specifically with multiple room environments in mind. PLP is also a profiling based approach, but it uses powerlines as the means to distribute RF signals throughout a building. Apart from the unconventional use of the powerline as an "antenna," PLP's approach is to apply a sufficiently large k for the nearest neighbors to localize and identify the room where the tag is located. The reference points from the identified room are then used with a smaller k to estimate the tag's location. We tested the same logic as PLP using our experimental data with k equal to six and four for, respectively, selecting the room and then estimating the location within the room. The results show that the estimation accuracy of PLP degrades as it subdivides the signal space according to the physical layout, and then it localizes within a selected space. There is apparent loss of information when we first identify the room and then localize strictly within that room; hence, the large error distance shown in Figure 2, even against LEMON without orientation information. LEMON outperforms PLP when we consider the three rooms as one signal space and use all reference points to perform the estimation. What this demonstrates is that the intuition of restricting the localization to a subset of reference points that might appear "closest" results in loss of information that could be crucial to localization, i.e., reference points in adjacent rooms *do matter*.

3.2   Effect of Reference Points

In the rest of this study, we compare LEMON against two well known general–purpose localization schemes, LANDMARC [5] and RADAR [1]. Specifically, we investigate the effect of the number of reference points and their arrangement on localization error. For this purpose, we gathered 49 reference points, i.e., from *every* grid point of a 7×7 grid, where the pegs were positioned at 16 locations (depicted as circles in Figure 3). When a reference point measurement was at the same location as a peg, the actual measurement was taken 6 cm away from the peg. We then removed reference points from the database in a regular fashion such that the density of reference points remained roughly the same across the

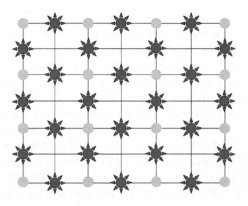

Fig. 3. Locations of pegs and of 24 reference points (Diamond)

grid. Figure 3 shows the layout of reference points after removing 25 points; the resulting setup is dubbed *Diamond* and consists of measurements from 24 reference points. By further removing points in the same fashion, i.e., one at a time, we arrive at two new layouts (see Figure 4) called, respectively, *Hexagon* (18 points) and *Nested-Hexagon-Diamond* (12 points). Starting with the Hexagon layout and eliminating more points, we produce two additional layouts dubbed *DoubleD* (14 points) and *Square* (12 points), respectively.

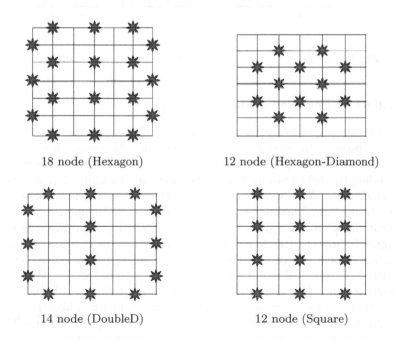

18 node (Hexagon)   12 node (Hexagon-Diamond)

14 node (DoubleD)   12 node (Square)

Fig. 4. Reference point location configurations

The average error in location estimation for LEMON, LANDMARC, and RADAR in each of the above layouts is shown in Figure 5. The average error for LEMON with all the initial 49 reference points was 0.43 m, but we could remove points in a regular fashion and end up with 12 points while the error distance was still less than 1 m. This indicates that it is possible to have a less complex deployment of LEMON without degrading the performance significantly (see, e.g., the case of Dia(24)). Overall, we have observed that the more the reference points the better the localization. Yet, the number of reference points alone is not sufficient, as their placement matters as well (see for example the case of DD(14) vs. Square(12)). Moreover, LEMON is less sensitive to the number of reference points and their layout compared to LANDMARC and RADAR. In fact, of all three, RADAR appears to be the one scheme that deteriorates the most when the number of reference points is reduced.

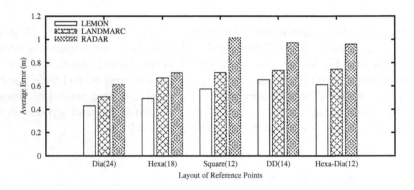

Fig. 5. Localization error for different reference point configurations

3.3 Effect of Peg Placement

The unique feature of LEMON is its flexibility of using as many pegs as needed without imposing any restriction on their number. However, there might exist better or worse ways of placing those pegs that may offer better or worse estimation. Finding the best placement of pegs for a target localization accuracy may pose a challenge. We may not need, say, 16 Pegs for 7×7 grid to keep the error distance below 1 m. Thus we conducted another experiment to try to produce configurations with fewer pegs that still provide a localization error less than 1 m. Figure 6 shows two such layouts with 8 Pegs and the corresponding average localization error is shown in Figure 7. As we can see, it is possible to eliminate half of the Pegs and still maintain error distance well below 1 m. We also found that removing just four pegs (the "center square" of circles in Figure 3) would leave 12 pegs (called "Square" in Figure 7) but somewhat higher localization error than the configurations with the fewer but "better placed" pegs. This may suggest that a clever design of placing pegs on the target area may help us reduce the error distance. Our observations indicate that removing pegs from inside the

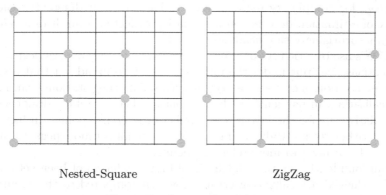

Nested-Square   ZigZag

Fig. 6. Peg layouts for eight pegs

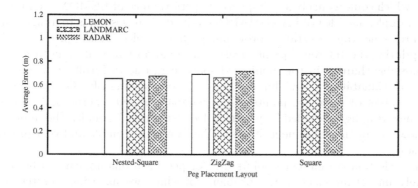

Fig. 7. Localization error for different peg layouts

grid may not be a good idea. Yet, what is even more interesting to note is that all three schemes (LEMON, LANDMARC, and RADAR) are relatively insensitive to the peg layout, compared to the impact that the layout (and number) of reference points could have. Additionally, LANDMARC appears to have a slightly better performance than the other two, but the statistical significance of the differences is very small. The lesson is that peg placement is not as critical an issue as the dense "sampling" of the space by many reference points. This is good news, as the placement of pegs is likely to be constrained or even dictated by external factors, e.g., placement of furniture, walls, etc.

4   Conclusions

Our proposed scheme, LEMON, can easily be applied to in-building localization using inexpensive devices and can provide an average localization error well below 1 m. Furthermore, we have found evidence that attempts at restricting the population of reference points to those in the suspected proximity of the target (e.g., for complexity reasons) tend to backfire. Put differently, the final accuracy of estimation depends on information collected from distant reference points in a surprisingly significant sort of way.

LEMON's use of a rich set of reference points, collected by pegs positioned at fixed locations, was put under the microscope and we found out that a smaller set of reference points or a smaller set of fixed points (peg stations) can provide good localization, in particular if their layout has been carefully designed. This is particularly true for the layout and number of reference points. In fact, faced with a choice between introducing more reference points or more pegs, the answer appears to be in favor of more reference points.

We can confidently say that the limits of LEMON have not been yet reached. LEMON achieved a localization error below 1 m compared to the accuracy of location estimation reported in [5], which was about 2 m using a *finer* grid than in our case. Yet, LEMON did not apply any of the additional techniques outlined in [5] which could conceivably improve its performance of LEMON even further. For example, one idea of LANDMARC and its derivatives [5], is to assign a fixed set of reference tags constantly providing profile samples. This approach can help by updating the reference point measurements almost in real–time to factor in the possible changes in the monitored space, e.g., due to furniture moving, the number of inhabitants etc. Thus one can think of adapting LEMON on-line, such that a subset of reference points in the database is being perpetually replaced with new readings provided by pre-installed "fixed tags." Finally, the uniformity and low cost of the equipment makes LEMON a highly viable and very practical solution.

The selection of reference points and/or peg locations appears to be an extremely important facet of the problem, of which we have just scratched the surface in this paper. We plan to define the placement problem formally as an optimization problem. Its solution may guide real-life deployments of pegs and the ways of collecting reference points (profiling) as to minimize the system's complexity (cost) for the required accuracy of localization. The analytical results can be verified with our experimental data.

References

1. Bahl, P., Padmanabhan, V.N.: RADAR: an in-building RF-based user location and tracking system. In: Nineteenth Annual Joint Conference of the IEEE Computer and Communications Societies (INFOCOM 2000), Israel (March 2000)
2. Gburzyński, P., Olesinski, W.: On a practical approach to low-cost ad hoc wireless networking. Journal of Telecommunications and Information Technology (1), 29–42 (January 2008)
3. Haque, I., Nikolaidis, I., Gburzyński, P.: A scheme for indoor localization through RF profiling. In: Proceedings of the Intl. Workshop on Synergies in Communications and Localization (SyCoLo 2009), Dresden, Germany (to appear) (June 2009)
4. Küpper, A.: Location–Based Services: Fundamentals and Operation. Wiley, Chichester (2005)
5. Ni, L.M., Liu, Y., Lau, Y.C., Patil, A.P.: LANDMARC: indoor location sensing using active RFID. Wirel. Netw. 10(6), 701–710 (2004)

6. Patel, S.N., Truong, K.N., Abowd, G.D.: PowerLine Positioning: A practical sub-room-level indoor location system for domestic use. In: Dourish, P., Friday, A. (eds.) UbiComp 2006. LNCS, vol. 4206, pp. 441–458. Springer, Heidelberg (2006)
7. United States Coast Guard Navigation Center. Global positioning system standard positioning service specification (June 1995)
8. Wallbaum, M., Diepolder, S.: Benchmarking wireless lan location systems. In: Proceedings of the 2005 Second IEEE International Workshop on Mobile Commerce and Services (WMCS 2005), Munich, Germany, July 2005, pp. 42–51 (2005)

Hyperbolic Position Location Estimation in the Multipath Propagation Environment

Jacek Stefański

Gdansk University of Technology
11/12 Narutowicza Str., PL-80233 Gdansk, Poland
jstef@eti.pg.gda.pl

Abstract. The efficiency analysis a hyperbolic position location estimation in the multipath propagation environment in the wideband code division multiple access (WCDMA) interface was presented. Four, the most popular methods: Chan's [1], Foy's [2], Fang's [3] and Friedlander's [4] were considered. These algorithms enable the calculation of the geographical position of a mobile station (MS) using the time differences of arrival (TDOA) between several base stations (BS) and MS. The simulation model is outlined and simulation results are presented.

1 Introduction

Hyperbolic position location estimation is accomplished in two stages. The first stage involves estimation of the time difference of arrival (TDOA) between transmitters through the use of time delay estimation techniques. The estimated TDOA's are then transformed into range difference measurements between base stations (BSs), resulting in a set of nonlinear hyperbolic equations. The second stage utilizes efficient algorithms to produce an unambiguous solution to these nonlinear hyperbolic equations.

Referring all TDOA's to the first base station, which is assumed to be the base station controlling, the distance between the i-th source (BS) and the receiver is given as

$$R_i = \sqrt{(X_i - x)^2 + (Y_i - y)^2} = \sqrt{X_i^2 + Y_i^2 - 2X_i x - 2Y_i y + x^2 + y^2} \qquad (1)$$

where (X_i, Y_i) and (x, y) are coordination of i-th base station and mobile station (MS) respectively. The range difference between base stations with respect to the first arriving base station is

$$R_{i,1} = v \cdot t_{i,1} = R_i - R_1 = \sqrt{(X_i - x)^2 + (Y_i - y)^2} - \sqrt{(X_1 - x)^2 + (Y_1 - y)^2} \qquad (2)$$

where v is the radio signal speed, $R_{i,1}$ is the range difference distance between the first base station and the i-th base station, R_1 is the distance between the first base station and the MS, and $t_{i,1}$ is the estimated TDOA between the first base station and the i-th base station. This defines the set of nonlinear hyperbolic equations whose solution gives 2-D coordinates of the mobile station.

This paper presents the efficiency analysis a hyperbolic position location estimation methods in the multipath propagation environment, especially Chan's, Foy's, Fang's and Friedlander's algorithms.

J. Wozniak et al. (Eds.): WMNC 2009, IFIP AICT 308, pp. 232–239, 2009.

2 Mathematical Model

Solving the nonlinear equations of (2) is difficult. Consequently, linearizing this set of equations is commonly performed. One way of linearizing these equations is through the use of Taylor-series expansion and retaining the first two terms [2]. An commonly used alternative method to the Taylor-series expansion method, presented in [1], [3] and [4], is to first transform the set of nonlinear equations in (2) into another set of equations. Rearranging the form of (2) into

$$R_i^2 = (R_{i,1} + R_1)^2 \tag{3}$$

Equation (1) can now be rewritten as

$$R_{i,1}^2 + 2R_{i,1}R_1 + R_1^2 = X_i^2 + Y_i^2 - 2X_i x - 2Y_i y + x^2 + y^2 \tag{4}$$

Subtracting (1) at $i=1$ from (4) results in

$$R_{i,1}^2 + 2R_{i,1}R_1 = X_i^2 + Y_i^2 - 2X_{i,1}x - 2Y_{i,1}y + X_1^2 + Y_1^2 \tag{5}$$

where $X_{i,1}$ and $Y_{i,1}$ are equal to $X_i - X$ and $Y_i - Y$ respectively. The set of equations in (5) are now linear with the source location (x, y) and the range of the first base station to the source R_1 as the unknowns, and are more easily handled.

In this paper a 2-D hyperbolic position location system using only three base station is considered.

2.1 Chan's Method

Following Chan's method [1], for a three base station system, producing two TDOA's, x and y can be solved in terms of R_1 from (5). The solution is in the form of

$$\begin{bmatrix} x \\ y \end{bmatrix} = - \begin{bmatrix} X_{2,1} & Y_{2,1} \\ X_{3,1} & Y_{3,1} \end{bmatrix}^{-1} \times \left\{ \begin{bmatrix} R_{2,1} \\ R_{3,1} \end{bmatrix} R_1 + \frac{1}{2} \begin{bmatrix} R_{2,1}^2 - K_2 + K_1 \\ R_{3,1}^2 - K_3 + K_1 \end{bmatrix} \right\} \tag{6}$$

where $K_1 = X_1^2 + Y_1^2$, $K_2 = X_2^2 + Y_2^2$, $K_3 = X_3^2 + Y_3^2$. When (6) is inserted into (1), with $i = 1$, a quadratic equation in terms of R_1 is produced

$$aR_1^2 + bR_1 + c = 0 \tag{7}$$

Substituting the positive root back into (6) results in the final solution. There may exist two positive roots from the quadratic equation that can produce two different solutions, resulting in an ambiguity. Simulations in this work have shown, that only the following root should be considered for cellular position location

$$R_1 = \frac{-b - \sqrt{b^2 - 4ac}}{2a} \tag{8}$$

2.2 Foy's Method

The Foy's method linearizes the set of equations in (2) by Taylor-series expansion then uses an iterative method to solve the system of linear equations. The iterative method begins with an initial guess and improves the estimate at each iteration by determining the local linear least square solution. With a set of TDOA estimates, the method starts with an initial guess (x_0, y_0) and computes the deviations $[\Delta x, \Delta y]^T$ of the position location estimation

$$\begin{bmatrix} \Delta x \\ \Delta y \end{bmatrix} = (\mathbf{G}^T \cdot \mathbf{G})^{-1} \cdot \mathbf{G}^T \cdot \mathbf{h} \tag{9}$$

where

$$\mathbf{h} = \begin{bmatrix} R_{2,1} - (R_2 - R_1) \\ R_{3,1} - (R_3 - R_1) \end{bmatrix} \tag{10}$$

$$\mathbf{G} = \begin{bmatrix} \dfrac{X_1 - x}{R_1} - \dfrac{X_2 - x}{R_2} & \dfrac{Y_1 - y}{R_1} - \dfrac{Y_2 - y}{R_2} \\ \dfrac{X_1 - x}{R_1} - \dfrac{X_3 - x}{R_3} & \dfrac{Y_1 - y}{R_1} - \dfrac{Y_3 - y}{R_3} \end{bmatrix} \tag{11}$$

The values R_1 and R_2 are computed from (1) with $x=x_0$ and $y=y_0$. In the next iteration, x_0 and y_0 are set to $x_0+\Delta x$ and $y_0+\Delta y$. The whole process is repeated until Δx and Δy are sufficiently small, resulting in the estimated position location of the (x, y). This Taylor-series method can provide accurate results, however, it requires a close initial guess (x_0, y_0) to guarantee convergence and can be very computationally intensive.

2.3 Fang's Method

Fang establishes a coordinate system so that the first base station is located at $(0, 0)$, the second BS at $(X_2, 0)$ and the third BS at (X_3, Y_3). The following relationships are simplified

$$R_i = \sqrt{(X_i - x)^2 + (Y_i - y)^2} = \sqrt{x^2 + y^2} \tag{12}$$

$$X_{i,1} = X_i - X_1 = X_i \tag{13}$$

$$Y_{i,1} = Y_i - Y_1 = Y_i \tag{14}$$

Using these relationships, the equation of (5) can be rewritten as

$$2R_{2,1}R_1 = -R_{2,1}^2 + X_2^2 - 2X_2 x \tag{15}$$

$$2R_{3,1}R_1 = -R_{3,1}^2 + X_3^2 + Y_3^2 - 2X_3 x - 2Y_3 y \tag{16}$$

Equating the two equations and simplifying results in

$$y = g \cdot x + h \tag{17}$$

where

$$g = \{R_{3,1} \cdot (X_2 / R_{2,1}) - X_3\}/Y_3 \tag{18}$$

$$h = \{X_3^2 + Y_3^2 - R_{3,1}^2 + R_{3,1} \cdot R_{2,1} \cdot (1 - (X_2 / R_{2,1})^2)\}/2 \cdot Y_3 \tag{19}$$

Substituting equation (17) into the equation (15) results in

$$d \cdot x^2 + e \cdot x + f = 0 \tag{20}$$

where

$$d = -\{1 - (X_2 / R_{2,1})^2 + g^2\} \tag{21}$$

$$e = X_2 \cdot \{(1 - (X_2 / R_{2,1})^2\} - 2gh \tag{22}$$

$$f = (R_{2,1}^2 / 4) \cdot \{1 - (X_2 / R_{2,1})^2\}^2 - h^2 \tag{23}$$

Solving the quadratic equation (20), we get two values for x. Using a priori information, one of the values is chosen and is used to find out y from (17). It has been found by simulations in this research that one of the roots of (20) is beyond the cell coverage area. Hence for position location in cellular systems we only need to evaluate the following root from (20)

$$x = \frac{-e - \sqrt{e^2 - 4df}}{2d} \tag{24}$$

As stated earlier, putting this value of x in (17) will give us the other coordinate of the mobile's position estimate.

2.4 Friedlander's Method

Friedlander's method [4] utilizes a least squares error criterion to solve for the position location. He first transforms the linear set of equations of (5) into

$$X_{i,1}x + Y_{i,1}y = 0.5 \cdot (X_i^2 + Y_i^2 - X_1^2 - Y_1^2 - R_{i,1}^2) - R_{i,1}R_1 \tag{25}$$

Then realizes this equation in matrix form as

$$S \cdot x = u - R_1 \cdot p \tag{26}$$

where

$$S = \begin{bmatrix} X_{2,1} & Y_{2,1} \\ X_{3,1} & Y_{3,1} \end{bmatrix} \tag{27}$$

$$x = [x \quad y]^T \tag{28}$$

$$\mathbf{u} = \begin{bmatrix} X_2^2 + Y_2^2 - X_1^2 - Y_1^2 - R_{2,1}^2 \\ X_3^2 + Y_3^2 - X_1^2 - Y_1^2 - R_{3,1}^2 \end{bmatrix} \tag{29}$$

$$\mathbf{p} = \begin{bmatrix} R_{2,1} & R_{3,1} \end{bmatrix}^T \tag{30}$$

In order to eliminate the second term of (26), which requires knowledge of the unknown term R_1, the equation in (26) is premultiplied by a matrix $\mathbf{N}$ which has $\mathbf{p}$ in its null-space. Matrix $\mathbf{N}$ is defined as

$$\mathbf{N} = (\mathbf{I} - \mathbf{Z}) \cdot \mathbf{D} \tag{31}$$

where

$$\mathbf{D} = (diag\{\mathbf{p}\})^{-1} = \begin{bmatrix} R_{2,1} & 0 \\ 0 & R_{3,1} \end{bmatrix}^{-1} \tag{32}$$

$$\mathbf{Z} = \begin{bmatrix} 0 & 1 \\ 1 & 0 \end{bmatrix}, \ \mathbf{I} = \begin{bmatrix} 1 & 0 \\ 0 & 1 \end{bmatrix}. \tag{33}$$

Closed form solution for the coordinates of the source is found by solving

$$\mathbf{N} \cdot \mathbf{S} \cdot \mathbf{x} = \mathbf{N} \cdot \mathbf{u} \tag{34}$$

The mobile station position can then be computed using the least square solution. A closed form solution which can be used is given by Friedlander as

$$\mathbf{x} = (\mathbf{S}^T \cdot \mathbf{N}^T \cdot \mathbf{N} \cdot \mathbf{S})^{-1} \cdot \mathbf{S}^T \cdot \mathbf{N}^T \cdot \mathbf{N} \cdot \mathbf{u} \tag{35}$$

3 Simulation Model

The experiments were carried out using the simulation model required for the UMTS [6]. This is a typical bad urban environmental model (Manhattan model). The area consisted 32 blocks with a total number of 21 base stations. The street width was 30 m and the distance between two street corners was 230 m (see Fig. 1). Base station antennas were placed 10 m above the mobile users but below rooftops. In our implementation, we covered the simulation area with a regular grid with resolution of 10 m. In the simulation model, the effect of a multipath propagation was implemented.

The time of radio signal arrival between the mobile station and base stations under the multipath environment was modeled by the sum of true value τ_0 and non-line of sight (NLOS) error τ_m [7]

$$\tau = \tau_0 + \tau_m \tag{36}$$

The variable τ_m is defined as mean excess delay and essentially correlated with the root mean squared delay spread τ_{rms} [8]

$$\tau_m \approx k \cdot \tau_{rms} = k \cdot T_1 \cdot d^\varepsilon \cdot y \tag{37}$$

where k is a constant proportional coefficient ($k=1$ in urban region), T_1 is the median value of τ_{rms} at $d=1$km (for urban environment $T_1 = 0.7$ μs), d is the distance between the mobile station and base station, ε is an exponent (for urban environment $\varepsilon=0.5$) and y is a lognormal variate. Specifically,

$$Y = 10 \cdot \lg y \tag{38}$$

is a Gaussian random variable over the terrain at the distance d, having zero mean and a standard deviation σ_y (for urban environment $\sigma_y = 4$dB).

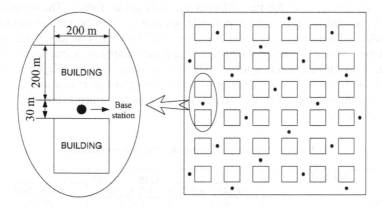

Fig. 1. Simulation model of the bad urban environment (Manhattan model)

4 Simulation Results

The most commonly used measure of positioning accuracy is the root mean square error metric. For a one dimensional case its value can be calculated as in [5]

$$RMS_x = \sqrt{\frac{\sum_{i=1}^{n}(x_i - x_t)^2}{n}}, \quad RMS_y = \sqrt{\frac{\sum_{i=1}^{n}(y_i - y_t)^2}{n}} \tag{39}$$

where x_t and y_t denote the true MS coordinate, x_i and y_i its estimation, and n ($n=10\,000$) represents a number of measurements. For the purpose of this article, however, the two dimensional metric is exploited instead. It can be easily derived by calculating the one dimensional metric for both coordinates, and than final value can be computed

$$RMS_{xy} = \sqrt{RMS_x^2 + RMS_y^2}. \tag{40}$$

Such a measure is very useful due to a possibility of representing it on a plane as a circle, which a center is the true target's position and radius' length is RMS_{xy} value. Within such a circle there are about 63% target's position estimations. Basing on the two dimensional root mean square error metric, another circle can be created. Its radius' length is two times longer than the previous one's. Within this circle there are about 98% target's position estimations. Moreover, the cumulative probability distribution functions (CDF) of the absolute position error were obtained from the simulation investigations. The absolute position error is defined as

$$\Delta d = \sqrt{(x - x_t)^2 + (y - y_t)^2} \qquad (41)$$

where x and y denote the estimated coordinates of a mobile station. All timing values have been assumed to be accurate within ± ½ of a WCDMA chip (the uniform random time error corresponds to a maximum distance error of about ± 38 m).

Comparisons of errors for the different methods are in Tab. 1. The simulation results are presented in Fig. 2. The hyperbolic position location estimation methods presented offer different accuracy's and complexities. The Chan's method offers a closed form solution, thus eliminating the need for an iteration approach, but requires a priori information to eliminate ambiguities. The Foy's method, using Taylor-series least square method offers accurate position location estimation at reasonable noise levels and is applicable to any number of range difference measurements, but can be computational intensive.

Table 1. Comparison of errors for the different methods

Method	Type of errors		
	RMS_x [m]	RMS_y [m]	RMS_{xy} [m]
Chan	82.13	134.05	157.21
Foy	107.99	103.60	149.65
Fang	78.19	108.61	133.83
Friedlander	240.79	233.35	335.31

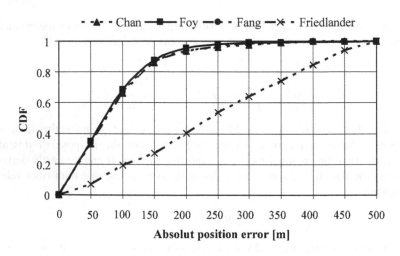

Fig. 2. The CDFs of absolute position error for the proposed methods

The Taylor-series method is iterative and has the risk of convergence to local minima. The Fang's method provides an optimal solution when the system of equations is consistent but does not make use of redundant measurements (only tree base stations are needed). The Friedlander's approach reduces the computational requirements for the solution but does is suboptimal because it eliminates a fundamental relationship and is difficult to implementation − inverse matrix with very low or high values.

5 Summary

To summarize Chan's method offers a closed form and is the best available option for solving hyperbolic equations (2). This method also requires a priori knowledge of the approximate location and distance between mobile station and first (serving) base station.

The optimal position location algorithm for a given situation depends on the geometrical configuration of the base stations, the number of coordinates of the source to be solved and range difference measurements utilized, computational requirements and complexity, assumptions on the statistical nature of the channel and desired accuracy.

An interesting observation that was made while studying the ambiguities in the Fang's and Chan's algorithms was that both these ambiguities are essentially the same. It was seen that if we make wrong choices in both algorithms for a given case then the wrong results given by both algorithms are identical.

Acknowledgments. This work was supported by the Ministry of Education and Science of the Republic of Poland, under grant No. O R00 0049 06.

References

[1] Chan, Y.T., Ho, K.C.: A Simple and Efficient Estimator for Hyperbolic Location. IEEE Trans. on Signal Proc. 42(8), 1905–1915 (1994)

[2] Foy, W.H.: Position-Location Solutions by Taylor-Series Estimation. IEEE Trans. on Aero. and Elec. Systems AES-12(2), 187–194 (1976)

[3] Fang, B.T.: Simple Solution for Hyperbolic and Related Position Fixes. IEEE Trans. on Aero. and Elec. Systems 26(5), 748–753 (1990)

[4] Friedlander, B.: A Passive Localization Algorithm and Its Accuracy Analysis. IEEE Jour. of Oceanic Engineer. OE-12(1), 234–245 (1987)

[5] Bronk, K., Stefanski, J.: Bad Geometry Effect in TDOA Systems. Polish Journal of Environmental Studies 16(4B), 11–13 (2007)

[6] ETSI, Universal Mobile Telecommunications System (UMTS); Selection procedures for the choice of radio transmission technologies of the UMTS, TR 101 112, ver. 3.2.0. (April 1998)

[7] Stefanski, J.: Method of Location of a Mobile Station in the WCDMA System without Knowledge of Relative Time Differences. In: Proc. of IEEE 65th Vehicular Technology Conference, pp. 674–678 (2007)

[8] Greenstain, L.J., Erceg, V., Yeh, Y.S., Clark, M.V.: A new path-gain/delay-spread propagation model for digital cellular channels. IEEE Trans. on Vehicular Techn. 46, 477–485 (1997)

Efficient Assignment of Multiple E-MBMS Sessions towards LTE

Antonios Alexiou[1], Christos Bouras[1,2], and Vasileios Kokkinos[1,2]

[1] Computer Engineering and Informatics Dept., Univ. of Patras, Greece
[2] Research Academic Computer Technology Institute, 26500 Rio, Patras, Greece
alexiua@ceid.upatras.gr, bouras@cti.gr, kokkinos@cti.gr

Abstract. One of the major prerequisites for Long Term Evolution (LTE) networks is the mass provision of multimedia services to mobile users. To this end, Evolved - Multimedia Broadcast/Multicast Service (E-MBMS) is envisaged to play an instrumental role during LTE standardization process and ensure LTE's proliferation in mobile market. E-MBMS targets at the economic delivery, in terms of power and spectral efficiency, of multimedia data from a single source entity to multiple destinations. This paper proposes a novel mechanism for efficient radio bearer selection during E-MBMS transmissions in LTE networks. The proposed mechanism is based on the concept of transport channels combination in any cell of the network. Most significantly, the mechanism manages to efficiently deliver multiple E-MBMS sessions. The performance of the proposed mechanism is evaluated and compared with several radio bearer selection mechanisms in order to highlight the enhancements that it provides.

1 Introduction

Nowadays, mobile industry rapidly evolves towards a multimedia-oriented model for providing rich services, such as Mobile TV and Mobile Streaming. LTE networks address this emerging trend, by shaping the future mobile landscape. However, the plethora of mobile multimedia services that are expected to face high penetration, poses the need for deploying a resource economic scheme.

In order to confront such high requirements for multimedia content, LTE networks rely on the E-MBMS framework. E-MBMS constitutes the evolutionary successor of MBMS, which was introduced in the Release 6 of Universal Mobile Telecommunication System (UMTS) [1], [2]. The main requirement in E-MBMS services is to make an efficient overall usage of radio and network resources. This necessity mainly translates into improved power control strategies, since the base stations' transmission power is the most limiting factor of downlink capacity. Under this prism, a critical aspect of E-MBMS performance is the selection of the most efficient radio bearer for the transmission of multimedia traffic.

In the frame of power control and transport channel selection during multimedia data delivery, several approaches have been proposed. The 3rd Generation Partnership Project (3GPP) TS 25.346 [3] and TR 25.922 [4], as well as work [5] are representative approaches. However, all of these works focus on MBMS, without considering E-MBMS and LTE requirements.

J. Wozniak et al. (Eds.): WMNC 2009, IFIP AICT 308, pp. 240–250, 2009.

In this paper, we propose a novel radio bearer selection mechanism for E-MBMS. The proposed scheme adopts downlink transmission power as the optimum criterion for radio bearer deployment and selects the transport channel combination that minimizes the transmission power of the base station. Therefore, Point-to-Point (PTP) and Point-to-Multipoint (PTM) transmission modes may be used separately or be combined and deployed in parallel. However, the most remarkable advantage of the proposed mechanism, that actually differentiates it from the above approaches, is that it may simultaneously serve multiple multimedia sessions. Our approach is compared with the 3GPP approaches in terms of power consumption and complexity so as to highlight its enhancements and the necessity for its incorporation in E-MBMS specifications.

The paper is structured as follows: In Section 2, we present the motivation behind our study and the related work in the specific field. Section 3 presents the proposed power control mechanism, while Section 4 is dedicated to the presentation of the results. Finally, the planned next steps as well as the concluding remarks are briefly described in Section 5.

2 Motivation and Related Work

The transmission of MBMS packets over the air interfaces may be performed on common (Forward Access Channel or FACH), dedicated (Dedicated Channel or DCH) or shared channels (High Speed-Downlink Shared Channel or HS-DSCH). Each channel has different power consumption characteristics [6], [7], [8].

The selection of the most efficient bearer is still an open issue in today's MBMS infrastructure, mainly due to its catalytic role in Radio Resource Management (RRM). The following paragraphs present the main radio bearer selection approaches existing in the bibliography.

2.1 MBMS Counting Mechanism (TS 25.346)

The 3GPP MBMS Counting Mechanism (TS 25.346) constitutes the prevailing approach of switching between PTP (multiple DCHs) and PTM (FACH) radio bearers, mainly due to its simplicity of implementation and function [3]. According to this mechanism, a single transport channel can be deployed in a cell at any given time. The decision on the threshold between PTP and PTM bearers is operator dependent, although it is proposed that it should be based on the number of served MBMS users. In other words, a switch from PTP to PTM resources should occur, when the number of users in a cell exceeds a predefined threshold. However, this mechanism provides a non realistic approach because mobility and current location of the mobile users are not taken into account. Moreover, this mechanism does not support FACH dynamic power setting. In other words, when FACH is employed, it has to cover the whole cell area that generally leads to unnecessary power wasting. Finally, TS 25.346 does not support the HSDPA technology, which could enrich MBMS with broadband characteristics [6].

2.2 MBMS PTP/PTM Switching Algorithm (TR 25.922)

3GPP TR 25.922 or MBMS PTP/PTM switching algorithm [4], assumes that a single transport channel can be deployed in a cell at any given time. Contrary to TS 25.346, it follows a power based approach when selecting the appropriate radio bearer, aiming at minimizing the base station's power requirements during MBMS transmissions. In TR 25.922, instead of using solely DCHs, HS-DSCH can also be transmitted. However, the restricted usage of either DCH or HS-DSCH in PTP mode may result to significant power losses. In both cases, the PTP (DCH or HS-DSCH, since the switching between HS-DSCH and DCH is not supported in this mechanism) and the PTM power levels are compared and the case with the lowest power requirements is selected. Even though TR 25.922 overcomes several inefficiencies of the TS 25.346, it does not support FACH dynamic setting.

2.3 Mechanism Proposed in 3GPP TSG RAN1 R1-02-1240

All the above mechanisms allow a single PTP or PTM transport channel deployment at any given time. On the other hand, the mechanism proposed in 3GPP TSG RAN1 R1-02-1240 [5], considers the mixed usage of DCHs and FACH, which can significantly decrease the base station's transmission power, depending on the number and the location of the users. According to this approach, the FACH channel only covers a dynamically selected inner area of a cell and provides the MBMS service to the users that are found in this part. The rest of the users are served using DCH to cover the remaining outer cell area.

However, none of the above MBMS power control mechanisms takes into account the ability of the base stations to support many simultaneous MBMS sessions. MBMS transmissions have increased power requirements and consume a large portion of the available power recourses of the base stations. Consequently, the number of parallel MBMS sessions that a base station could support is limited. Therefore, the selection of the appropriate radio bearer for a MBMS service should be done with respect to other existing MBMS sessions in the corresponding cell. The number of parallel MBMS sessions that a base station could support depends on many parameters. We could classify these parameters in three categories: user related parameters, MBMS session related parameters and provider related parameters. User related parameters are parameters such as UEs' (User Equipment) distances from the base stations and UEs' Quality of Service (QoS) parameters. The number of active MBMS sessions per cell, the number of UEs per MBMS session per cell and the bit rates of the MBMS services are some of the MBMS session related parameters. Finally, the portion of the available power recourses of base stations that could be used for MBMS transmissions is a provider related parameter. All these parameters should be considered in the RRM of MBMS so as to have efficient power control.

3   Proposed Mechanism for PTP and PTM Bearers Combination

This section presents the architecture and the functionality of the MBMS session assignment mechanism. The proposed mechanism incorporates all the basic functionalities of the standardized 3GPP approaches and furthermore, it integrates several enhancements (power based transport channel selection, combined usage of transport channels, parallel MBMS sessions and user mobility support).

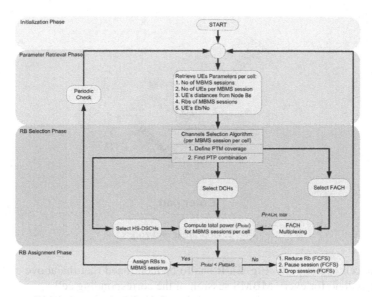

Fig. 1. Block diagram of the mechanism

The block diagram of the mechanism is illustrated in Fig. 1. According to Fig. 1, the mechanism consists of four distinct operation phases. These are: the initialization phase, the parameter retrieval phase, the radio bearer (RB) selection phase and the RB assignment phase.

The initialization phase (Fig. 1) launches the mechanism when one user expresses his interest in receiving a MBMS service (i.e. the mechanism begins when the first user requests the first MBMS service).

The parameter retrieval phase is responsible for retrieving the parameters of the existing MBMS users and services in each cell. In this phase, the mechanism requires the two of the three types of parameters, mentioned in the previous section: the user related parameters and the MBMS session related parameters. Regarding the latter type of parameters, the mechanism requires information about the number of active MBMS sessions per cell, the number of UEs per MBMS session per cell and the bit rates of the MBMS sessions. This information is retrieved from the Broadcast Multicast – Service Center (BM - SC). On the other hand, user related parameters are retrieved from the UEs through uplink channels.

The RB selection phase is dedicated to the selection of the transport channels for the MBMS sessions in any cell of the network. The most critical operations of the phase are executed by the Channels Selection Algorithm block (Fig. 1). The algorithm executed in this block selects the combination of PTP and PTM bearers that minimizes the downlink base station's transmission power in any cell of the network that multicast users are residing. In particular, the algorithm is executed in two steps. In the first step (Define PTM coverage) the algorithm estimates the optimum coverage of FACH for the users' distribution of any MBMS session in the cell. This coverage area is called inner part of the cell as illustrated in Fig. 2. In the second step (Find PTP combination), the mechanism decides which PTP bearer(s) will cover the rest

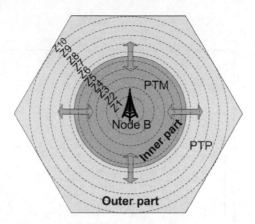

Fig. 2. Cell areas and zones

part of the cell (outer part - Fig. 2). It has to be mentioned that the above cell characterization is done for every MBMS session of the corresponding cell.

In order to estimate the optimum coverage of FACH (for any MBMS session in the
cell) in Define PTM coverage step (Fig. 1), the algorithm initially divides the cell in
ten zones (Z1 to Z10) according to the FACH dynamic power setting technique. Each
zone Zi refers to a circle with radius equal to 10i% of cell radius. Afterwards, the
algorithm scans all the zones and calculates the total base station's transmission
power for the following 21 transport Channel Configurations (CC):

– CC1: No FACH used. All users are covered by DCHs.
– CC2: No FACH used. All users are covered by HS-DSCHs.
– CC3: FACH for UEs up to Z1. All the rest UEs covered by DCHs.
– CC4: FACH for UEs up to Z1. All the rest UEs covered by HS-DSCHs.
– ....
– CC19: FACH for UEs up to Z9. All the rest UEs covered by DCHs.
– CC20: FACH for UEs up to Z9. All the rest UEs covered by HS-DSCHs.
– CC21: FACH for all UEs (up to Z10). DCHs and HS-DSCHs are not used.

The CC that consumes less power indicates the coverage of the FACH and determines the inner part of the cell. The same procedure is executed for any MBMS session in the cell. The output of the Define PTM coverage step is the coverage of the
FACH for any MBMS session in the examined cell.

Once the appropriate FACH coverage for any MBMS session in the cell is defined,
the algorithm enters the Find PTP combination step (see Fig. 1), which determines the
appropriate PTP radio bearer(s) that will cover the MBMS users residing in the outer
part of the cell for any MBMS session. The procedure is similar to the procedure
described in the Define PTM coverage step. The algorithm scans all the zones in the
outer part of the cell and calculates the total base station's transmission power in
order to cover all the outer part MBMS users only with PTP bearers. The first zone of
the outer part is Z(inner part+1), therefore the algorithm will have to scan the following PTP transport Channel Configurations (PTP_CC):

- PTP_CC1: DCHs for outer part UEs up to Z(inner part+1). All the rest outer part UEs (up to Z10) covered by HS-DSCHs.
- PTP_CC2: DCHs for outer part UEs up to Z(inner part+2). All the rest outer part UEs (up to Z10) covered by HS-DSCHs.
- ....
- PTP_CC(10-inner part): All MBMS users in the outer part cell are covered by DCHs. HS-DSCHs are not used.
- PTP_CC(10-inner part+1): HS-DSCHs for outer part UEs up to Z(inner part+1). All the rest outer part UEs (up to Z10) covered by DCHs.
- PTP_CC(10-inner part+2): HS-DSCHs for outer part UEs up to Z(inner part+2). All the rest outer part UEs (up to Z10) covered by DCHs.
- ....
- PTP_CC(2*(10-inner part)): All MBMS users in the outer part cell for the specific session are covered by HS-DSCHs. DCHs are not used.

After these calculations, the different PTP_CCs are compared and the PTP_CC with the lowest power requirements determines the PTP transport channel configuration for the outer part MBMS UEs of the specific MBMS session in the cell. The procedure is recursively executed for any MBMS session in the cell.

Generally, the output of the Channels Selection Algorithm block is the combination of PTM and PTP transport channels that consumes the lowest power resources between all possible combinations in the corresponding cell for any MBMS session running in it.

In the case of FACH there is another block in the mechanism's block diagram named FACH Multiplexing. When the number of MBMS sessions requiring FACH in cell is greater than one, these FACHs should be multiplexed onto a Secondary Common Control Physical Channel (S-CCPCH) [9], [10]. After the multiplexing procedure, the capacity of the S-CCPCH is calculated and based on this, the total power required for the common channels ($P_{FACH,total}$) in the corresponding base station is estimated. In this paper we consider a one to one mapping between MBMS sessions and FACHs.

The last action performed in the RB selection phase is the computation of the total base station's power (P_{total}) required to support all the MBMS sessions in each cell. However, we have to mention that the selected radio bearers are not yet assigned to the MBMS sessions. This action is performed in the following phase.

During the RB assignment phase, the P_{total} is compared with the available power assigned by the network provider to MBMS sessions in each base station (P_{MBMS}). Obviously, the P_{MBMS} constitutes the third type of parameters mentioned in the previous section, known as provider related parameter. If P_{total} is smaller than P_{MBMS}, the selected from the RB selection phase transport channels are assigned to MBMS sessions and the MBMS data transfer phase begins. In case when P_{total} is bigger than P_{MBMS}, a session reconfiguration procedure should occur due to the fact that there are no available radio resources to the base station so as to serve all the MBMS sessions in the examined cell. In this paper, we propose three possible reconfiguration events that could be used in such a case. The first is the reduction of the transmission rate of a MBMS session, the second is the pause of a MBMS session for a short time period and the last is the service cancellation.

The simplest policy in order to perform the three above reconfiguration events, is a First Come First Served (FCFS) policy. Following the FCFS policy and considering the available power, the mechanism performs the optimum event to the most recent MBMS sessions.

The above description refers to a dynamic model, in the sense that the UEs are assumed to be moving throughout the topology and the number of MBMS sessions varies. The parameter retrieval phase is triggered at regular time intervals so as to take into account the user related parameters, the MBMS session related parameters and the operator related parameters. This periodic computation inserts a further complexity as this information is carried in through uplink channels. This entails that a certain bandwidth fraction must be allocated for the transmission of this information in uplink channels, thus resulting to a system's capacity reduction. A further complexity is inserted due to the fact that the mechanism is executed many times for each cell in the topology.

4   Performance Evaluation

In this section, analytical simulation results for the evaluation of the mechanism are presented. In particular, through two different scenarios we examine the following key aspects of the mechanism (efficiency, comparison with current 3GPP approaches, handling of multiple parallel MBMS sessions).

The main assumptions of our simulations are presented in Table 1 and refer to a macrocell environment [8], [10]. In addition, Block Error Rate (BLER) target is set to 1% and no Space Time Transmit Diversity (STTD) is assumed.

Table 1. Simulation assumptions

Parameter	Value
Cellular layout	18 hexagonal grid cells
Sectorization	3 sectors/cell
Site-to-site distance	1 Km
Maximum BS Tx power	20 W (43 dBm)
Other BS Tx power	5 W (37 dBm)
CPICH power	2 W
Common channel power	1 W (30 dBm)
Propagation model	Okumura Hata
Multipath channel	Vehicular A (3km/h)
Orthogonality factor	0.5
E_b/N_0 target	5 dB

4.1   Comparison with 3GPP Approaches

The first scenario lasts for 200 sec and can be divided into four time periods, depending on the number of MBMS users. According to this scenario, a 64 Kbps service should be delivered to a group of users, whose initial position at each time period is presented in Table 2. For example, for the period 0 to 50 sec, 25 UEs receive the service at distance 50% and 7 UEs at distance 80% of the cell radius.

Table 2. UE Number, Coverage per time period

Time (sec)	UEs No	Coverage (%)	Best Performance
0-50	25	50	Our Mechanism
	7	80	
51-100	25	50	R1-02-1240, Our Mechanism
	2	80	
101-150	17	50	TR 25.922 (HS-DSCH), Our Mechanism
151-200	4	50	All except TR 25.922 (HS-DSCH)

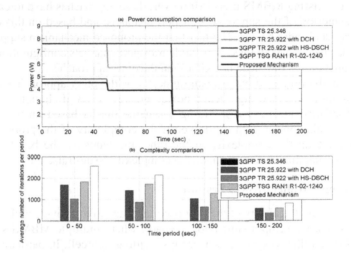

Fig. 3. (a) Power consumption and (b) complexity comparison

Fig. 3a depicts the power levels of the examined approaches. For example, for the period 0 50 sec, the total number of users in the cell is 32. By assuming that the threshold for switching between DCH and FACH (HS-DSCH is not supported) in TS 25.346 is 8 UEs (a value proposed in the majority of research works), TS 25.346 will deploy a FACH with 100% cell coverage (requiring 7.6 W).

The high initial users' population favors the deployment of FACH in order to serve all the UEs in 3GPP TR 25.922. However, as TS 25.346, TR 25.922 does not support FACH dynamic setting. This is the reason why TR 25.922 has the same power requirements with TS 25.346 (7.6 W) for the time period 0-50 sec.

The mechanism proposed in RAN1 R1-02-1240 allows the mixed usage of DCHs and FACH and supports FACH dynamic power setting. As shown in Fig. 3a, this mechanism requires 4.8 W in order to serve all the users, for the first time period. More specifically, this mechanism will deploy only a FACH with 80% coverage, since the user with the worst path loss resides at the borders of zone Z8.

Finally, Fig. 3a depicts the power requirements of the proposed mechanism. For the time period 0-50 sec, the output of the Channels Selection Algorithm block (Fig. 1) specifies that the users up to Z5 should be served by a FACH. Moreover, the most efficient combination of PTP bearers for the outer part MBMS users is to serve the remaining 7 users in zone Z8 with HS-DSCH. Therefore, 4.5 W in total are required in order to serve all the MBMS users with this mechanism. Obviously, a significant power budget, ranging from 0.3 to 3.1 W, may be saved for the period 0-50 sec compared with the other approaches.

On the other hand, Fig. 3b presents the computational overhead that each mechanisms inserts (number of iterations required to calculate the power and assign the ideal channel), based on the above scenario. In general, TS 25.346 inserts the lowest computational overhead, since it requires only the number of served MBMS users in order to assign the appropriate transport channel. The other approaches have higher computational overhead due to the fact that these mechanisms have to periodically retrieve the parameters of existing MBMS users. Moreover, these approaches have to calculate the power consumption of the supported transport channels; and based on this calculation to assign the ideal radio bearer. The fact that the proposed mechanism supports all the available channels and examines all possible transport channels configurations explains why its computational overhead is higher than the other approaches.

To sum up, the proposed mechanism outperforms the other approaches in terms of power consumption, since significant power budget is saved. It puts together the benefits of all mechanisms by providing a scheme that is based on the concept of transport channels combination. And even if the complexity of the proposed mechanism is higher than the complexity of the other mechanisms, the benefits from the optimal power planning counterbalance the complexity issues raised.

4.2 Managing Parallel MBMS Sessions

In order to evaluate the ability of the proposed mechanism to manage multiple parallel MBMS sessions, we setup a simulation scenario where multiple MBMS services are transmitted in parallel to several user groups residing in a cell. In particular, we suppose that four user groups receive four distinct MBMS services with characteristics presented in Table 3. Moreover, Table 3 presents the appropriate transport channel (with respect to power consumption) to serve each group at each time interval. Fig. 4 depicts the power consumption of each MBMS session as well as the total, aggregative power required to support the transmission of all services to the multicast users in the corresponding cell.

Users of the 1st MBMS session are served with a HS-DSCH channel, due to the small population, throughout the whole service time. At simulation time 50 sec, MBMS service 2 is initiated (Fig. 4). At this time instant, the mechanism, through the

Table 3. Scenario parameters

MBMS No.	Duration (sec)	Bit Rate	UEs Number	Maximum Coverage	Channel
1	0-600	64	10	80%	HS-DSCH
2	50-600	64	22	60%	FACH
3	100-150	64	2-13	60%	DCH
	151-300	64	14-19	60%	HS-DSCH
	301-600	64	20-27	60%	FACH
4	150-560	64	7	70%	DCH
	561-600	32	7	80%	DCH

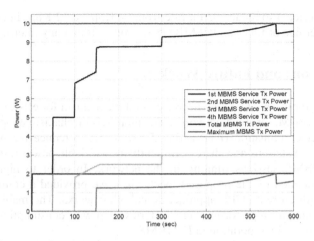

Fig. 4. Power levels of the MBMS sessions

RB selection phase, selects FACH as the most efficient transport channel for the transmission of the second MBMS traffic, since MBMS session 2 is delivered to a large number of users (22 UEs).

MBMS service 3 starts at simulation time 100 sec. At this time the 3rd multicast group consists of only two UEs; thus, the mechanism selects multiple DCHs for this MBMS service. The number of users receiving the service successively increases (join requests), reaching 13 UEs at simulation time 150 sec, 19 at 300 sec and 27 at the end of the simulation. The increasing number of users in the group forces the mechanism to perform a channel switching from DCH to HS-DSCH at simulation time 151 sec and another one from HS-DSCH to FACH at simulation time 301 sec, securing the efficient resource utilization.

At this point we have to mention that from time 300 sec until the end of the simulation, MBMS services 2 and 3 employ FACHs for the transmission of the MBMS data (see Table 3). During this time interval, the deployment of two parallel FACHs forces the mechanism to perform a FACH multiplexing procedure in the RB selection phase. Consequently, a single S-CCPCH with bit rate of 128 Kbps is used to deliver services 2 and 3. Additionally, P_{total} is lower than P_{MBMS}, which translates into efficient provision of the three parallel MBMS sessions.

At simulation time 150 sec, the MBMS service 4 is initiated and is targeted to a multicast group consisting of seven members. Multiple DCHs are selected by the mechanism to deliver the MBMS content to the 4th multicast group. Additionally, at the same time instance, P_{total} still remains smaller than P_{MBMS}, which means that the MBMS service 4 is accepted for transmission in the system. From simulation time 150 until the end of the simulation, four parallel MBMS sessions running in the system and our mechanism handles them in an efficient way.

Due to the fact that the users of the 4th multicast group are moving towards the cell edge an increase in P_{total} occurs and at simulation time 560 sec; and P_{total} exceeds P_{MBMS} value (Fig. 4). Thus, a session reconfiguration procedure is performed, forcing the MBMS service 4 to reduce its bit rate from 64 Kbps to 32 Kbps in order to ensure the efficient service of four parallel MBMS sessions without any interruption.

5 Conclusions and Future Work

In this paper we presented a novel power control mechanism for efficient radio bearer selection in E-MBMS enabled networks. The proposed mechanism adopts the concept of radio bearer combination (PTP and/or PTM) so as to reduce the power require-ments of the base stations and shares efficiently the available power resources of base stations to MBMS sessions running in the network. In order to highlight the en-hancements obtained by the proposed mechanism, we provided a comparison of the mechanism with current 3GPP approaches and other works. The main conclusion is that our mechanism outperforms them in terms of power consumption, underlining the necessity for its incorporation in E-MBMS.

The steps that follow this work could be at a first level the evaluation of the mechanism through additional simulation scenarios so as to measure other parameters such as delays in air interfaces during MBMS transmissions. At a second level, we plan to improve the functionality of our mechanism by incorporating the enhance-ments obtained from the use of multiple-input multiple-output (MIMO) antennas in High Speed Downlink Packet Access (HSDPA).

References

1. 3GPP TS 22.146 V8.3.0. Technical Specification Group Services and System Aspects; Multimedia Broadcast/Multicast Service; Stage 1 (Release 8) (2007)
2. 3GPP TR 23.846 V6.1.0. Technical Specification Group Services and System Aspects; Multimedia Broadcast/Multicast Service; Architecture and functional description (Release 6) (2002)
3. 3GPP TS 25.346 V8.1.0. Technical Specification Group Radio Access Network; Introduc-tion of the Multimedia Broadcast Multicast Service (MBMS) in the Radio Access Network (RAN); Stage 2, (Release 8) (2008)
4. 3GPP TR 25.922 V7.1.0. Technical Specification Group Radio Access Network; Radio re-source management strategies (Release 7) (2007)
5. 3GPP TSG-RAN WG1#28 R1-02-1240. Power Usage for Mixed FACH and DCH for MBMS, Lucent Technologies (2002)
6. Holma, H., Toskala, A.: HSDPA/HSUPA for UMTS: High Speed Radio Access for Mo-bile Communications. John Wiley & Sons, Chichester (2006)
7. Perez-Romero, J., Sallent, O., Agusti, R., Diaz-Guerra, M.: Radio Resource Management Strategies in UMTS. John Wiley & Sons, Chichester (2005)
8. 3GPP TR 25.803 V6.0.0. Technical Specification Group Radio Access Network; S-CCPCH performance for MBMS (Release 6) (2005)
9. 3GPP TS 25.211 V7.4.0. Physical channels and mapping of transport channels onto physi-cal channels (FDD) (Release 7) (2007)
10. 3GPP TS 25.212 V8.0.0. Multiplexing and channel coding (FDD) (Release 8) (2007)
11. 3GPP TR 101.102 V3.2.0. Universal Mobile Telecommunications System (UMTS); Selec-tion procedures for the choice of radio transmission technologies of the UMTS (UMTS 30.03 version 3.2.0)

CATS: Context-Aware Triggering System for Next Generation Networks

José Simoes[1], João Goncalves[2], Telma Mota[2], and Thomas Magedanz[1]

[1] Fraunhofer Institute FOKUS, Kaiserin-Augusta-Allee 31,
10589 Berlin, Germany
[2] PT Inovacao, Rua Eng. José Ferreira Pinto Basto,
3810-106 Aveiro, Portugal
{jose.simoes,thomas.magedanz}@fokus.fraunhofer.de,
{joao-m-goncalves,telma}@ptinovacao.pt

Abstract. Considering context information, namely location, in order to create new services has become a commercial trend. The innovative approaches that context-aware mechanisms make possible are being targeted by service providers of diverse areas. On the other hand, Service Oriented Architectures play a central role in allowing component reuse and low cost service creation. Together with IP Multimedia Subsystem enable the convergence of telecommunications and web services, allowing the network transport technologies to be abstracted from the services above. By integrating these three technologies, a number of synergies can be explored. Existing services can be easily enriched with context information, made available on a variety of networks and new services can be composed using previously existing building blocks. This paper explains how this integration can be achieved, and demonstrates the potentialities of this architectural paradigm with a prototype service.

1 Introduction

With the rapid advance in technology, it is becoming increasingly feasible for people to take advantage of the devices and services in the surrounding environment to remain "connected" and continuously enjoy the activity they are engaged in, be it sports, entertainment, or work. Such a ubiquitous computing environment will allow everyone permanent access to the internet anytime, anywhere and anyhow. Nevertheless, mobility is just one of the innumerous aspects that will play a preponderant role in the services of the future. In fact, being aware and able to communicate context is a key part of human interaction. Context is a much richer and more powerful concept, particularly for mobile users and can make network services more personalized, adaptable, interactive and therefore useful. Harvesting of context to reason and learn about user behavior will enhance the "internet of services" or "cloud computing" vision, allowing services to be composed and customized according to user context [1]. Combining both concepts, will allow services to be tailored to customer needs, providing richer experiences, thus enabling a whole new set of sensations, improving the user overall Quality of Experience (QoE). By QoE the authors refer to the subjective measure from a user's perspective of the overall value of the service provided.

J. Wozniak et al. (Eds.): WMNC 2009, IFIP AICT 308, pp. 251–262, 2009.

To support such vision, it is important that services are deployed under Next Generation Networks (NGNs), empowering cross-fertilization scenarios among different access networks. The IP Multimedia Subsystem (IMS) represents a natural response for this dilemma by combining traditional telecommunications concepts and internet service technologies. Furthermore, IMS can be considered to be an overlay control subsystem over heterogeneous networks [2], which makes it a natural choice for converging different access networks. On the service creation point of view, we need an architectural model that technically supports the previous described concepts. Service Oriented Architecture (SOA) is considered as the philosophy of encapsulating application logic in services with uniformly defined interfaces and making these publicly available via discovery mechanisms. Not only the notion of complexity-hiding and reuse but also the idea of loosely coupling services, are part of such ideology.

Based on these principles, we introduce the Context-aware Triggering System (CATS), a service created by the composition of different enablers, over an IMS environment, focusing not only in the technical structure of such application, but also on the business models that can arise from such architectural innovation. The ability to trigger a reaction to a specific occurrence or a set of events is a key functionality in context-aware systems. It can be seen as a standalone or distributed service build over a myriad of variables. In recent applications, it is usually implemented as a notification in a variety of ways: sound, light, messaging, video, etc. A concise survey of the current approaches to provide context-based services over mobile devices will be addressed in section 2. Then, we present in section 3 the proposed architecture and the concepts that allow ubiquitous context-aware service creation. A special focus over the architecture and concept validation is covered in section 4. Finally, section 5 concludes the paper by discussing future work.

2   Related Work

In the past few years, the interest for context-aware services (CAS) has raised increasingly, not only from the industry but also for the research perspective. Google mail was one of the first companies to provide such services by offering targeted advertising and Really Simple Syndication (RSS) feeds according to user demographic information and e-mail content type [3]. From the mobile marketing market, AcuityMobile developed and patented Spot Relevance [TM], a technology with the ability to deliver the right marketing content, to the right person, at the right time, in the right location.

The panorama for research institutes is not that different. VTT Information Technology (Finland) conducted a study to evaluate user needs for location-aware mobile devices [4] where they present user needs under five main themes: topical and comprehensive contents, smooth user interaction, personal and user-generated contents, seamless service entities and privacy issues. This study helped to identify key issues that should be addressed, but didn't propose any improvement. In order to manage various context data and make the best use of this data for application services in ubiquitous environments, Yoon-Ae Ahn [5] proposes a mobile object data management framework for location enhanced applications, which consists of a data collector, a context manager, a knowledge base, an inference engine and a mobile object database. Despite the validity of such an approach, it does not define interfaces outside a closed environment and does not target other types of context, namely presence, weather, time, etc. The work done by Kukhun et al. [6] presents a location-aware geographical pervasive system that provides mobile users with a service that corresponds not only to their requests but also to their preferences, location and querying time. Although this vision is going towards the authors' ideas, the work only establishes initial ground steps and does not specify any kind of architecture.

By providing an analysis of the requirements of a middleware for context-aware systems, focusing on mobility and privacy issues, the authors of [7] present PACE middleware, whose intent was the design of context-aware applications and solutions for modeling and managing context information. Focusing on mobile environments, the Web, semantics and their convergence with a shared communications world, project SPICE [8] presents a platform, which is capable of context acquisition, context representation, context enabling and use in mobile services. Although the work developed is very interesting in what concerns contexts, it does not target any notification or distribution mechanisms.

In fact, notification mechanisms are getting more prominent, especially with the advent of social networking (communities), as many see it as the new trend in the landscape of communications. The first success example of this interaction model that transmits an "always connected" sensation to the users was RSS, even if the underlying technology is actually based on polling.

In order to meet the demand for popularized application, socialized service of geographical information and solve the sharing, interoperation and integrated application of geographical information in network environments, paper [9] introduces a Service Oriented Architecture (SOA) model, which can conduct distributed deployment, by combining loosely-coupled and coarse-grained application components on the internet according to the demand and effectively support the development of a geographical information service. On the other hand, there are approaches such as [10] which use IP Multimedia Subsystem (IMS) [11] as an architectural framework to provide multimedia services over a packet based next generation network (NGN), where they use IMS application servers and enablers for service delivery. In fact, the idea of extending IMS management to SOA based next generation networks has already been acknowledged at [12].

By combining concepts from the previously enunciated approaches, we present an architecture based on IMS as infrastructure technology and SOA as a software architectural model, which will enable the provision, creation and dissemination of context-aware services for next generation networks.

3 Architectural Approach

The Context-aware Triggering System may be described as a multi context notification service, enhanced with user preferences and social activities. It allows the originating party to create one or more multimedia content items that can be distributed in a myriad of forms according to the recipient(s) set of contexts available, over an unicast or multicast overlay network. This means that the same trigger may be experienced in different ways depending on the user device, presence status, user preferences or other external factors/contexts. In other words, the main contribution focuses on a framework, which has the capability of casting content into context, allowing personalized, contextualized, adapted and interactive multimedia distribution across heterogeneous access networks.

3.1 General Overview

In order to achieve such service, the proposed architecture adopts a SOA model and it is implemented using an IMS infrastructure in order to provide a network agnostic control layer, facilitating at the end, seamless service delivery across a variety of heterogeneous networks. This approach will provide a way to have a flexible and modular architecture, making it possible to extend in the future by adding extra enablers and consequently add more functionalities. Figure 1 depicts a high level overview of the used components as well as indicates the protocols involved to communicate between themselves.

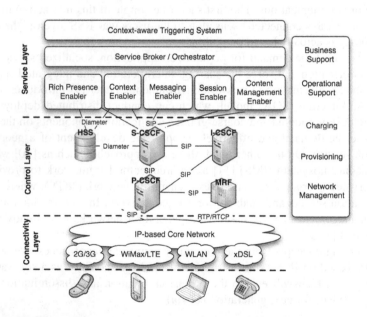

Fig. 1. General overview of the Context-aware Triggering System architecture

3.2 Functional Architecture

In the previous section we introduced the main components, which interact with the CATS; however, it is important to know how each one of them work and how they communicate with each other. Figure 2 details the functional architecture of CATS, focusing only the service layer, where our contributions are most notable.

Although all enablers interact directly with the application server, from figure 2 it is possible to identify that some of them are triggered through the service broker (the Rich Presence and Context enablers). The reason why not all components are directly induced by the service broker is related to the fact that at this stage the Messaging, Content Management and Session enablers specifications still aren't mature enough to allow reuse for any service. Consequently, at this point it makes no sense to provide means for these enablers to be correctly registered and exposed in order to be discoverable and orchestrable, as it would be usual in a SOA environment. Nevertheless, the further specification of these enablers is one of the goals of the authors' research, maximizing the simplicity of the application by exploiting the advantages of a service-oriented architecture. One of these advantages is the possibility that different service providers implement different parts of the system.

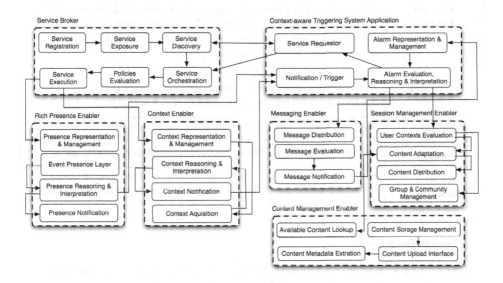

Fig. 2. Functional architecture of the Context-aware Triggering System

In the preceding scenario, we assume that the capabilities, ontologies, semantics and policies of the specified enablers where previously registered and exposed at the service broker. That means that the service broker is able to know how to execute every of the functions of the other components, as well as make them available whenever other components request so. To simplify, in this section we abstract provisioning and charging mechanisms, as well as business and operational support systems, focusing them later on, when convenient, on section 3.3. In order to understand the role of each enabler, table 1 provides a small description of the basic functionalities of each component. A full description of the functionalities and characteristics of these components can be found in [13].

Table 1. Description of the components inside the Context-aware Triggering System

Component	Description
CATS Application	Responsible for managing and representing the triggers defined by the end user.
Service Broker / Orchestrator	Coordinates a group of services that communicate with each other. In this context, services comprise intrinsically unassociated units of functionality that have no calls to each other embedded in them.
Rich Presence Enabler	Offers the basic presence functionalities as well as location-based information.
Context Enabler	Decouples context from application. In general it deals with context acquisition as a mechanism to obtain context from diverse sources, context fusion for merging correlated contextual information, context discovery as a mechanism to locate and access context sources, context diffusion/dissemination for efficiently propagating the context while ensuring availability and reliability [14].
Messaging Enabler	Two main functions: send messages across multi platforms to one or multiple users and allow two-way communication, authorizing users to use the same channels to create, modify or delete triggers
Session Management Enabler	Manages all the user-to-content and, vice versa, the content-to-user relationships (Content Distribution module). In fact, it provides the necessary signaling to deliver a specific content to its consumers.
Content Management Enabler	Responsible for the content provisioning, storage and management. When uploaded, the content may be tagged with metadata supplied by the user or content provider.

3.3 Design Decisions

Although the research in what concerns context brokers, enablers or frameworks is very active, it usually does not cover the entire life chain of context acquisition, reasoning and distribution. Our approach on the other hand focus on the whole cycle, giving particular emphasis to the efficient matching of content to context, as well as distribution mechanisms (context-aware multiparty delivery). In order to better understand our decisions, we separated the problem analysis into four different topics:

Context View

A generic context-aware infrastructure that effectively decouples context from the application and supports many different types of context-aware services requires a flexible model. Such model shall allow diverse context processing components, probably operated by different service providers, to be integrated, enabling the applications to handle only the highest level context. A generic model that satisfies some of these constraints is a producer-consumer, publish-subscribe, broker model [15]. This is the approach used in the CATS framework and its design principles are documented in [16]. All context-processing entities can either be CxC, CxP or a combination of both. In this sense, we can identify two main parts in our context-aware system:

- Context Management subsystem concerned with context acquisition and dissemination.
- Context modeling concerned with manipulation, representation, recognizing and reasoning about context and situations.

Service View

Considering a SOA and IMS approach allows for different service providers and network providers to share responsibilities in the service execution. IMS puts forward a well defined set of interfaces towards the upper layers. The network providers, depending on their business model, may allow third party enabler implementations to interact directly with their IMS infrastucture, or for certain key enablers only allow a SOA bus to be set up on top of them. Either way, the synergic possibilities between different service providers and network providers, relying on different business models, are obvious.

One of the consequences of context adaptation, particularly when using SOA, is that adaptation can be partly satisfied by the ability to compose and orchestrate services on the fly, based on user and/or environmental context. As service composition is a dynamic and flexible process, which allows for reconfiguration as the context changes, this actuation will represent an indispensable part of what it means to be context aware. A step toward meeting this challenge was to design a clear separation between the context management supporting architecture and service architecture and to clearly understand the interaction points between context and services. Previous context-aware architectures have shown that incorporating context change decision triggering within the application, service or service enabler is a poor performing design.

Network View

From the network perspective, one key objective of CATS is to investigate the dynamic optimization of content delivery to a group of users based on the group's context; i.e. context-aware multiparty delivery, with QoS guaranteed. The rationale for this work and the associated network's view of the CATS architecture enabling it are presented in detail in [17]. As efficient network support of real-time group communications requires dedicated multiparty networking technologies, such as IP multicast associated with QoS dynamic control, to send the same content simultaneously to multiple receivers, with intermediate routers duplicating packets and allocating network resources (e.g., bandwidth) as needed, CATS takes this features into consideration.

Moreover, the heterogeneity of group members, e.g. in terms of link characteristics, access to network resources, devices used, physical mobility and environment, makes it almost impossible to deliver content to all group members in the same manner. Dynamic adaptation of the multiparty delivery is thus needed to optimize group communications and maximize satisfaction of each individual group member. Hence, although the same content is to be sent to all group members, its delivery needs to be optimized for each user based on his/her context. In that respect, optimizations need to be considered at all levels of the protocol stack, including session, transport and network layers. Furthermore, the context considered to drive the dynamic adaptation of the multiparty delivery needs to encompass not only the networking context of each group member (e.g. link quality and characteristics) but also its environmental context (device capabilities, physical location, speed, etc.). Naturally, different adaptations can be applied simultaneously for different members of a group receiving the same content, with adaptations for each user being driven by its own networking and environmental contexts. Figure 3 illustrates the network architecture for context-aware multiparty delivery, including context detection and delivery.

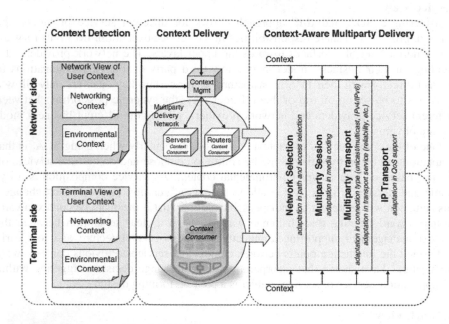

Fig. 3. Context-aware Multiparty delivery

Content View

Our view focuses on four different stages: production, processing, management and consumption. The first stage includes content identification, targeting, selection, and classification. Here, content is carefully prepared, monitored, and very importantly, related content metadata is generated. The subsequent phase includes adaptation of the content and its preparation for mobile use. In order to push the content into the platform, there is a preliminary step to pre-process the content in order to make it suitable for the platform's infrastructure. Then the item, and its metadata are stored in a database, which is part of the content management system. The last step is content consumption. After a group of people, an individual user or a context driven event trigger the consumption of the content, it will be delivered to the end-user terminals for final consumption, via multicast or unicast.

4 Validation

4.1 Use Case: Targeted Advertising

One of the most common use cases is advertising. Using CATS, advertising companies can send targeted campaigns towards their customers or even aiming at new clients that are willing to receive commercials in exchange for some other benefits (discounts, free goods or even free SMS's or minutes). This brings opportunities not only for operators but also to developers which build applications under other platforms such as social networks. By correlating this new hype (social networks) with the CATS application, the possibilities and amount of collectable contexts are huge, not to mention the increasing target audience that can arise from concepts such as "Friend of a Friend" (FOAF) [1], widely available inside these environments.

A typical scenario would present users in proximity of certain sell-points with information about the goods that are available on saving prices. Moreover, we will assume that there is not a "one-to-one" relationship between the advertising company and the shop, but a "one-to-many". This means that, although the trigger will be triggered by the location, the information targeted for customers will not only take rich presence information and user preferences into consideration, but also other contexts, providing the user with a new type of experience. Although the matching between content and context is still in a premature phase, exploring this topic will enable us to increase the relevancy of a specific trigger towards a specific user (this is done during the inference process).

When the session enabler receives a notification to start a media session with the user, it verifies the user's device settings and network conditions to check whether the content needs to be adapted. If not, the advertised content and probably a discount coupon are sent to the user. Upon entering the shop, the user uses his coupon to get a 20% discount on a brand article. Later on, when the coupon number is inserted in the system, the user's history profile is updated accordingly, leveraging this information for future reference. Despite the fact this architecture can be used for advertising purposes, it would have to be improved to fulfill both advertisers and customers' needs. Moreover, in a world in which the line between users, providers and advertisers is becoming increasingly blurry, a trend best exemplified by the Web 2.0 phenomena, avoiding intrusion and advertisement spam has become very pertinent. Work towards this issue will be addressed by future work.

4.2 CATS Concept Validation

In order to test the concept proposed on this paper, a prototype of the Context-aware Triggering System was developed. The application was developed using JAVA and deployed on a BEA Weblogic SIP Server[2], using both SIP and Hypertext Transfer Protocol (HTTP) protocols. At this stage, it was not possible to interconnect this application with any Broker or Orchestration server, therefore, the discovery process was bypassed in our trials and the service orchestration hard coded on the CATS application to enable service execution between all the enablers.

The validation scenario was setup under FOKUS Open SOA Telco Playground[3] where the following components where used: OpenIMSCore[4], FOKUS Presence

---

[1] Friend of a Friend, FOAF Vocabulary Specification 0.91, http://xmlns.com/foaf/spec/

[2] BEA WebLogic® SIP Server, Converging Internet, SOA and next-generation telecom services, www.bea.com/sip

[3] FOKUS Open SOA Telco Playground, http://www.opensoaplayground.org

[4] Open IMS Playground, www.open-ims.org

Server[5] and Converged Open Messaging Server (COMS)[6]. Once again, no Context enabler was available and consequently, the CATS application was responsible for acquiring the contexts. Session Management Enabler and Content Management Enabler were also not tested at this stage. By using a proper configuration webpage, a user could select which capabilities the trigger should have.

On our tests, we decided that a user should be notified every time a specific friend is in a 1Km distance of his house. Furthermore, the trigger should only be triggered if the user was also in the same region and it is not raining. As a notification mode, the user defined to be alerted via message: via SIP message when online and via SMS otherwise. Moreover, we enabled trigger modification via SMS, so that the trigger would consider not only raining but also snowing into its evaluation. When the user's friend finally moves inside the targeted area, the trigger evaluates the user location, the current weather conditions and when all conditions are verified, its checks the current user presence status. As it was offline, the user was notified via SMS that his friend was in the surroundings. Using such a testing environment enabled us to see that the concept works and can be useful. Nevertheless, this was not performing in a dynamic way like presented in the architecture. Therefore, although the concept was tested, there is still a need to validate the architecture.

5   Conclusions and Future Work

Service creation paradigms are changing as users demand for adapted, contextualized and personalized services, allowing interactivity and easy integration with their social environments, enabling them to access and consume content/services anytime, anywhere, anyhow. Moreover, users want to be in control of their services which require an adaptation of today's business models and technologies in order to support this desire. In fact, to achieve this vision, we need an architectural model that technically supports the previous described concepts. SOA is considered as the philosophy of encapsulating application logic in services with uniformly defined interfaces and making these publicly available via discovery mechanisms. Based on such premises, CATS architecture enables users to design complex and personalized notification mechanisms using a simple and clear interface. Moreover, this architecture allows two-way communications, introducing interactivity into the scenario, enabling users to add, edit or delete a trigger across a set of different platforms (e-mail, SMS, SIP message, XMPP, HTTP, etc.). By using IMS as a control technology, the architecture becomes access network independent and therefore easy to integrate with a different set of devices using multiple technologies.

Furthermore, exploring such an architectural approach enabled the identification of interesting topics that could improve the functionalities, efficiency and scopes of the CATS application. Firstly, the context to content matching, which will improve the relevance of the content recommended or targeted to the end user, depending on the user's current context and profile. Both these dimensions will be relevant to choose content based on its semantic metadata and on its format. The effectiveness of such

---

[5] FOKUS Presence Server, www.open-ims.org/presence
[6] FOKUS Converged Open Messaging Server, www.open-ims.org/coms

improvement is believed to be directly impacted by the completeness of content metadata, the availability of context information, and the existing mechanisms to perceive user preferences. Secondly, the integration with social networks will be addressed under the context enabler; not only to improve context-gathering mechanisms but also to cover business related issues. Thirdly, it showed how interesting this architecture can be for advertising. Correlating this fact with industry reports [18], [19] which indicate advertising will be the business model of the future, makes all sense to explore this idea in the future.

As future work, this architecture will be extended under two parallel branches. The first will be under the European Project C-Cast where a framework to collect data, manage context groups, enable context driven content creation, reason contexts, distribute and efficiently manage context aware multiparty and multicast transport is defined. The second, inside the FOKUS Open SOA Telco Playground where an advertising enabler will be deployed to create a unified multi channel (TV, Mobile and WEB) advertising solution capable of targeting users according to their preferences and needs, concerning intrusion and advertisement spam avoidance by introducing the concept of buddy list, already existent and well developed in the context of social networks, generalizing it for advertising companies.

To sum up, this paper presented an application and architecture which pretend to act as a starting point for future context aware next generation network architectures (supporting both unicast and multicast distribution mechanisms) improving at the end the users quality of experience by addressing topics such as personalization, contextualization, adaptation, interactivity and mobility.

References

1. Baker, N., Zafar, M., Moltchanov, B., Knappmeyer, M.: Context-Aware Systems and Implications for Future Internet. In: Future Internet Conference and Technical Workshops, Prague, Czech Republic (May 2009)
2. Camarillo, G., Garcia-Martin, M.: The 3G IP Multimedia Subsystem: Merging the Internet and the Cellular Worlds. John Wiley & Sons, Chichester (2004)
3. Google Mail Privacy Policies: Targeted Ads in Google Mail (2009),
   http://mail.google.com/mail/help/intl/en_GB/about_privacy.html
4. Kaasinen, E.: User needs for location-aware mobile services. In: Personal and Ubiquitous Computing. Springer, Heidelberg (2003)
5. Ahn, Y.: Design of a Mobile Object Data Management Framework for Location enhanced Applications. In: International Conference on Convergence and Hybrid Information Technology, pp. 270–273 (2008)
6. Kukhun, D., Soukkarieh, B., Lopes-Ornelas, E., Sedes, F.: LA-GPS: A location-aware geographical pervasive system. In: 24th IEEE International Conference on Data Engineering Workshop, Cancun, Mexico, pp. 160–163 (2008)
7. Hightower, J., LaMarca, A., Smith, I.: Practical Lessons from Place Lab. IEEE Pervasive Computing Magazine (2006)
8. Zhdanova, A., Zoric, J., et al.: Context Acquisition, Representation and Employment in Mobile Service Platforms. In: 15th IST Mobile and Wireless Communications Summit, Mykonos, Greece, June 4-8, 2006, pp. 64–68 (2006)

9. Shujun, D., Liang, L., Chengqi, C.: Research on Geographical Information Service Based on SOA. In: Proceedings of the IEEE International Conference on Automation and Logistics, Qingdao, China (2008)

10. Brajdic, A., Lapcevic, O., Matijasevic, M.: Service Composition in IMS: A Location Based Service Example. In: 3rd International Symposium on Wireless Pervasive Computing, pp. 208–212. IEEE, Santorini (2008)

11. 3GPP TS 23.228: Service requirements for the Internet Protocol (IP) multimedia core network subsystem (IMS); Stage 1. Release 9, 2008-12-19

12. Blum, N., Magedanz, T., Schreiner, F., Wahle, S.: From IMS Management to SOA based NGN. Journal of Network and Systems Management 17 (2009) (accepted for publication), ISSN 1064-7570

13. full description of the functionalities and characteristics of these components

14. Requirements and concepts for context detection and context-aware multiparty transport. In: Deliverable 6-C-CAST: Content Casting, European Project ICT-2007-216462 (2008)

15. Chen, H.: An Intelligent Broker Architecture for Pervasive Context-Aware Systems, PhD Thesis, University of Maryland, Baltimore County (2004)

16. Goncalves, J., Moltchanov, B., et al.: Context Management Architecture for Future Internet Services. In: ICT Mobile Summit 2009, Spain (June 2009)

17. Antoniou, J., Simoes, J., et al.: Context-Aware Multiparty Networking. In: ICT Mobile Summit 2009, Spain (June 2009)

18. Blackshaw, P.: The Global Village, Virtually Realized: Social networking engages users. In: Nielsen Online Customer Insight (May 2008)

19. Global Mobile Forecasts to 2013 Worldwide Market Analysis, Strategic Outlook & ed-Forecasts to 2013. In: Informa telecoms & media Reports/Forecasts (December 2008)

Performance Analysis of Uplink Packet Schedulers in Cellular Networks with Relaying

Desislava C. Dimitrova[1], Hans van den Berg[1,2], and Geert Heijenk[1]

[1] University of Twente, Postbus 217, 7500 AE Enschede, The Netherlands
{d.c.dimitrova,geert.heijenk}@ewi.utwente.nl
[2] TNO ICT, The Netherlands
J.L.vandenBerg@ewi.utwente.nl

Abstract. Deployment of intermediate relay nodes in cellular networks, e.g. UMTS/ HSPA, has been proposed for service enhancement, which is of particular importance for uplink users at the cell edge suffering from low power capacity and relatively poor channel conditions. In this paper, we propose and investigate a number of uplink packet scheduling schemes deploying the relay functionality in different ways. Using a combined packet and flow level analysis capturing the specifics of the scheduling schemes and the random behavior of the users (initiation and completion of flow transfers), the performance of the various schemes is evaluated and compared to a reference scenario where relaying is not used. The main performance measures considered in our study are the instantaneous data rate, the energy consumption and the mean flow transfer time. Interestingly, considering flow transfer times, it is found that the use of relay nodes is not only particularly beneficial for users at the cell edge but also has a strongly positive effect on the performance of users at locations close to the base station.

1 Introduction

In the last decade UMTS (Universal Mobile Telecommunications System) evolved from an on-paper standard to a full scale technology adopted by many operators around the globe. The system continues to advance towards improved radio resource efficiency and higher data rates with the HSPA (High Speed Packet Access) technologies being the latest upgrade. For the downlink, from base station to mobile, the HSDPA (High-Speed Downlink Packet Access) technology has been standardized by 3GPP in [1]. Alternatively, [2] introduces the EUL (Enhanced Uplink) for the uplink, from mobile to base station. In EUL, the channel resource is shared among all active users. Channel access is organized by the base station (BS) according to a particular scheduling scheme operating on a time scale of 2 ms (TTI - Transmission Time Interval).

The key resource in EUL is the total power budget at the base station which is shared among the mobile stations (MSs). A mobile station is an EUL compatible device which often operates on batteries and as such has limited transmit power. Depending on its distance to the BS a MS can or cannot use the total available resource on its own. If it can, system throughput is optimized by single transmission during a time slot (TTI) [8]. Alternatively, if a MS cannot fully use the total available resource, simultaneous transmissions by multiple MSs are a better choice [6].

J. Wozniak et al. (Eds.): WMNC 2009, IFIP AICT 308, pp. 263–273, 2009.

In recent years many studies concentrated on EUL performance evaluation of various scheduling schemes for EUL, [7,4] to mention some. In principle, independently of the scheme, remote MSs suffer from high path loss and thus low data rates. Some authors, see [10], propose service in parallel due to users' limited power capacity. However, this only improves resource utilization but not the offered service. Placing a relay station (RS) can positively influence data rates by breaking a long communication path MS-BS into two shorter paths, MS-RS and RS-BS namely. The idea of relaying is rather attractive, however its implementation is trivial since it poses new implementation decisions such as where to set the relay or what scheduler to apply. As each decision has several possible outcomes the number of potential implementations rapidly grows. Selecting an optimal solution requires identifying main implementation choices as well as their advantages and disadvantages. Most importantly, before any research, we need to show that relaying can improve performance.

The application of relaying has been researched for a range of cellular and computer networks. [12] discusses a downlink scenario with relays placed at 2/3 of the cell radius. In the presence of interference the author shows, for a single MS, that sending via the relay is beneficial in terms of packet errors and delays. The authors of [9] have chosen for a LTE (Long Term Evolution) network with changing number of MSs but evaluate only outage probability. Furthermore, the paper benefits from considering scheduling which adapts to channel conditions but lacks realism since unlimited flow sizes are assumed. Other studies which show the advantages of relaying for OFDMA networks are [11,14], while [13] and [3] concentrate on WiMAX networks.

In this paper we show the potential performance benefits of relaying for EUL, with focus on the resulting modifications on packet scheduling. In particular, we discuss four relay-enabled Round Robin schedulers and a reference scheme which does not use relaying. The schemes are compared with respect to a range of performance measures such as received powers, energy consumption, instantaneous rates and mean flow transfer times. Additionally, we show how RS characteristics such as transmit power and position influence MS performance. In summary, the most prominent contributions of our research are: incorporating the impact of flow level dynamics in terms of randomly changing number of active users; diversity of schedulers with relaying as opposed to the few options discussed by other authors; and combined assessment of both MS performance and RS specifics.

The paper continues as follows. First, in Section 2, we briefly discuss the relaying concept and describe the scheduling schemes considered in this paper. The model description and analysis appear in Section 3 and Secion 4, respectively. Section 5 presents our findings of the performance evaluation. Finally, Section 6 summarizes our work.

2   Relay-Enabled Round Robin Scheduling Schemes

We first briefly recall the idea of relaying in a cellular system and introduce the required notation. Next, the scheduling schemes considered in this paper are described.

Relaying allows us to reduce signal degradation due to path loss by breaking up a long path into two short ones. However, these gains are against the cost of an additional transmission of the data at the relay node. In a relay-enabled system a MS chooses

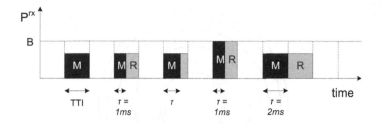

Fig. 1. Scheduling schemes: OBO, SOBO, SoptOBO, SidOBO and ExOBO

between direct communication to the BS (direct path MS-BS) and sending via the relay (relay path MS-RS-BS), depending on the data rates that can be realized on these paths. Note that the relay path consists of two sub-paths - MS-RS and RS-BS, and the data rate realized on the relay path depends on the achievable data rates on the two sub-paths.

Each (sub-)path is characterized by a set of transmission parameters: the distance between the communicating devices d_{zz}, the path loss L_{zz}, the transmit power P_{zz}^{tx}, the duration of a transmission opportunity τ_{zz} and the realized data rate r_{zz} during a transmission opportunity. The index zz refers to the specific (sub-)path, i.e. ms for the direct path from MS to BS, mr for the sub-path from MS to RS, and rs for the sub-path from RS to BS. The transmission times τ_{mr} and τ_{rs} are scheduler specific; their sum is denoted by $\tau = \tau_{mr} + \tau_{rs}$. The relations between the transmission parameters are further discussed in Section 4. We will now continue to introduce the specific scheduling schemes considered in our study.

Scheduling Schemes
Scheduling decisions are taken by the BS for which a relay behaves as a stationary MS that needs to be scheduled. Several approaches towards incorporating relay transmissions into scheduling are possible. We consider only Round Robin (RR) type of schedulers where mobile users are served one-by-one (OBO), independently of their channel conditions. Studies among which [10,8] show that OBO is inefficient in resource utilization when remote users with limited power capacity are served. However, a relay is closer to the MSs and it is more likely that a single MS is capable to fully use the available resources. In our study we consider four variants of relay-aware OBO schedulers: *SOBO, SoptOBO, SidOBO* and *ExOBO*, see Figure 1. In addition, plain *OBO* is considered as a reference scheme in which a MS always transmits directly to the base station independently of its location in the cell (and the duration of a single transmission opportunity equals one TTI, i.e. $\tau_{ms} = 2ms$. In the other four schemes a MS can select the direct or the relay path, depending on its location relative to the BS and the RS.

Extended OBO (ExOBO) assigns (in the case an indirect path is chosen) a single TTI to the transmissions on both sub-paths, i.e. $\tau_{mr} = \tau_{rs} = 2ms$. Therefore, for the service of a particular mobile station in a single round of the RR scheme the base station sets aside 2 TTIs meaning $\tau = 4ms$. Transmissions on a direct path are assigned one TTI.

In the *shared OBO* (SOBO) scheme transmissions on direct and indirect paths are assigned one TTI, i.e. $\tau = 2ms$. In the case of an indirect path, the single TTI is divided

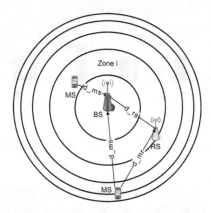

Fig. 2. Single cell model with cell division in zones. The distance between MS-BS, MS-RS and RS-BS are denoted as d_{ms}, d_{mr} and d_{rs}, respectively.

into two equal intervals of $1ms$ and MS and RS both receive one interval to transmit, i.e. $\tau_{mr} = \tau_{rs} = 1ms$. Assigning fixed-length transmission times of one TTI for direct and indirect paths eases implementation, but the static subdivision for indirect paths is inefficient when the achievable rates on the sub-paths differ.

Optimized SOBO (SoptOBO) also uses one TTI to serve a single MS, both for direct and indirect paths (i.e. $\tau = 2ms$). However, for indirect paths, it selects the transmission times on the sub-paths MS-RS and RS-BS such that the same amount of data is transferred, i.e. $\tau_{mr}r_{mr} = \tau_{rs}r_{rs}$. Although SoptOBO maximizes the resource utilization it is rather challenging for implementation. The selection of the specific transmission times for the sub-paths requires complex functionality in the base station.

Ideal SOBO (SidOBO) is the last relay-aware scheduling scheme we consider in our study. It is like SOBO but assumes ideal forwarding conditions, namely unlimited RS transmit capacity and interference budget. Under these assumptions, the relay is eliminated as a potential bottleneck. The scheme is more of a theoretical design rather than a practical solution, and provides us some notion of best-case performance.

3   Model

Our model consists of a single cell with EUL users (MSs) generating data traffic. In order to differentiate in the location of MSs we divide the cell in K zones with equal area, see Figure 2. Each zone is characterized by its distance to the base station d_i, measured from the outer edge of the zone, and a corresponding path loss L_i, $i = 1 \cdots K$. All MSs within the same zone i are assumed to have the same distance d_i, $i = 1 \cdots K$ to the base station. The distribution of (active) MSs over the zones is represented by the system state vector $\underline{n} \equiv (n_1, n_2, \cdots n_i, \cdots n_K)$, where n_i is the number of currently active users (MSs) located in zone i.

Mobile stations become active, at uniformly distributed positions in the cell, according to a Poisson process with rate λ, and have to transfer a file with mean size F (the file size has an exponential distribution). Hence, due to the equal area zones, all zones have the same file transfer initiation rate (or: arrival rate) $\lambda_i = \lambda/K$. Active MSs keep their position during the file transfer. A MS chooses between direct transmission or via the relay, depending on its location. All mobile stations have the same maximum transmit power capacity P_{max}^{tx}. Depending on their location either the maximum transmit power or a lower power is used, i.e. $P^{tx} \leq P_{max}^{tx}$, such that a MS maximizes its utilization of the available budget B.

In our study a relay is a device with one basic functionality, namely resending data from MS to BS. Such simple design keeps the cost down and still provide improved performance to the MSs.

In this paper we assume multiple relay stations on a concentric circle around the base station, such that BS, RS and MS are on a straight line[1]. A fixed path selection policy based on distance is adopted according to which all MSs located beyond the relay, i.e. $d_{ms} > d_{rs}$, transmit via the indirect path and in all other cases, i.e. $d_{ms} \leq d_{rs}$, direct transmission is chosen. An ideal switching mechanism between transmitting and receiving at the RS is assumed as the same frequency band is used.

At both BS and RS limited channel resource, budget B, is assumed. Given noise rise η and constant thermal noise N, the shared budget B at the BS is derived: $B = \eta \times N$. The same budget B is set also for the RS, with the exception of SidOBO for which the RS budget is unlimited. The other interference component we account for is the self-interference, i.e. the own signal parts which travel via secondary (reflected) paths. Intentionally disregarding important factors such as inter- and intra-cell interference allows us to identify the effect relaying has on performance. However, we realize that such factors may have negative effects on the performance gains.

4 Analysis

Relaying has two opposite effects on performance (excluding effects outside the scope of this paper): on the one hand it enables a higher received power level (increasing the data rate), but on the other hand forwarding at the RS requires an additional transmission of the data (decreasing effective data rates). Apart from that performance is influenced by the varying number of active users in the cell. In order to account for these effects we apply a combined packet and flow level analysis. For each scheduler we start with calculating received powers from which subsequently instantaneous data rates can be derived. Next we consider the flow level dynamics - flow initiation and flow completion - and model the system at flow level by a continuous time Markov chains (CTMC) with transmission rates derived from the instantaneous data rates and the system state $\underline{n}$. From the steady state distribution of the Markov chain the mean flow transfer time of a MS in a particular zone can be obtained. Our combined approach exhibits several advantages above other existing approaches. Starting at the packet level allows us to catch

---

[1] In fact, this assumption requires unlimited number of relays on the concentric circle but it can be shown that this ideal situation can be very well approximated by 8 or 10 RS. In principle, the analysis in this paper can easily be extended to any other assumption concerning the number and position of the RSs. However, for the purpose of the present paper, this specific choice suffices.

specifics of scheduling scheme and environment. Applying Markov models supports the modeling of a real network where the number of active users constantly changes. The approach is also rather scalable since changing the scheduler or the environment only ask for recalculation of the instantaneous rate at the packet level.

4.1 Received Powers

According to the propagation law, the received power P^{rx} is a direct result of applied transmit power $P^{tx} \leq P^{tx}_{max}$ and path loss $L(d)$. Furthermore, the maximum possible received power of any MS is limited by the available budget B resulting in: $P^{rx} = min(P^{tx}/L(d); B)$. The assumed path loss model, taken from the Okumura-Hata model, is given by $L(d) = 123.2 + 35.2log10(d)$ (in dB) with d the distance in kilometer.

In Section 2 we explained how each scheduling scheme organizes its transmissions. Furthermore, when relay is used, we should differentiate between the received powers $P^{rx}(BS)$ and $P^{rx}(RS)$ at the base station and the relay station, respectively.

When a relay is used in SOBO, SidOBO and ExOBO the slower sub-path limits the transmission. Therefore, the transmit powers at MS and RS are chosen such that the received powers at RS and BS are equal ($P^{rx}(RS) = P^{rx}(BS)$) and determined by the minimum of the achievable received powers on both sub-paths. In the case of SoptOBO $P^{rx}(RS)$ is in general not equal to $P^{rx}(BS)$ since the scheme deploys different data rates on the two relay sub-paths. Note that SidOBO differs from the rest in the higher maximum transmission powers of RS it uses. The scheme-specific received powers at the BS can be given as:

$$P_i^{rx}(BS) = \begin{cases} P^{rx}_{i,ms} & \text{for OBO and direct path in other schedulers} \\ min(P^{rx}_{i,mr}, P^{rx}_{rs}) & \text{for relay path in SOBO, SidOBO and ExOBO} \\ P^{rx}_{i,rs} & \text{for relay path in SoptOBO} \end{cases} \quad (1)$$

4.2 Instantaneous Rate

The instantaneous rate r of a particular MS represents its realized data rate when the MS is scheduled. Hence, r can be defined as the amount of unique data transferred during the period τ when the BS serves that MS. Note that how much unique data is transferred is determined by how long the MS sends, i.e. τ_{mr}. The dependency of the instantaneous rate on the other transmission parameters is thus given by:

$$r_i = \frac{r_{chip}}{E_b/N_0} * \frac{P_i^{rx}}{N + \omega P_i^{rx}} * \frac{\tau_{i,mr}}{\tau}, \quad (2)$$

In Equation 2 $i = 1 \cdots K$ indicates zone number, r_{chip} is the system chip rate and E_b/N_0 is the energy-per-bit to noise ratio. The parameter ω is used to account for reflected signals.

Each scheduler uses different policy for the selection of the total transmission time τ and the MS transmission times $\tau_{i,mr}$, see Section 2. After substitution the expressions of the scheduler specific instantaneous rates become:

$$r_i = \begin{cases} r_{i,ms} & \text{for OBO and direct path in others} \\ min(r_{i,mr}, r_{i,rs}) * \frac{1}{2} & \text{for relay path in SOBO, SidOBO and ExOBO} \\ \frac{r_{i,mr} * \tau_{i,mr}}{\tau} & \text{for relay path in SoptOBO} \end{cases} \quad (3)$$

In OBO and for all MSs selecting direct path $\tau_{mr} = \tau$ holds. SOBO, SidOBO and ExOBO assign to the mobile station only half of the total time τ, since the factor 1/2 in Equation (3). Finally, the more generic expression for SoptOBO follows from the strategy to adapt the transmission opportunities $\tau_{i,mr}$ according to the MS's zone.

4.3 Flow Dynamics

The actual rate a MS receives is lower or equal to the instantaneous rate r_i and depends on the state $\underline{n}$ of the system and on the number of users n in particular. This new rate we term state-dependent throughput $R_i(\underline{n})$ and it differs per scheduler. In all schemes but ExOBO the state-dependent throughput can be derived from $R_i(\underline{n}) - r_i/n$. The difference for ExOBO originates in the different number of TTIs assigned to MSs using the direct and the relay path - one and two TTIs respectively. Hence, for $R_i(\underline{n})$ we can write:

$$R(\underline{n}) = \begin{cases} \frac{r_i}{n+n_{relay}} & \text{for MSs using direct path} \\ \frac{2r_i}{n+n_{relay}} & \text{for MSs using relay path} \end{cases} \quad (4)$$

where n_{relay} is the number of MSs which make use of the relay station. Note that r_i is difference for each path according to Equation (3).

The changes in the system state $\underline{n}$ (due to flow initiations and completions) are described by CTMC of K dimensions with transition rates λ_i and $R(\underline{n})$. The particular form of the Markov chain is scheduler specific via $R(\underline{n})$. A more detailed description of a related CTMC can be found in [4]. All schemes but ExOBO are a multi-class M/G/1 processor sharing (PS) model whose steady state distribution is readily available, e.g. [5]. For the ExOBO model we implemented a Markov chain simulator which based on large number of state transitions, i.e. 1 million, provides us with the steady state distribution. Subsequently, we can derive performance measures such as mean flow transfer times T_i:

$$T_i = \frac{1}{r_i F} * \frac{1}{1 - \rho} \quad (5)$$

where $i, i = 1 \cdots K$ is zone number, ρ the system load and F the mean file size.

4.4 Energy Consumption

The retransmission of data on the relay path suggests higher energy consumption E compared to direct transmission, given both have the same total duration τ. However, E is an absolute measure and does not account for the fact that different schemes might transfer different amount of data. To compensate we use the energy-per-bit E_{bit} measure which incorporates the details on data traffic and is given by:

$$E_{bit,i} = \frac{E_i}{r_{i,mr} \tau_{mr}} = \frac{E_i}{r_{rs} \tau_{rs}} \quad (6)$$

where E_i, $i = [1 \cdots K]$, is the energy consumption of a MS in zone i. The product $\tau_{mr} *$ $r_{i,mr}$ gives the unique amount of data transferred on the relay path, which is a scheduler specific value.

5   Numerical Results on Relaying

In this section we present a quantitative evaluation of performance at both packet and flow level for the discussed scheduling schemes. We are particularly interested in comparing performance based on measures such as received powers, energy consumption, instantaneous rates and mean flow transfer times. Special attention is given to the impact of flow level dynamics. Most results are generated from mathematical analysis with small part coming from simulations, i.e. flow level analysis of ExOBO. Simulations are carried out with a generic simulator developed in Matlab for deriving the steady-state distribution of multi-dimensional Markov chains.

5.1   Parameter Settings

In the numerical experiments we apply system chip rate r_{chip} of 3840 kchips/s, thermal noise level N of -105.66 dBm and noise rise target η of 6 dB, see [6]. For SidOBO the available budget B at the RS is unlimited. Self-interference of 10% is considered, i.e. $\omega = 0.1$.

The cell is divided in $K = 10$ zones with equal area[2], see Figure 2. Given a cell radius of 2 km we applied straightforward link budget calculations to determine the zone radii and the path losses. By default relay stations are located in zone 3 with some experiments explicitly saying when this changes. Both MS and RS has maximum transmit power $P_{max}^{tx} = 0.125$ Watt. EUL flows are taken with an E_b/N_0 target of 5 dB [6] and mean file size is $F = 1000$ kbit. The call arrival rate is set to $\lambda = 0.4$ calls/sec.

5.2   Received Power Levels

In order to build up good understanding of the impact relaying has on schedulers, we confine the discussion of received powers to OBO, SOBO and ExOBO. The received powers at the BS as a function of the distance are presented in Figure 3(a) for a range of positions of the RSs. As expected, relaying increases received powers but the position of the RS has strong impact on the gain. Moving the RS towards the edge of the cell lowers the gain - effect caused by increased distance d_{rs} between RS and BS and fewer MSs using the relay.

Recall that transmission on the relay path can be limited by both relay and mobile station, which is illustrated by the graphs for RSs in zone 1 and 2 in Figure 3(a). Typically signal degrades with distance as we can see from the graph of direct transmission, i.e. OBO. However, when the RS forms the bottleneck on the relay path, all MSs served by it are characterized by the same received powers at the BS which explains the unchanging graphs. Note that with RS in zone 3 and further this is always the case. If the MSs are the bottleneck and the received powers decrease in distance and differ per zone.

---

[2] Extensive numerical experiments showed that this granularity is sufficient for our purposes.

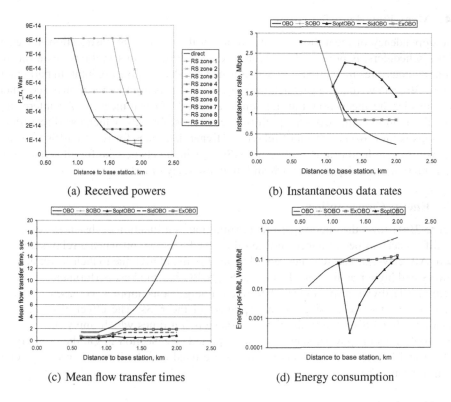

(a) Received powers

(b) Instantaneous data rates

(c) Mean flow transfer times

(d) Energy consumption

Fig. 3. Performance evaluation based on (a) received powers; (d) energy-per-bit consumption; (b) instantaneous rates; and (c) mean flow transfer times (flow level)

5.3 Instantaneous Data Rates

The instantaneous data rates generated by each of the five scheduling schemes are presented in Figure 3(b). Fixed path selection policy is applied. In all cases when direct transmission is used all schemes show the same performance. However, beyond zone 3 (after relay is used) the relay-enabled schemes outperform OBO meeting our expectations. The graphs of SOBO and ExOBO coincide, since both schemes have the same received powers at BS and use only half the transmission time τ to transfer unique data, see Equation (3). Contrary to our expectations not SidOBO is the best performing scheme but SoptOBO. Even if the RS does not limit transmissions, a MS can use only the available budget B at the BS for the fixed time τ_{mr}. SoptOBO circumvents the limitation by using adaptive transmission times.

The disadvantage of relay due to forwarding is well illustrated by the graphs of SOBO and ExOBO. Given our fixed path selection policy, MSs close behind the relay would be better if they could select the direct path. Furthermore, in OBO and SoptOBO rate degrades in distance since both schemes differentiate in users from different zones. Alternatively, given a relay in zone 3, all transmissions on the relay path in SOBO, SidOBO and ExOBO are limited by the relay which explains the flat graphs.

5.4   Mean Flow Transfer Times

The dependency of mean flow transfer times of the distance to base station, for each of the schedulers, is presented in Figure 3(c). Some of the observations made for the instantaneous rates in Section 5.3 are confirmed but also new ones exhibit. In particular, Figure 3(c) shows that the gain achieved by using RSs is even more emphasized when we consider the mean flow transfer times as a performance measure. An interesting new observation is that even the performance of close-by MSs which use direct transmissions is improved as result of relaying. Higher data rates for remote MSs translate to lower system load (number of active MSs n), which is beneficial for all active MSs, see Equation (4), independently whether they use relay or not.

5.5   Energy Consumption

Figure 3(d) shows the energy-per-bit consumption as a function of distance. As so far, the differences in performance between the schedulers exhibit when a relay station is used. All relay-enabled schemes record lower E_{bit} than OBO due to their lower applied powers and higher achievable data rates. SoptOBO has the best performance since its optimal TTI division maximizes data rates. Energy consumption increases with distance as result of increasing transmit powers of the MSs. The effect of distance is largest for SoptOBO due to the flexible transmission times which imply that stations with high transmit powers send for a long period, see Section 2. On the contrary, in SOBO and ExOBO the transmission times are fixed and the increase in energy consumption smoother.

6   Conclusion

We have compared the performance of four Round Robin packet schedulers for uplink transmission in a cellular network with relaying. As performance measures we used received powers at the base station, energy consumption, instantaneous data rates and mean flow transfer times. Analysis of the flow transfer times (taking into account that the number of active users varies over time) is enabled by a combined packet- and flow level analysis. Our study reveals the pros and cons of the four relay-aware scheduling schemes, and makes a comparison to the performance achieved in the case without relaying. As expected, relaying significantly improves the service offered to remote users with poor channel conditions. Interestingly, it is also beneficial for users which do not use the relay - an effect that can be seen only at the flow level. The latter observation supports our claim that the flow level dynamics have crucial influence on scheduling performance, and needs to be taken into account in performance comparisons.

As topics for further study we suggest investigating alternative relay-aware schedulers (e.g. with parallel transmissions), and examining the mutual relay influence in a more complex scenario with multiple cells.

Acknowledgments

The authors would like to thank Remco Litjens for the prolific discussions and for providing us with challenging new ideas.

References

1. 3GPP TS 25.308. High Speed Downlink Packet Access (HSDPA); Overall Description
2. 3GPP TS 25.309. FDD Enhanced Uplink; Overall Description
3. Bel, A., Seco-Granados, G., Vicario, J.L.: The benefits of relay selection in wimax networks. In: ICT-MobileSummit 2008 (2008)
4. Dimitrova, D.C., van den Berg, H., Heijenk, G., Litjens, R.: Flow-level performance comparison of packet scheduling schemes for UMTS EUL. In: Harju, J., Heijenk, G., Langendörfer, P., Siris, V.A. (eds.) WWIC 2008. LNCS, vol. 5031, pp. 27–40. Springer, Heidelberg (2008)
5. Haverkort, B.: Performance of computer communication systems. John Wiley & Sons Ltd., Chichester (1999)
6. Holma, H., Toskala, A.: HSDPA/HSUPA for UMTS. John Wiley & Sons Ltd., Chichester (2006)
7. Liu, T., Mäder, A., Staehle, D., Everitt, D.: Analytic modeling of the UMTS enhanced uplink in multi-cell environments with volume-based best-effort traffic. In: IEEE ISCIT 2007, Sydney, Australia (October 2007)
8. Ramakrishna, S., Holtzman, J.M.: A scheme for throughput maximization in a dual-class CDMA system. In: ICUPC 1997, San Diego, USA (1997)
9. Reetz, E., Hockmann, R., Tonjes, R.: Performance study on cooperative relaying topologies in beyond 3g systems. In: ICT-MobileSummit 2008 (2008)
10. Rosa, C., Outes, J., Sorensen, T.B., Wigard, J., Mogensen, P.E.: Combined time and code division scheduling for enhanced uplink packet access in WCDMA. In: IEEE VTC 2004, Los Angeles, USA (Fall 2004)
11. Schoenen, R., Halfmann, R., Walke, B.H.: An fdd multihop cellular network for 3gpp-lte, May 2008, pp. 1990–1994 (2008)
12. Umlauft, M.: Relay devices in umts networks - effects on. In: Proceedings of the Fifth Annual Mediterranean Ad Hoc Networking Workshop, Med-Hoc-Net 2006 (2006)
13. Vidal, J., Munoz, O., Agustin, A., Calvo, E., Alcon, A.: Enhancing 802.16 networks through infrastructure-based relays. In: ICT-MobileSummit 2008 (2008)
14. Wei, H.Y., Ganguly, S., Izmailov, R.: Ad hoc relay network planning for improving cellular data coverage. In: Personal, Indoor and Mobile Radio Communications, 2004. PIMRC 2004, September 2004, vol. 2, pp. 769–773 (2004)

A Method of the UMTS-FDD Network Design Based on Universal Load Characteristics

Slawomir Gajewski

Gdansk University of Technology,
11/12 Narutowicza Str., PL-80-233 Gdansk, Poland
slagaj@eti.pg.gda.pl

Abstract. In the paper an original method of the UMTS radio network design was presented. The method is based on simple way of capacity-coverage trade-off estimation for WCDMA/FDD radio interface. This trade-off is estimated by using universal load characteristics and normalized coverage characteristics. The characteristics are useful for any propagation environment as well as for any service performance requirements. The practical applications of these characteristics on radio network planning and maintenance were described.

1 Introduction

The WCDMA/FDD radio interface provides various ways of its resources utilization for UMTS radio network planning [1,2,3]. The most effective utilization of these resources depends however, on the knowledge of relationships among all the techniques used in that interface and their impact on the transmission performance, system capacity and coverage in real conditions. The coverage of mobile and base stations depends in turn on interface load that has influence on system capacity [4]. It is well known that system capacity is dependent on transmission performance and traffic characteristics required for different services. Capacity also depends on propagation environment and speed of terminals [3,5]. In addition, all these characteristics change over time.

Methods for estimating a coverage using characteristics defining the maximum allowable pathloss in the radio channel, allowing the performance of services are well known. A novel approach described in this paper relies on the use of the family of load characteristics to determine the capacity-coverage trade-off. It permits the quantitative assessment of benefits and losses arising from the use of this phenomenon during the design and operation of the UMTS networks.

The capacity-coverage trade-off estimation is facilitated by the designation of a single parameter of the system. This parameter can be easily calculated on the basis of link budget or can be estimated using the method described in [6].

2 WCDMA/FDD Radio Interface Load

The load of WCDMA/FDD radio interface in a given cell depends on multiple access interference coming from that cell and surrounding cells. As known, excessively high value of load cannot be accepted in the system. Coverage of the mobile and base stations can be reduced significantly for heavy-load and many services may be inaccessible. Accordingly, when calculating the link budget we have to take into account the interference margin (usually 3 dB) [2].

In the case of uplink transmission the load factor η_{UL} can be found [2, 3] according to the following expression

$$\eta_{UL} = \left(1 + \bar{\xi}_{UL}\right) \sum_{j=1}^{M} \frac{v_j}{d_{pc,j}} \frac{\left(E_b/N_t\right)_{n,j}}{G_j} \tag{1}$$

where M is the number of connections at a given time, v_j is the activity factor of signal source (for j-th connection), $d_{pc,j}$ is the factor of dynamic power control efficiency for j-th connection, $\bar{\xi}_{UL}$ is the average value of other-cell to own-cell interference ratio in uplink, G_j is the processing gain for j-th connection, $(E_b/N_t)_{n,j}$ is the required (nominal) bit energy per interference and noise power density ratio for j-th connection (for the service implemented in this connection).

Maximum allowable pathloss $L_{max,UL,j}$ for the uplink j-th connection is related to the load factor η_{UL} by

$$L_{max,UL,j} = \frac{P_{MS,j}}{F_{Tx} F_{Rx}} \frac{G_{Tx} G_{Rx} G_{SHO}}{N} \frac{G_j}{\left(\dfrac{E_b}{N_t}\right)_{n,j}} (1 - \eta_{UL}) \tag{2}$$

where $P_{MS,j}$ is the power of signals transmitted by the mobile station for j-th connection, G_{Tx} is the transmitter antenna gain, G_{Rx} is the receiver antenna gain, F_{Tx} is the signal loss of transmitter antenna feeder, F_{Rx} is the signal loss of receiver antenna feeder, G_{SHO} is a soft handover gain, N is the power of thermal noise (including the receiver noise figure).

Expression (2) is called the uplink load characteristic [3] of j-th connection. This characteristic can be used to determine the range of a mobile stations.

Additionally, the interface load η_{DL} for downlink can be defined on similar principle as for uplink including average value of downlink orthogonality factor $\bar{\alpha}$ [3]. It is given by

$$\eta_{DL} = \left(1 - \bar{\alpha} + \bar{\xi}_{DL}\right) \sum_{j=1}^{M} \frac{v_j}{d_{pc,j}} \frac{\left(E_b/N_t\right)_{n,j}}{G_j} \tag{3}$$

where $\bar{\alpha}$ represents the average orthogonality factor, $\bar{\xi}_{DL}$ is the average value of other-cell to own-cell interference ratio in downlink.

In this case the maximum allowable pathloss $L_{max,DL,j}$ for j-th connection is related to the load factor η_{DL} by the following expression

$$L_{max,DL,j} = \frac{P_{BS,j}}{F_{Tx} F_{Rx}} \frac{G_{Tx} G_{Rx} G_{SHO}}{N} \frac{G_j}{\left(\dfrac{E_b}{N_t}\right)_{n,j}} (1 - \eta_{DL}) \tag{4}$$

when $P_{BS,j}$ is the power of signals transmitted by the base station over the traffic channel in *j-th* connection [3,6,7]. Expression (4) is called downlink load characteristic of *j-th* type of service.

It was proved in [3,6], that the maximum allowable pathloss $L_{max,DL}$ for all connections (and all types of implemented services) not always depend on the type of service and it can be calculated from expression

$$L_{max,DL} = \frac{P_{BS,max}}{N} \frac{G_{Tx}G_{Rx}G_{SHO}}{F_{Tx}F_{Rx}} \left(1 - \overline{\alpha} + \overline{\xi}_{DL}\right)\left(\frac{1}{\eta_{DL}} - 1\right) \tag{5}$$

when $P_{BS,max}$ is the maximum power of signals transmitted by the base station to all active mobile stations.

The expression (5) is called the generalized downlink load characteristic [3] of the WCDMA/FDD radio interface. This characteristic is valid for different services having various performance requirements and data rates with the limitations characterized in [3]. Expression (5) can be used to estimate the coverage and capacity-coverage trade-off for mixed services of any type.

3 Family of Universal Load Characteristics in Uplink

As we can see in expression (2), maximum allowable uplink pathloss $L_{max,UL,j}$ for the *j-th* connection can be calculated on the basis of link budget. We can introduce the parameter $A_{UL,j}$ resulting from uplink link budget, in form

$$A_{UL,j} = \frac{P_{MS,j}G_{Tx}G_{Rx}G_{SHO}G_j}{F_{Tx}F_{Rx}\left(\frac{E_b}{N_t}\right)_{n,j} N} \tag{6}$$

Then we can describe the load characteristic, which is simplified to the form

$$L_{max,UL,j} = A_{UL,j}\left(1 - \eta_{UL}\right) \tag{7}$$

Now, the relationship between $L_{maxUL,j}$ and η_{UL} depends only on single parameter $A_{UL,j}$. On the other hand, the $A_{UL,j}$ parameter of the uplink load characteristic depends on the mobile station transmitter power and another radio link budget parameters. The power of transmitted signals can change and the value of the parameter $A_{UL,j}$ can also be changeable in the time.

If we have a realistic estimate of possible values for the parameter $A_{UL,j}$ then we can draw the family of uplink load characteristics. The family reflects the relationship between $L_{max,UL,j}$ and η_{UL} for different parameters of radio link budget. The family is shown in Fig. 1.

In this simple case, through the use of the load characteristics we can determine the scope of variation of maximum allowable pathloss ($L_{max,UL,j}$) for any services and for a specific load. Also, we can determine the pathloss variability when changing load, or we can estimate the extent of the coverage variation.

As we can see, the family of load characteristics can be used to cellular network design, as explained in section 6 and 7.

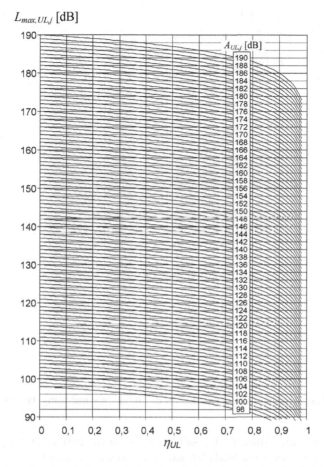

Fig. 1. Family of load characteristics for uplink designed using the link budget

4 Families of Universal Load Characteristics in Downlink

In downlink, maximum allowable pathloss $L_{max,DL,j}$ for *j-th* connection, we can calculate from expression (4) (on the basis of link budget) and then we have

$$L_{\max,DL,j} = A_{DL,j}\left(1 - \eta_{DL}\right) \tag{8}$$

where parameter $A_{DL,j}$ has the form

$$A_{DL,j} = \frac{P_{BS,j}G_{Tx}G_{Rx}G_{SHO}G_j}{F_{Tx}F_{Rx}\left(\dfrac{E_b}{N_t}\right)_{n,j} N} \tag{9}$$

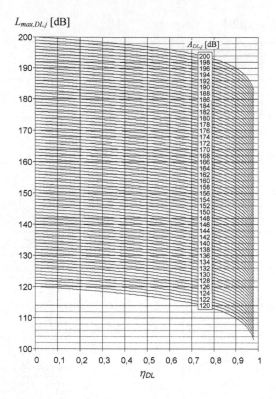

Fig. 2. Family of load characteristics for downlink designed using the link budget

If we have a realistic estimate of possible values for the parameter $A_{DL,j}$ then we can draw the family of downlink load characteristics, as we can see in Fig. 2.

We can also use another method for calculation of the downlink load characteristics, which is presented by details in [6]. Then the family of downlink load characteristics we can calculate using formula (5).

Then the maximum allowable pathloss $L_{max,DL}$ [3,6] for the service of any type (implemented in downlink connection) is given by

$$L_{max, DL} = A_{DL}\left(1/\eta_{DL} - 1\right) \tag{10}$$

where

$$A_{DL} = \frac{P_{BS,max}\left(1 - \overline{\alpha} + \overline{\xi}_{DL}\right)G_{Tx}G_{Rx}G_{SHO}}{F_{Tx}F_{Rx}N} \tag{11}$$

The graph of the family of downlink load characteristics calculated using the expression (5) is presented in Fig. 3.

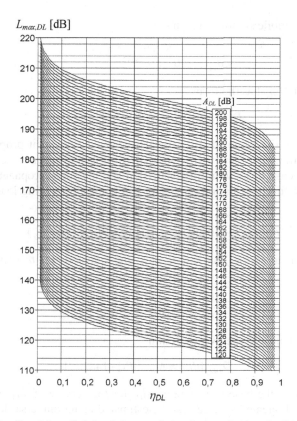

Fig. 3. Family of downlink load characteristics designed using the formula (5)

5 Normalized Coverage Characteristics

In order to evaluate the coverage on the basis of universal load characteristics, the propagation loss characteristic is needed. For that purpose, we can use any empirical propagation loss model. For example, we can use standardized ITU-R propagation models published in [8].

In particular for microcell in UMTS i.e. for the Outdoor to Indoor & Pedestrian (OIP) propagation environment the propagation loss L_{prop} can be estimated as

$$L_{prop}\ [dB] = 148 + 40\lg(d) \tag{12}$$

where the range d is given in [km].

For example, in uplink we can calculate the range of mobile station using the expression (2) for the service implemented in j-th connection. If we compare the maximum allowable pathloss with the propagation loss of a given microcell propagation environment $L_{max,UL,j} = L_{prop}$ then we get

$$A_{UL,j}[dB] + 10\lg(1 - \eta_{UL}) = 148[dB] + 40\lg(d_{UL,j}) \tag{13}$$

where $d_{UL,j}$ is the range of mobile station for the service implemented in j-th connection.

The range of mobile station is given by

$$d_{UL,j} = 10^{\frac{D_{UL,j}[dB]+10\lg(1-\eta_{UL})}{40}} \qquad (14)$$

where parameter $D_{UL,j}$ has the form

$$D_{UL,j}[dB] = A_{UL,j}[dB] - 148\,[dB] \qquad (15)$$

Parameter $D_{UL,j}$ is the value of $A_{UL,j}$ normalized by the constant propagation loss in a given propagation environment, i.e. normalized by 148 dB as in our case.

Also, for the case of downlink transmission and the same propagation loss model we can calculate the base station range (for *j-th* connection) using load characteristic given by (4). Then we have

$$A_{DL,j}[dB] + 10\lg(1 - \eta_{DL}) = 148\,[dB] + 40\lg(d_{DL,j}) \qquad (16)$$

The base station range $d_{DL,j}$ for the service implemented in *j-th* connection can be calculated from

$$d_{DL,j} = 10^{\frac{D_{DL,j}[dB]+10\lg(1-\eta_{DL})}{40}} \qquad (17)$$

Parameter $D_{DL,j}$ is now given by expression

$$D_{DL,j}[dB] = A_{DL,j}[dB] - 148\,[dB] \qquad (18)$$

On the other hand, the base station range d_{DL} for the service of any type and for the mixed set of implemented services can be estimated using universal load characteristic given by the formula (5). Then we get

$$A_{DL}[dB] + 10\lg\left(\frac{1}{\eta_{DL}} - 1\right) = 148\,[dB] + 40\lg(d_{DL}) \qquad (19)$$

In this case, the base station range d_{DL} for the service of any type has the form

$$d_{DL} = 10^{\frac{D_{DL}[dB]+10\lg\left(\frac{1}{\eta_{DL}} - 1\right)}{40}} \qquad (20)$$

Then the parameter D_{DL} is given by

$$D_{DL}[dB] = A_{DL}[dB] - 148\,[dB] \qquad (21)$$

Using expressions (14), (17) and (20) we can draw the families of normalized coverage characteristics for a given propagation model for both the uplink and downlink transmission. Sample graph of the family of normalized coverage characteristics for uplink is presented in Fig. 4 (shown in section 6).

6   Example Network Design Procedures

For example, let $D_{UL,j} = 9$ dB and the value of uplink load $\eta_{l,UL} = 0.5$ for a given connection *j* provided the same service for all connections is used. Then the mobile station range $d_{UL,j}$ is approximately 1.3 km, as we can see on the relevant coverage characteristic in Fig. 6.1.

For the cell radius decreased from 1.3 km to 1 km the value of $D_{UL,j}$ is decreased to 6 dB. We may obtain the same result by increasing the load factor to $\eta_{UL,2} \approx 0.87$ as is shown in Fig. 4. It gives the capacity growth ΔM in proportion to the ratio of discussed load factors i.e.

$$\Delta M = \frac{\eta_{UL,2}}{\eta_{UL,1}} = \frac{0.87}{0.5} = 1.74 \tag{22}$$

for that type of service. This means that the number of connections can be increased by 74 % in comparison with those ones for $\eta_{l,UL} = 0.5$ and a specific type of service. Note that, the mobile station range is reduced by 300 m for that service. So, if the coverage may be reduced in a cell, the capacity can be greater.

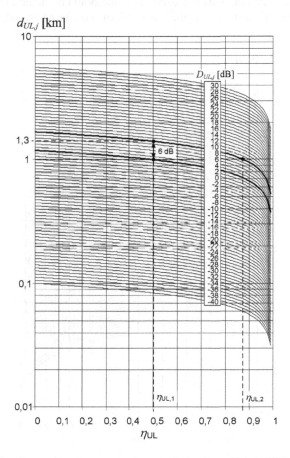

Fig. 4. Family of normalized coverage characteristics for uplink in OIP environment

As we can see in Fig. 4, normalized coverage characteristics can be used to:

– e.g. if we would like to reduce the cell radius from 1.3 km to 1 km, it can reduce the value of $D_{UL,j}$ by 6 dB. If the network designer wants to do it, it can use lower antenna gain in the base station or it can decrease the mobile station transmitter power,
– result will be the same, if we not change the parameters of the radio interface, but allow an increase in load factor to 0.87 (instead of the previous 0.50).

Presented consideration shows, that the normalized coverage characteristics can be used to simple way of capacity-coverage trade-off estimation for the service implemented in j-th connection, for both the uplink and downlink. In addition, by using the characteristics (5) the capacity-coverage trade-off can be estimated for mixed services of any type in downlink.

Let's consider another example. Suppose that calculated for downlink the value of the parameter D_{DL} = 23 dB. This corresponds to the bold characteristic in Fig. 5, passing through points "1" and "2". At the point "1" of the characteristic, the base station

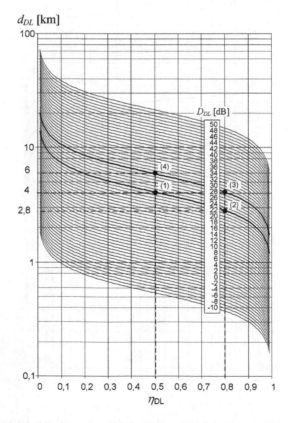

Fig. 5. Family of normalized coverage characteristics for the services of any type implemented in downlink (OIP environment)

range is approximately 4 km, with the assumed load equal to 0.5. As we can see, by increasing the allowable load to about 0.8 coverage achieved will decrease to approximately 2.8 km (point "2"). If this value of the load is acceptable for a designer, it can allow the maximum load of 0.8.

From the other side, if we would like to provide coverage at the same level as at the beginning of the analysis, we can proceed as follows. Increasing the values for D_{DL} to 29 dB will allow us to obtain a range of approximately 4 km, with an assumed load of 0.8 (point "3"), as we can see in Fig. 5. It can be achieved e.g. by increasing the base station transmitter power of about 6 dB or use the antenna with higher gain. We can therefore make the appropriate changes to the base station. Of course, with such changes is increasing coverage (for the load of 0.5), which in this case would be about 5.9 km (point "4"). Note, that a similar analysis also for the uplink must be done.

We can see that by measuring the load we are able to control capacity and coverage in various operational conditions of the radio interface. Using this method in the planning and operation of the network can simplify the radio resource management. Also, we can increase the efficiency of the use radio resources. The presented method of analysis can be performed for different types of services implemented in any propagation environment. Moreover, described method can be used to reconfigure the network, which may allow a temporary increase in capacity [3].

7 Applications of the Load and Coverage Characteristics

For design the 3G radio network both the load and coverage characteristics can be used. They may be applied for the verification of propagation and coverage conditions, both at network startup, as well as during its operation. By using the characteristics we can rapidly assess the necessary changes in the link budget parameters and thus, increase the efficiency of the use of radio resources. The presented method can be used for the simulation and measurement investigations.

It should be noted, that the use of load characteristics may significantly speed up the process of network design, because the consequences of changes in the value placed on the radio link parameters are directly visible. In addition, clearly visible is the effect of changes of these radio link parameters on another characteristics of the network: e.g. load, capacity, coverage etc.

Among the many applications of universal load characteristics we can mention:

- analysis of the allowable pathloss value for any configuration of link budget parameters,
- estimation of pathloss variation depending on the variability of the load,
- calculation of allowable pathloss for different values of the load and miscellaneous services of any type,
- determination of the extent of transmitter power variation in different operating conditions, for both the base stations and mobile stations.

Note that, the normalized coverage characteristics in any propagation environment and for any propagation loss model can be used. Furthermore, all of the listed applications of universal load characteristics can be used as an application of normalized coverage characteristics.

Among the applications of normalized coverage characteristics we can mention in particular:

- analysis of the capacity-coverage trade-off for a given set of implemented services,
- calculation of coverage in different operating conditions, for various values of the load and services of any type,
- defining the coverage changes resulting from the modification of link budget parameters.

Of course, the primary purpose of the use of coverage characteristics is to designate the coverage, for both the base stations and mobile stations. In spite of the significant complexity of the WCDMA/FDD radio interface, the designation of capacity-coverage trade-off is very simple, when presented method is used.

8 Summary

The method of the estimation of capacity-coverage tradeoff presented in this paper is useful for both the planning and maintenance of the UMTS network. As seen, we can use the phenomenon of the trade-off to determine the necessary changes in the value of radio link parameters, depending on the volatility of the operating conditions in a radio network.

In the network planning process we can shape the cell coverage at the cost of capacity for the service of any type and for arbitrary propagation environment in order to utilize the radio resources in the most efficient way.

In the maintenance process we can easily evaluate the coverage when the traffic is increasing in some periods of time and borrow the resources (capacity) needed from neighbouring cells. Ultimately, if the traffic is increased permanently we can redesign the local network appropriately.

The presented method can be used for simple calculations, when the values of link budget parameters are modified in a real network. Also, to simulate the UMTS network in the expert systems it can be used.

References

1. Laiho-Steffens, J., Wacker, A., Aikio, P.: The Impact of the Radio Network Planning and Site Configuration on the WCDMA Network Capacity and Quality of Service. In: Proc. of IEEE Vehicular Technology Conference, VTC 2000 Fall (2000)
2. Laiho, J., Wacker, A., Novosad, T. (eds.): Radio Network Planning and Optimisation for UMTS, 2nd edn. Wiley & Sons, Chichester (2006)
3. Gajewski, S.: Performance Analysis of the WCDMA /FDD Technique in the UMTS Terrestrial Radio Access Network, doctoral dissertation (in Polish), Gdansk University of Technology, Gdansk, Poland (2004)
4. Veeravalli, V., Sendonaris, A.: The Coverage – Capacity Tradeoff in Cellular CDMA Systems. IEEE Transactions on Vehicular Technology 48(5) (September 2002)
5. Gajewska, M., Gajewski, S.: Simulation Results of Dynamic Capacity Reallocation in Hierarchical Cell Structure of UMTS Network. Polish Journal on Environmental Studies 16(4B) (2007)

6. Gajewski, S., Gajewska, M.: Downlink Capacity-Coverage Trade-off Estimation Based on Measurement of WCDMA/FDD Interface Load. Polish Journal of Environmental Studies 16(4B) (2007)
7. Shikh-Bahaei, M.R., Aghvami, A.H.: Joint Planning of Soft-capacity and Soft-coverage for 3G WCDMA Systems, 3G Mobile Communication Technologies, Conference Publication No. 477, IEE 2001 (2001)
8. ITU-R, Guidelines for Evaluation of Radio Transmission Technologies for IMT-2000, Recommendation ITU-R M.1225 (1997)

Dynamic Localized Broadcast Incremental Power Protocol and Lifetime in Wireless Ad Hoc and Sensor Networks

Julien Champ, Anne-Elisabeth Baert, and Vincent Boudet

LIRMM - University of Montpellier II - CNRS, 161 rue Ada,
F-34392 Montpellier Cedex 5, France
{champ,baert,boudet}@lirmm.fr

Abstract. This paper deals with the problem of broadcasting messages and lifetime in wireless ad-hoc and sensor networks. The study is based on the best known localized algorithm, namely *LBIP*, which is based on a centralized one, *BIP*, whose principle consists in constructing a broadcast tree rooted on the source node, taking into account the specificities of wireless networks. Even if *LBIP* has excellent performances regarding energy consumption, it selects for each broadcast the same nodes to retransmit the message; if the source of the broadcast, the base station in a sensor network, is always the same, this will lead to deplete quickly the energy of relay nodes. In this paper, we propose *DLBIP*, a new localized broadcast algorithm based on *LBIP*, which dynamically changes the broadcast tree to balance energy consumption on nodes without any additional messages. We show that proposed strategy can significantly increase the number of broadcasts before the network failure. We provide simulations results that clearly demonstrates the lifetime enhancement due to our optimization.

1 Introduction

As a consequence of recent advancements in miniaturization and wireless communications, a new kind of network has come to the fore: *Wireless Sensor Networks* (WSN). In those networks, nodes can gather information from their environment, such as temperature, gas leak, etc. They can also communicate, thanks to their wireless communication device, with other nodes in their transmission range. WSN are also composed of at least one special node, called the base-station, or sink, the purpose of which is either to centralize collected data from the WSN, send queries in the network, or connect the WSN to other networks. WSN recently attracted a lot of attention because of their wide range of applications. They can be used in a many different fields, monitoring tasks either for the military, or the environment, security, health-care, and habitat automation [1].

When broadcasting, the source node needs to send a message to all the nodes in the network. Many applications need to broadcast messages to the whole network: so as to send a query to all the nodes, to broadcast an information, or to do some route discovery ... The broadcasting task occurs more frequently in such networks. Proposed methods need to be designed for wireless sensor networks: sensor nodes are small objects working thanks to a tiny battery and communicating thanks to their wireless communication device.

J. Wozniak et al. (Eds.): WMNC 2009, IFIP AICT 308, pp. 286–296, 2009.

Due to the limited battery power, these networks are power constrained, and as communication ranges are limited, an important set of nodes needs to retransmit the message in order to cover the whole network. The easiest way to broadcast a message to all sensors in the area is called *Blind Flooding* and it works as follows: each node relay once the message, and if there exists a path between the broadcast source and any node in the network, all nodes will receive the message properly. But this method implies a lot of redundant messages.

We can find in the literature various broadcast algorithms used to save energy consumption in the WSN: sometimes nodes can adjust their transmission power in order to save energy and obtain better results, sometimes it is only possible to reduce the number of retransmitting nodes to achieve a full coverage.

Nevertheless, reducing the energy consumption is always realized for the same purpose: to increase the lifetime of the network. It is not sufficient to analyze the energy consumption for one broadcast; it is more interesting to study the lifetime network after several broadcasts. The notion of network lifetime is not clearly defined for ad-hoc and sensor networks in the literature and is clearly application dependent [2]. This point is discussed in section 2.3.

In this paper, we try to guarantee as long as possible the reception of broadcast messages in the network, *i.e.*, over 90% of the sensors have to receive the broadcast messages. We based our work on the *Localized Broadcast Incremental Power Protocol (LBIP)* [3], the best known localized algorithm regarding to energy consumption when transmission range adjustment is possible, and we propose the *Dynamic Localized Broadcast Incremental Power Protocol (DLBIP)* a new localized broadcast protocol whose principle is to use dynamic broadcast trees to improve lifetime. We provide simulation results demonstrating its efficiency regarding to lifetime.

This paper is organized as follows: we introduce the network and the energy consumption models, and also definitions of network lifetime in Section 2. Section 3 is dedicated to a brief overview of existing broadcasting algorithms. In section 4, we introduce our protocol *DLBIP*. Section 5 presents simulation results comparing *DLBIP* to *LBIP*. Section 6 concludes this article and presents future directions.

2 Preliminaries

2.1 Network Model

We represent a WSN using the widely adopted Unit Disk Graph Model, denoted UDG. An UDG is defined by $G = (V, E)$ where V is the set of nodes (sensors), and E the set of edges representing available communications. Let R be the maximum communication range for all nodes. There is an edge $e = (u, v) \in E$ if and only if the Euclidean distance between u and v, denoted $d(u, v)$ is less or equal R:

$$E = \{(u, v) \in V^2 | d(u, v) \leq R\} \tag{1}$$

Two nodes that can communicate are considered to be neighbors. The hop-distance between nodes i and j is the minimum number of edges to cross to reach j from i. The k-hop neighborhood for node i, is defined as the set of nodes reachable within at most k-hops of node i. Nodes can adjust their transmission range, so as to consume less energy, while chosen transmission range r is less or equal maximum transmission range R. We assume that nodes are equipped with omnidirectional antennas: if a node i transmits a message with its transmission range set to x, all its neighbors j with $d(i, j) \leq x$ receive it.

2.2 Energy Consumption Model

In networks where nodes are not able to adjust their transmission range, one easy way to measure the energy consumption of a broadcasting algorithm is to count the number of nodes which retransmit the message. In this paper, we consider networks where sensors can adjust their transmission range to reduce their energy consumption. Thus, we use the most commonly used energy consumption model where energy consumption is given according to the chosen transmission range. If a node broadcasts a message with a transmission range equals to r the energy consumption will be:

$$energy(r) = \begin{cases} r^\alpha + c & \text{if } r > 0 \\ 0 & \text{else} \end{cases} \tag{2}$$

The most used values for this model are given by Rodoplu and Meng in [4]. They propose to use it with $\alpha = 4$ representing the signal attenuation and $c = 10^8$ an overhead due to signal processing.

2.3 Lifetime Definition

Many broadcasting algorithms proposed in the literature are designed to reduce global energy consumption. But energy consumption reduction is made to extend network lifetime. It seems insufficient to consider only energy consumption of only one broadcast; it is more suitable to analyze the behaviour of the network after more broadcasts.

Lifetime in WSN is still not very well defined whereas choosing a good lifetime criterion is very important when designing new protocols [2]. A good lifetime metric is needed to analyze exactly protocol's behaviour, and to optimize your protocol regarding requirements. There is no definition of lifetime suitable to all kind of applications in Wireless Sensor Networks. The choice of one or another criterion depends clearly on your application requirements.

We can find in the litterature communication algorithms using the Time To First Failure (TTFF), or number of tasks done before one failure to define network lifetime [5]. But, unless if the failure of only one node is a disastrous state regarding our application requirements, TTFF criterion seems to be insufficient. In many applications, the network is not considered as being faulty if one node runs out of its battery energy, or if one node do not receive a broadcast.

It is important to note, that some nodes may be considered as *more important* than other nodes in WSN. For example, a node can run out of its battery energy and partition the network : but, it is not always the First Failure which partition the network into two components. When our application needs that all sensors have to be alive, TTFF is suitable, else we should analyze the number of broadcasts (or time) until less than X% of nodes receive the message, with X depending on application requirements.

3   Related Work

3.1   Broadcasting without Range Adjustment

The easiest way to reduce energy consumption consists in reducing the number of nodes which retransmit the message to achieve the broadcast; many solutions have been proposed to minimize the number of communications.

We can find clustering algorithms constructing connected dominating sets, providing a backbone for communications, so as to reduce the number of nodes used to retransmit messages. Probabilistic protocols have also been proposed such as [6,7].

In the Neighbor Elimination Scheme (*NES*) [1][8], when node *i* receives the message, it monitors if its neighbors have received the message, until a timeout. If all nodes seems to be covered, node *i* does not rebroadcast r the message; else node *i* needs to retransmit the message. This leads to a significant elimination of redundant messages. It is important to note that *NES* can often be used as an additional mechanism over another protocol. In the Multipoint Relay protocol (*MPR*) proposed in [9], 2-hop neighborhood knowledge is needed. The source node selects a subset of its 1-hop neighbors to relay the message in order to cover all its 2-hop neighbors, and sends the message, including its choice in the packet so as to propagate them to its 1-hop neighbors. Finding such a minimal set of nodes is a NP-complete problem, however an interesting greedy heuristic is proposed in [9].

3.2   Broadcasting with Range Adjustment

When nodes are able to adjust their transmission range so as reduce their energy consumption, it is not sufficient to reduce the number of nodes retransmitting messages.

RBOP, LBOP and TR-LBOP: Broadcast Oriented Protocols. *RBOP*, *LBOP* and then *TR-LBOP* [10] have been proposed to improve the efficiency of already existing protocols, taking into account the possibility to adjust transmission range. Their general principle is to use *NES* on a subset of their neighbors defined with respectively the Relative Neighborhood Graph (RNG), the Local Minimum Spanning Tree (LMST), and LMST computed with a target radius. These protocols, specially *TR-LBOP*, offer goods performances regarding to energy consumption, but suffers of latency due to the use of *NES*.

BIP: Broadcast Incremental Power Protocol. In [11], authors proposed a centralized algorithm, named Broadcast Incremental Power protocol (*BIP*), allowing transmission range adjustment, and providing interesting results regarding to energy consumption. The main idea consists in proposing a variant of Prim's minimum spanning tree algorithm, taking into account the wireless multicast advantage: when a node sends a message using its maximal transmission range, all nodes inside its transmission range (its neighbors) receive the message. The principle of *BIP* is the following:

- Initially, the broadcast tree is empty, and the source node is marked.
- All nodes start with their transmission range set to 0.

---

[1] Also known as Wait & See protocol.

– At each step and until all nodes are covered, *BIP* selects the pair (i, j), with i a marked node and j an unmarked node, and such that the *additional power* needed to reach j is minimum; then i sets its transmission range so as to reach j.

The *additional power* is defined as being the cost for i to reach j, minus the cost of its already selected transmission range.

LBIP: Localized Broadcast Incremental Power Protocol. The Localized *BIP* (*LBIP*) protocol, in [3], is the localized version of *BIP*. Its main principle is the following; the source node applies the *BIP* algorithm in its k-hop neighborhood, and forwards its instructions with the message, as with *MPR* protocol. When node i receives the message, if there is instructions inside the packet i applies the *BIP* algorithm on its k-hop neighborhood using previously received instructions. If there is no instruction in the packet, this means that i neighborhood is already entirely covered, and thus i can drop the packet.

Simulation results show that *LBIP* outperforms other distributed protocols regarding to energy consumption, and is only slightly more energy consumming than centralized *BIP* protocol. Authors shows that the *NES* could be additionally used in order to guarantee a complete coverage. Else, without *NES*, conflicting decisions lead to reach nearly 98% of the nodes in the network (which is in most cases sufficient). They also show that using *LBIP* with $k = 2$ is the best compromise, providing excellent results while required knowledge is not too important.

Lifetime oriented centralized algorithms. In [12], authors consider the problem of broadcasting the maximum number of messages until a node do not receive a broadcast. They propose interesting heuristics but proposed algorithms are centralized and they consider a different source for each broadcast. Another interesting centralized work can be found in [13].

4   Dynamic Localized Broadcast Incremental Power Protocol

4.1   Principle

We consider here a static wireless ad hoc or sensor network, where always the same node (the base station in a WSN) broadcasts a message to all nodes. Our purpose is to maximize the number of broadcast until the reachability goes below than 90%. The reachability is defined as being the number of receiving nodes divided by the total number of nodes.

The general principle of our optimization consists in balancing energy consumption between nodes, by changing relay nodes according to remaining energy. The main part of the protocol is the same used in *LBIP*: each sensor computes its broadcast tree on its k-hop neighborhood, and according to received instructions.

In our protocol, we change the weight metrics to compute dynamic broadcast trees, *i.e.*, which can change at each broadcast. Weights used to compute the broadcast tree are not only computed according to the energy consumption of the communication but also regarding to remaining amount of energy. Let B_i be the remaining amount of energy on node i. If we denoted $C(i \rightarrow j)$ the weight for a communication from i to j, our new weight is computed as follows:

$$C'(i \rightarrow j) = \frac{C(i \rightarrow j)}{B_i} \tag{3}$$

Using this new weight to compute the broadcast tree, nodes with lower remaining energy are less selected to retransmit the broadcast, or are asked to communicate on a shorter distance. This leads to a better balance of energy consumption between nodes. This solution extend network lifetime : the broadcast tree, contrary to *LBIP*, is now not unique, it changes according to remaining energy. As our optimization is based on a dynamic broadcast tree, and is based on *LBIP*, we name it *DLBIP* for Dynamic *LBIP*.

Fig. 1 illustrates our dynamic algorithm. On the left we can see the unique broadcast tree computed by *LBIP* ; this is also at least the first broadcast tree computed by *DLBIP* (if initially $B_i = B_j$ for every node i and j). On the right we can see one of the other broadcast trees computed by *DLBIP*.

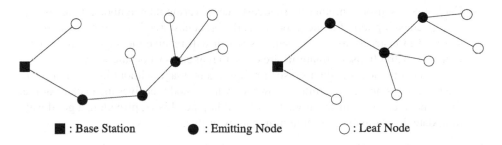

■ : Base Station   ● : Emitting Node   ○ : Leaf Node

Fig. 1. A small network

In the following example, we compare the weightings used to compute broadcast trees, with *LBIP* and *DLBIP*. We consider three nodes i, j and k such that $d(i,j) < d(i,k)$. Node i wants to communicate the message to j and k.

With *LBIP* algorithm, the choice is made by the following inequality:

$$C(i \rightarrow k) < C(i \rightarrow j) + C(j \rightarrow k) \tag{4}$$

If previous inequality is true, i sets its transmission range to communicate with both nodes ; else i sets its transmission range to communicate with j and asks it to retransmit the message to k.

As described before, with *DLBIP* remaining energy on nodes i and j change previous inequality:

$$C(i \rightarrow k)/B_i < C(i \rightarrow j)/B_i + C(j \rightarrow k)/B_j$$
$$\Leftrightarrow \quad C(i \rightarrow k) < C(i \rightarrow j) + C(j \rightarrow k) \times B_i/B_j \tag{5}$$

The choice is related both to communication costs and remaining energy. If j remaining energy is still high while i remaining energy has decreased, j is more easily chosen as a relay node. On the other hand, if $B_i > B_j$ node i will probably send the message to both nodes.

Remark: In this paper, we have decided to consider only remaining energy of sending nodes to compute the dynamic broadcast trees. But other ways to weight edges can be used to balance energy consumption in the network: for example, critical nodes whose failure lead to partition the network can be detected. We can use this to limit their energy consumption, so as to partition the network as late as possible.

4.2   Energy Updates

However, so as to use *DLBIP*, each sensor needs to know or assess remaining energy of each its k-hop neighbors. We propose here a method which can be divided in two parts:

Approximate calculations
Each sensor tries to estimate remaining energy on nodes in its neighborhood according to its knowledge:

- Each time a sensor computes a broadcast tree, covering its neighbors, it can assess energy consumption of its neighbors. As the broadcast tree is computed according to received instructions, and as sensors which will receive messages will obey to included instructions, estimations are close to real energy consumption.
- For a given broadcast, when sensor i receives a message without instruction for i, the message is not retransmitted. However, when a node receives the packet for the first time, it can read instructions for all its k-hop neighbors, providing a good way to update the estimation of remaining energy of its neighbors.

Accurate updates: When sensor i computes the broadcast tree needed to cover its k-hop neighbors, and includes instructions in the message, it can also include a field with its remaining energy B_i. This message is sent to its neighbor, which can accurately update their estimation of remaining energy on node i. These updates can be done until the message reaches node outside i neighborhood, then the field can be removed.

Remark: to avoid an excessive packet size increase, it is not needed to include remaining energy for each broadcast, this can be done once in a while.

5   Performances Evaluation

5.1   Simulation Parameters

In order to evaluate performances of *DLBIP*, we present simulation results in this section using WSNET simulator [14]. We compared our protocol to *LBIP*, because our protocol is based on it and as it outperforms other localized protocols regarding to energy consumption. We have chosen nearly the same parameters used for the evaluation of *LBIP* in [3]. As said in Section 2 we used the Unit Disk Graph to model available communications. We consider a static network, composed of 500 nodes randomly deployed using an uniform distribution inside a square area. The size of the area is computed according to chosen network density. The maximum transmission range is fixed to 250.

All nodes have initially the same amount of energy, except the base station which can transmit as many messages as needed. The energy consumption model is the one presented in section 2 equation 2 with $\alpha = 4$ and $c = 10^8$. As in [3], *LBIP* and *DLBIP* have been implemented with an ideal MAC layer: two nodes can transmit a message at the same moment, without collisions. *LBIP* and *DLBIP* compute their broadcast tree within their 2-hop neighborhood as simulations in [3] shows that $k = 2$ seems to be the best compromise.

To resolve conflicting decisions in *LBIP* protocol, the authors proposed in [3], to use the Neighbor Elimination Scheme to reach a total coverage. Without the additional *NES* over *LBIP*, the coverage is still enough high for most applications. As our algorithm is based on *LBIP*, the same conflicting decisions may appear, and can also be solved using the *NES*. Thus, so as to measure precisely the impact of our proposition we have decided to analyze simulation results without the use of the *NES* in *LBIP* and in *DLBIP*.

We have decided to consider that the network is faulty when less than 90% of nodes receive the broadcast, so we study the number of broadcasting tasks that can be done until reachability goes below 90%. Simulation results are similar, whatever the initial energy on sensor is; that's why we do not compare results according to the initial amount of energy.

5.2 Simulation Results

In Fig. 2 we give for *LBIP* and *DLBIP* the percentage of receiving nodes, regarding to the number of broadcasts done. With *LBIP* protocol, the chosen broadcast tree is always the same, thus always the same node are selected as relay nodes. In density 20 networks, this obviously lead to a achieve less than 90% of covered nodes after 210 broadcasting tasks. Contrary to *LBIP*, *DLBIP* uses a dynamic broadcast tree, relay nodes and transmission ranges are selected according to both communication costs and remaining energy: this leads to a better lifetime of the network. Indeed, 600 broadcast are achieved before less than 90% nodes are covered. This means an increase of nearly 185% until 10% of nodes are not covered in density 20 networks.

The percentage of transmitting nodes for density 20 networks, given in Fig. 3, is obviously initially the same for *LBIP* and *DLBIP*, but as the number of realized broadcast

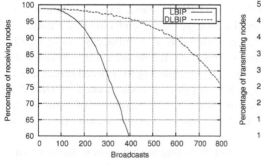 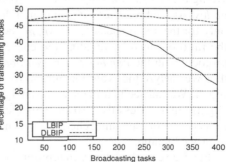

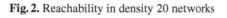

Fig. 2. Reachability in density 20 networks   **Fig. 3.** Percentage of transmitting nodes

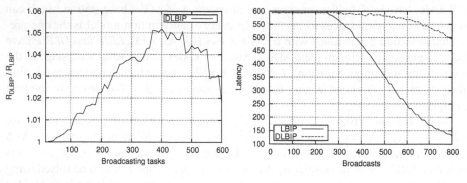

Fig. 4. Ratio R_{DLBIP}/R_{LBIP} **Fig. 5.** Broadcasts latency

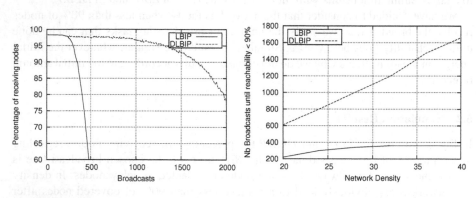

Fig. 6. Reachability in density 40 networks   **Fig. 7.** Lifetime regarding to density

increases, *DLBIP* use a few more emitters. This can be explained: if node *A* was used to cover some nodes, when its remaining energy decreases too much, there is two possible alternatives. Either another node increases its transmission range to cover its neighbors, or two other nodes communicate the message, leading to an higher number of emitters.

Let R_{LBIP} be the average energy consumed by a node, *i.e.* the ratio obtained by dividing the energy consumption for a broadcast using *LBIP* protocol with the *number of nodes*. As the broadcast tree computed thanks to *LBIP* is static, we compute this ratio only for the first broadcast. Let R_{DLBIP} be the average energy consumed by a node, *i.e.* the ratio obtained by dividing the energy consumption for a broadcast using *DLBIP* protocol with the *number of covered nodes*. This time, the broadcast tree is dynamic, thus we analyze the evolution of ratio R_{DLBIP}/R_{LBIP} regarding to realized broadcasts in Fig. 4. We can see that the average energy consumption by node, needed by *DLBIP* is slightly higher (less than 10%). This increase is due to the weighting of communication costs with remaining energy, and is needed to balance energy consumption so as to improve lifetime.

Fig 5 give us the latency in arbitrary units of the broadcasting tasks regarding to the number of broadcast realized. The latency is computed as being the elapsed time for all sensors to receive a broadcast. There is no significant increase of the latency after several broadcast when using *DLBIP* compared to *LBIP*. The latency is only linked to the number of remaining alive nodes and to the broadcast tree depth which is nearly the same for both protocols.

We provide in Fig. 6 reachability, *i.e.* the percentage of receiving nodes, in a network with a higher density than in Fig. 2; this is a network with density equals to 40. We can note that our algorithm is scalable: indeed, in density 40 networks *DLBIP* perform more than 350% more broadcast than *LBIP* until reachability goes below 90%. This is confirmed by Fig 7, which gives the average number of broadcasts done before the reachability goes below 90%. We can note that when the density is between 30 and 40, *LBIP* achieve less than 400 broadcasts, while *DLBIP* goes on increasing the number of broadcasting tasks. The higher the density is, the more choices *DLBIP* has to balance energy consumption.

6 Conclusion and Future Work

In this paper, we deal with optimizing the number of broadcasts until 10% of sensors failed receiving the message. We argue that only measuring the energy consumption of one broadcast is not sufficient to make a good performance analysis of the lifetime of a sensor network. Thus we propose a new protocol, *DLBIP*, which is a new dynamic broadcast algorithm based on the *LBIP* algorithm.

Actually, in the case of a static wireless ad hoc or sensor network, where always the same node wants to broadcast a message to all nodes, computing the energy consumption of one broadcast is not sufficient. Our method consists in taking into account the remaining energy of nodes in the network, to obtain in time a dynamic broadcasting tree which optimizes the lifetime of network. The main property of *DLBIP* is that it does not need additional messages to work. Simulations show the efficiency of the protocol in term of lifetime comparatively to *LBIP*, even if *DLBIP* is slightly more energy consuming.

Further research should address other ways to balance energy consumption, such as detecting critical nodes, or trying other weightings to compute *DLBIP* broadcast trees, to see the most efficient trade-off between energy consumption and load balance. Other simulations made with different broadcast sources for each broadcasting task should also be done to analyze the impact of such a protocol when the source is different for each broadcast. As for *LBIP* protocol, it should be interesting to analyze our protocol using an energy consumption model which considers reception costs and also with more realistic physical layer.

References

1. Akyildiz, I.F., Su, W., Sankarasubramaniam, Y., Cayirci, E.: Wireless sensor networks: a survey. Computer Networks 38(4), 393–422 (2002)
2. Champ, J., Baert, A.E., Saad, C.: Lifetime in wireless sensor networks. In: CISIS: Conference on Complex, Intelligent and Software Intensive Systems, Fukuoka, Japan (2009)
3. Ingelrest, F., Simplot-Ryl, D.: Localized broadcast incremental power protocol for wireless ad hoc networks. Wirel. Netw. 14(3), 309–319 (2008)

4. Rodoplu, V., Meng, T.: Minimum energy mobile wireless networks. In: Proc. of IEEE International Conference on Communications, ICC (1998)
5. Chang, J.H., Tassiulas, L.: Energy conserving routing in wireless ad-hoc networks. In: INFOCOM, pp. 22–31 (2000)
6. Tseng, Y.C., Ni, S.Y., Chen, Y.S., Sheu, J.P.: The broadcast storm problem in a mobile ad hoc network. Wirel. Netw. 8(2/3), 153–167 (2002)
7. Cartigny, J., Simplot, D.: Border node retransmission based probabilistic broadcast protocols in ad-hoc networks. In: Telecommunication Systems, pp. 189–204 (2003)
8. Peng, W., Lu, X.C.: On the reduction of broadcast redundancy in mobile ad hoc networks. In: MobiHoc '00: Proceedings of the 1st ACM international symposium on Mobile ad hoc networking & computing, Piscataway, NJ, USA, pp. 129–130. IEEE Press, Los Alamitos (2000)
9. Qayyum, A., Viennot, L., Laouiti, A.: Multipoint relaying for flooding broadcast messages in mobile wireless networks. In: HICSS 2002: the 35th Annual Hawaii International Conference on System Sciences (HICSS 2002), vol. 9 (2002)
10. Ingelrest, F., Simplot-Ryl, D., Stojmenovic, I.: Optimal transmission radius for energy efficient broadcasting protocols in ad hoc and sensor networks. IEEE Trans. Parallel Distrib. Syst. 17(6), 536–547 (2006)
11. Wieselthier, J.E., Nguyen, G.D., Ephremides, A.: Energy-efficient broadcast and multicast trees in wireless networks. Mob. Netw. Appl. 7(6), 481–492 (2002)
12. Park, J., Sahni, S.: Maximum lifetime broadcasting in wireless networks. IEEE Trans. Comput. 54(9), 1081–1090 (2005)
13. Elkin, M., Lando, Y., Nutov, Z., Segal, M., Shpungin, H.: Novel algorithms for the network lifetime problem in wireless settings. In: Coudert, D., Simplot-Ryl, D., Stojmenovic, I. (eds.) ADHOC-NOW 2008. LNCS, vol. 5198, pp. 425–438. Springer, Heidelberg (2008)
14. Fraboulet, A., Chelius, G., Fleury, E.: Worldsens: development and prototyping tools for application specific wireless sensors networks. In: IPSN 2007: Proceedings of the 6th international conference on Information processing in sensor networks, pp. 176–185. ACM, New York (2007)

Information-Importance Based Communication for Large-Scale WSN Data Processing

Ahmad Sardouk, Rana Rahim-Amoud, Leïla Merghem-Boulahia,
and Dominique Gaïti

ICD/ERA, FRE CNRS 2848, Université de Technologies de Troyes,
12 rue Marie Curie, 10010 Troyes Cedex, France
{ahmad.sardouk,rana.amoud,leila.boulahia,dominique.gaiti}@utt.fr

Abstract. Gathering information in an energy-efficient and scalable manner from a wireless sensor network is always a basic need. In this work, we use the multi-agent approach in order to build an Information-Importance Based Communication for large scale wireless sensor network data processing. The principal goal of our proposition is to tackle the problem of network density and scalability in an energy efficient manner. Simulation results are provided to illustrate the efficiency of our proposition.

1 Introduction

Due to advances in wireless communications and electronics over the last few years, the development of networks of low-cost, low-power, and multifunctional sensors has received increasing attention. These sensors are small in size and integrating the capabilities of sensing, computing and communication. Large amounts of these small-size and low cost sensor nodes can be rapidly deployed in a region of interest and form a loosely coupled distributed networking system called Wireless Sensor Network (WSN). The main goal of this network is to collect information from the target environment. The WSN is generally composed of a large number of dense, randomly deployed and energy limited nodes. To the best of our knowledge, processing the information locally in the sensor nodes is very cost effective comparing to its communication. This is especially due to the fact that a lot of the sensed information could be redundant or not important.

In this paper, we propose a novel scheme for data gathering in wireless sensor networks. This scheme does not only reduce heavily the power consumption but it also tackles the density of the network and the scalability problems. This new scheme is based on the multi-agent approach in order to build a Communication structure Based on the Importance of the Information (IIBC). In addition, a multi-agent cooperation is used to gather the processed information from multiple sensor nodes with an inter-sensor-nodes redundancy elimination.

The paper is organized as follows. In section 2, we give a brief description of the multi-agent systems. In section 3, we review the related work. Our solution is described in section 4. Next, in section 5, we present our simulation setup

J. Wozniak et al. (Eds.): WMNC 2009, IFIP AICT 308, pp. 297–308, 2009.

parameters and the performance criteria. Then, in section 6, we evaluate and analyze the performance of our proposition. Finally, the conclusion is given in section 7.

2   Multi-Agent Systems: Some Definitions

According to [1], an agent is a physical (robot) or virtual (real time embedded software) entity having trends and resources, able to perceive its environment, to act on it and to acquire a partial representation of it (called the local view of the agent). An agent is also able to communicate with other peers and devices and has a behavior that fits its objectives according to its knowledge and capabilities. Furthermore, an agent can learn, plan future tasks and is able to react and to change its behavior according to the changes in its environment. A multi-agent system (MAS) is a group of agents able to interact and to cooperate in order to reach a specific objective. Agents are characterized by their properties that determine their capabilities. Different properties are defined like autonomy, proactiveness, flexibility, adaptability, ability to collaborate and to coordinate tasks and mobility. According to its role within its environment, the agent acquires some of these properties. Multi-agent approach is well suited to control distributed systems. WSN are good examples of such distributed systems. This explains partly the considerable contribution of agent technology when introduced in this area.

3   Related Works

The basic role of sensor nodes in a WSN is to gather information from the environment. This gathering should respect the finite battery of the sensor node to maintain the lifetime longevity of such network, the density of the network, and its scalability. The traditional and base model of data gathering is the client/server (C/S) communication architecture. In this architecture, when the sensors of a sensor node perceive information from their environment, they send it directly as it is (raw not processed) to the sink to be processed. Moreover, to send information to the sink, the communication passes through a multi hop communication. That means, intermediate nodes relay farther nodes information. Consequently, this relaying causes supplementary power consumption for this node. Several works have been done to optimize this traditional architecture.

In [2], the authors proposed to merge the information of a maximum number of nodes. Thus, they proposed a serial incremental fusion which can be described as follows. When a node sends its data to the sink, intermediate nodes merge their data with the first node data. As the information of multiple sensor nodes are merged into one message (one overhead instead of many ones), this solution saves some energy. However, intermediate nodes do not have always important information to send and they do not eliminate the unimportant or redundant information. Furthermore, this solution is only suitable to some scenarios. A scenario, showing that this solution is not scalable, is when a node, which is at

100 hops from the sink, sends it a message. In this case, the size of the message will be incremented, in each hop, by the size of the current message hop. For example, in the middle of the network, the size will be approximately equal to the original message size multiplied by 50. Such situation (message with huge size) may consume a lot of power and causes a lot of loss in term of packets.

The authors of [3] present an ant colony based data aggregation for wireless sensor networks. They are trying to tackle the problem of constructing an aggregation tree for a group of source nodes within the WSN to send sensory data to a single sink node. The proposed mechanism assigned artificial ants to source nodes to establish low-latency paths between the source nodes and the sink. They suppose that every ant will explore all possible paths from the source node to the sink. The data aggregation tree is constructed by the accumulated pheromones. By exploring all possible paths to sink, each ant consumes extra power that could be eliminated. Furthermore, the construction of the appropriate tree depends heavily on the nodes' deployment, which is generally random. Such tree construction consumes an important amount of power.

In [4], the authors propose an adaptive data aggregation (ADA) scheme for clustered WSN. The authors define temporal and spatial aggregation degrees, which are controlled by the reporting frequency at sensor nodes and the aggregation ratio at cluster heads respectively. The temporal and spatial aggregation degrees are determined by the current scheme state according to the observed reliability. Their work focuses principally on the incorporation of an adaptive behavior into protocols in such dynamic network. ADA is based on the cluster heading paradigm, which needs an expensive construction in term of energy. Furthermore, in the implementation of this scheme, the authors did not address the needed power consumption which we suppose important. In addition, authors neglected the importance of scalability of such kind of networks.

The authors of [5] have been oriented towards the agent technologies to gather the information. The central work of these authors was based on the use of Mobile Agents (MAs) in WSN for an energy-efficiency data gathering. In these propositions, the MA is defined as a message which contains an application code, a list of source nodes (predefined by the network administrator or the sink) and an empty field to put the gathered data. The message contains also an important header including the required fields (such as the next destination and next hop fields) to route it in the network following the list of source nodes and the header. The MA gathers the information from the source nodes. At each source node, the MA processes the collected data locally and concatenates it with previous source nodes' data. This solution reduces the power consumption in cases of low bandwidth and power constrained networks such as WSN. However, the size of the mobile agent message is large enough to waste an important part of this reduced power, when sending it on the WSN. Another drawback of this type of solutions is the difficulty to create the source nodes' list and to define the starting time of data gathering. Indeed, the sink node is not able to decide by itself when the nodes have important data to send. Furthermore, this solution appears to be so far from the scalability issue imposed in a lot of WSN applications. This lake appears when measuring the required time for a mobile agent to gather the information from far regions and the number of source nodes' lists and mobile agents needed to gather information from the entire network. Another limitation is the definition of the region which will be treated by the MA.

After analyzing the previously presented solutions, we can see clearly that there is a serious problem in term of scalability for all of them. In addition, we can deduce that there is still a lot of work in term of energy-efficiency with paying attention to the packet delivery ratio and latency. Moreover, the needed solution should be independent from the network topology. It should be efficient for a randomly deployed network.

4   A Multi-Agent Cooperation for Energy-Efficiency Data Gathering

According to [6], a sensor node expends a maximum energy in data communication where the energy expenditure in data processing is much less compared to it. The energy cost for the transmission of 1KB on a 100m distance is approximately the same as that for the execution of 3 million instructions by a 100 million instructions per second processor. Hence, we have defined three main points to save the power of each node and so to extend the lifetime of the network:

1. The Information Importance Based Communication: its main role is to reduce the number of communications. Thus, by estimating the importance of the information locally (in the sensor node), it is possible to prohibit an important number of communications corresponding to non important or redundant information. A good example of this kind of information could be seen in the case where we monitor a stable environment. Hence, the sensor node may not detect new values or events during a long time. Consequently, the difference between the gathered values will not be important.
2. The Elimination of Non Important inter-sensor-nodes Information: generally, sensor nodes are deployed randomly (e.g., a plane throws them in hazards zones). Thus, two or more sensor nodes can cover the same point and trivially they will give always the same information (inter-sensor-nodes redundancy).
3. The Data Concatenation: Due to protocol overheads, the communication cost, in term of energy, of sending a long message is usually less than the one of sending the same amount of data using many short messages.

These three points will be discussed within the system design of our proposal.

4.1   System Design

A sensor node has generally a battery, a processing unit, a memory and a radio entity. To convert a sensor node to a fully autonomous entity, we should empower it by some capacities. A sensor node should be able to make decisions to estimate the importance of the sensed information. It should be also able to cooperate with other sensor nodes in order to eliminate the inter-sensor-nodes redundancy and/or to concatenate data. Therefore, we have chosen to use an MAS to bring up the full autonomy to WSNs. As we presented in section 2, an agent is by definition capable of making decisions and of cooperating with other agents to execute tasks. To illustrate our proposal, we use the architecture presented in fig. 1, where we associate an agent to each sensor node. This agent processes the sensed data locally, and estimates its importance. It also makes decision to communicate this information and/or to cooperate with other neighbor agents in order to eliminate the inter-sensor-nodes redundancy and concatenate the processed information of other sensor nodes. In our previous work [7] [8], we have defined some mechanisms to process the information and to estimate its importance. These mechanisms will be enhanced and used in our present work. Our agent is implemented in the application layer; however it could be seen as a cross-layer entity as it uses the information of the routing level to build its dedicated view. In general, an agent is interested in the events occurring in its neighborhood. That is why, in our proposition, the local view of the agent will be restricted to its one hop neighbors and the first node on its path to the sink. An agent is also able to know if it is the first hop on the path to the sink for other nodes. We would like to underline here that an agent will never know the whole WSN architecture as this is cost expensive in term of communication and non important to achieve the agent goal.

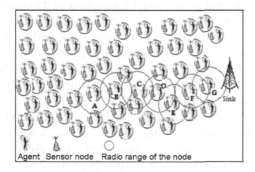

Fig. 1. Network topology

4.2   Agent Knowledge Base

One of the basic attributes of an agent is to be situated (situadness [9]). That is, an agent is a part of its environment. Its decisions are based on what it perceives of this environment and on its current state [10]. The situated view of an agent is then composed of the information obtained from its local observation and the information exchanged with its neighbors. For that, we should define carefully the required information for the agent to achieve its mission. This information will be stored in a data base, which we call a knowledge base. The knowledge base contains the list of the one hop neighbor agents, the first agent in its path to sink and also if the agent is a first agent on the path to the sink for other agents.

This information could be found through the routing level. At this level, our proposal is based on the Dynamic Source Routing protocol [11]. This protocol uses flooding technique to discover the route from a source to a destination. During the route discovery, network nodes build their routing tables from the information existing in the route request messages. For a computer network or a non limited resource network, there is no problem to save all of this information. However, in the case of WSN, we should pay attention to the memory size. In our proposal, as we said previously, the agent needs only the knowledge based on its situated view. Therefore, the agent filters the information received from the route request messages, saves only what interests it and keeps out the others.

4.3   Gathering Session Scenario

The information gathering session starts when an agent (sensor node) detects important information. This agent invites its one hop neighbors to cooperate in order to gather the maximum possible of information and to create a cooperation message resuming these collected information. However, the neighbor agent, who is at the same time the first hop on the path to the sink for the agent in question (source node), will not response the cooperation request. Indeed, once the co-operation message is ready, this neighbor agent (called intermediate agent) will receive the message and will invite its one hop neighbors' agents to cooperate. The intermediate agent will gather the information of its one-hop neighbors and extend the initial cooperating message. This message will be then sent to the next intermediate agent. The new intermediate agent will, in its turn, repeat the same scenario. This scenario will be repeated until reaching the sink node.

In the following, we present an example to illustrate the main operation of the agents implemented into the sensor nodes during a gathering session. We consider in this example the network shown in fig. 1.

We suppose that the sensors of node A detect information. The information is sent directly to the corresponding agent (agent A) to be processed. After processing, we consider that agent A estimated this information as important. Consequently, agent A sends a cooperation request to its one hop neighbors. The communication of this request passes by a one hop broadcast. Indeed, the one hop neighbors will be programmed to not re-broadcast it.

By sending the cooperation request, agent A invites its one hop neighbor agents to join cooperation for a data gathering session. A neighbor agent decides to cooperate if it has also important information.

After taking the appropriate decision, each cooperating agent responses by sending its processed data. This data will be concatenated within a coopera-tion message after an inter-sensor-nodes redundancy elimination. A model of this cooperation message has been previously presented in one of our previous works [8]. This message contains two main parts. The first one is for the sensor nodes' addresses and the second one is reserved for the correspondent data.

Agent A sends the cooperation message to its first hop on the path to the sink which is agent B (fig. 1). As agent B is a one hop neighbor for agent A, it has previously received the cooperation request sent by A. Agent B did not response this request as it knows that it is an intermediate agent (on the path to the sink for agent A). When receiving the cooperation message, agent B sends its cooperation request to its one hop neighbors to gather their data. If agent A and agent B have common one hop neighbor(s) agent(s), the common agents receive two cooperation requests but answer only the first request and neglect the second one.

Agent B concatenates its cooperation message with the initial message received from agent A. Then it sends it to its first hop to the sink, which is the agent C. Finally, agent C and all the intermediate agents repeat the same procedure as agent B until reaching the sink node.

5   Simulation Setup

In order to evaluate the relevance of our proposal (IIBC), we have carried out a set of simulation tests. In this section, we define the simulation parameters in order to demonstrate the performance of IIBC by comparing it to the traditional client/server (C/S) communication architecture. We have implemented these two approaches on GlomoSim [12] which is a scalable simulation environment for wireless and wired network systems.

We have run our simulation over a 1000mx1000m square with a random distribution of nodes during 1000 seconds. We have limited the radio range and the data rate of each node to 87 meters and 1Mbps respectively as suggested in [13]. The size of sensed data is 24 byte per node and the sensed data interval is 10 seconds. The transmission and reception powers parameters, which influence directly the radio range, have been chosen carefully from the ranges defined in the sun spot system technical document [14].

In order to test the scalability and the relevance of IIBC across different network densities, the simulations are done for a number of nodes varying from 100 to 900 nodes with an interval of 200. The local processing time is inspired from the work realized in [5], where the processing code is put in a message (mobile agent) sent by the sink. Indeed, transferring this code from the message to the node and placing it in the appropriate place of the memory will take some time. We have estimated this time to 10 ms. The authors of [5] have fixed the processing time to 50 ms which means that 40 ms will be sufficient in IIBC.

5.1   Performance Criteria

In this section, we present the main performance criteria and the base of their evaluation through our simulations:

1. Energy consumption is the parameter that defines the life duration of a sensor node and consequently of the concerned wireless sensor network. Therefore, we consider this parameter as the most important criterion to evaluate the performance of IIBC. In our simulations, we compute the average value of power consumed by each node which is composed of three main parts:

- The communication entity of the sensor node, which is the most energy-intensive function in the node. In order to compute the amount of energy consumed in communication, we use the equation (1) defined in [13]. E_{TX} is the power consumed during transmission and E_{RX} is the power consumed during the reception. Both of them are computed following the data length and distance of transmission (radio range of the node) (l,d);

$$E_{TX}(l, d) = lE_c + led^s \qquad \text{where} \quad e = \{ \begin{matrix} e_1 & s = 2, & d < d_{cr} \\ e_2 & s = 4, & d > d_{cr} \end{matrix} \tag{1}$$
$$E_{RX}(l, d) = lE_c$$

Where E_c is the base energy required to run the transmitter or receiver circuitry. A typical value of E_c is 50nJ/bit for a 1-Mbps transceiver; d_{cr} is the crossover distance, and its typical value is 87m; e_1 (or e_2) is the unit energy required for the transmitter amplifier when d < d_{cr} (or d > d_{cr}). Typical values of e_1 and e_2 are 10pJ/bit.m^2 and 0.0013pJ/bit.m^4 respectively.

- The energy consumed by the CPU: to compute this energy, authors in [13] have defined a rule based on the number of instructions and the frequency of the processor. In IIBC, we use the processor defined in the sun spot technical document [14], which defines the processor frequency of their sensor nodes to 180 Mhz. According to [13], a processor with such frequency consumes approximately 0.8 nJ per instruction. For this parameter, we would like to underline that for the C/S approach the power consumption of the CPU is neglected as the information processing occurred in the sink and not in the sensor node;

- The consumption of the sensing action: we neglected this consumption as it is supposed to be the same for IIBC and C/S approach.

2. The average end to end delay is an important criterion. This parameter represents the average latency needed to carry a message from a source sensor node till the sink. This delay computation is applicable as it to the C/S approach as all the processing is done in the sink node. However, in IIBC, the average end to end delay includes the local processing time (needed for estimating the importance of data and cooperating with neighbor agents);

3. The packet delivery ratio is the ratio of data packets received by the sink node to the number of packets generated by the source nodes. In IIBC, the source nodes are the nodes that start a data gathering cooperation, which means that the number of messages sent by source nodes is equal to the number of data gathering cooperation sessions;

4. The saved overhead: this parameter emphasizes the importance of the data concatenation. It defines the average number of saved messages' headers needed to carry out the information of n-source nodes. In the traditional C/S architecture, we need one message header to send the information of one node to the sink, while in IIBC; we need one message header to carry out the information of n-cooperating nodes.

6 Results and Analysis

In this section, we present the simulation results to highlight the relevance of IIBC. We show the advantages of IIBC by comparing it to the traditional C/S approach. We have chose to compare IIBC to the C/S approach as it is the main base for all the propositions and actually it is the most deployed one. We focus mainly on the efficiency of IIBC in terms of power consumption and scalability in different network densities. As presented in the simulation setup section.

In fig. 2 a, we can observe that IIBC highly decreases the power consumption. In addition, it is clear that the saved power is more important for the higher number of nodes. These results prove that IIBC is significantly better designed for scalable or dense networks than the C/S approach. Indeed, for a number of nodes varying from 100 till 900, the power consumption obtained by using IIBC is in average reduced by a factor of 7.11. In fig. 2 b, we show the average end-to-end delay in the network. We see that the values when IIBC is employed are higher than those given by the C/S approach.

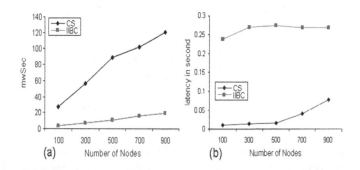

Fig. 2. a) Average Power Consumption per Node, b) Average End-to-End Delay

This gives the false impression that the performance of IIBC in term of end-to-end delay is bad. However, this is not correct, as the end-to-end delay in the C/S approach does not include the processing time. If we add this processing time to the end-to-end delay, we think the difference with IIBC will be really reduced. In fact, by observing the results, we can see that the different values, obtained by using IIBC and the C/S approach, are lower than 0.3 second. This means only the applications, which are very sensitive to latency and require less than 0.3 second precision, could be influenced. In the other hand, it is clear that the average end-to-end delay is approximately stable and follows a straight curve for a number of nodes varying from 300 to 900. That means the network density does not influence the end-to-end delay and the scalability could be supported easily in term of latency. Moreover, this difference could be easily explained when discussing later the saved overhead results (fig. 4).

In Fig. 3, we plot the packet delivery ratio (percentage of received packets). We see that the packet delivery ratio decreases when the density of the network increases. This is attributed to the fact that the main source of loosing packets is the collision. Furthermore, the number of messages sent in the network and the probability of collision are higher for the denser networks. Nevertheless, the packet delivery ratio in IIBC is always close to 100%. This high level of packet delivery ratio in IIBC could be explained by the fact that we have just one hop communications. Indeed, the initial (or intermediate) node sends a cooperation request to only its one hop neighbors. Next, the neighbors respond to the initial (or intermediate) by a one hop communication. Finally, the initial (or intermediate) node sends the cooperation message to its first node on the path to the sink. Thus, we eliminated the problems of multi hop communication in term of packets loss.

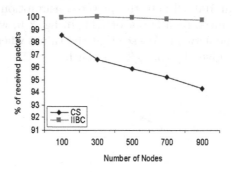

Fig. 3. Percentage of received packets

This result means that IIBC is scalable in term of packet delivery ratio. However, the C/S approach is sensitive to the density of the network as the values of packet delivery ratio decrease when the number of nodes increases. Thus, the C/S approach is not scalable regarding the packet delivery ratio. It should be noted here that the fact of having high values in both solutions is due to the use of a relatively big sensing interval (10s), which is the case of monitoring applications.

The last performance criterion is the saved overhead. This criterion has a special particularity as it helps to better explain the results obtained in term of power consumption and end-to-end delay. As we can observe in fig. 4, for a number of nodes equal to 100 (non dense network), we have around 17 sensor-nodes information sent in only one cooperation message. In term of power consumption, this result means that in IIBC, we send one message with one header instead of sending 17 messages with 17 headers in the C/S approach. By referring to fig. 2 a, we can notice that the power consumption in the C/S approach and in IIBC was around 27 mw and 3 mw respectively, which is a ratio of 9. This ratio means that for a non dense network, IIBC allows us to send the same amount of significant data with a power consumption divided by 9 comparing to the C/S approach. For the denser networks, the values obtained show a real gain in term of energy.

The saved overhead explains also the end-to-end delay occurred in IIBC. By looking to the values obtained in fig. 2 b, we can consider that IIBC sends the information of 17 sensor nodes processed in around 250 ms, while the C/S approach requires 10 ms to send one message of non processed information. By comparing the time that the sink node takes to compute 17 messages and their communication time to the 250 ms required in IIBC, we could estimate that there is not a real latency difference between IIBC and the C/S approach.

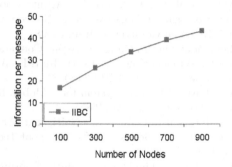

Fig. 4. Saved overhead for each cooperation

7   Conclusion and Future Works

In this paper, we have presented a new information importance based communication, scalable and energy efficient approach for data gathering in wireless sensor networks (IIBC). This approach is based on a multi-agent cooperation solution, in order to gather the maximum possible of important information in each data gathering session. By cooperating, the agents eliminate the inter-sensor-nodes redundancy and concatenate all the processed information of one gathering session into one message. That means a gain in the amount of information and the overhead needed to send them. We have proved through successive simulations that IIBC is very scalable comparing to the client server approach. Indeed, IIBC consumes little amount of energy, justified latency by taking into consideration the gain in overhead, and has a very high packet delivery ratio even if the network is dense. In the future work, we are intended to define an agent strategy to make a more appropriate decision to cooperate or not taking in consideration several important parameters (the available power in the node, its position within the network, etc.)

References

1. Ferber, J.: Multi-Agent System: An Introduction to Distributed Artificial Intelligence. Addison Wesley Longman, Harlow (1999)
2. Das, S.R., Nasipuri, A., Patil, S.: Serial Data Fusion Using Space-filling Curves in Wireless Sensor Networks. In: IEEE (ed.) First Annual IEEE Communications Society Conference on Sensor and Ad Hoc Communications and Networks, pp. 182–190 (2004)

3. Liao, W.H., Kao, Y., Fan, C.M.: Data aggregation in wireless sensor networks using ant colony algorithm. J. Netw. Comput. Appl. 31(4), 387–401 (2008)
4. Chen, H., Mineno, H., Mizuno, T.: Adaptive data aggregation scheme in clustered wireless sensor networks. Comput. Commun. 31(15), 3579–3585 (2008)
5. Chen, M., Kwon, T., Yuan, Y., Leung, V.C.M.: Mobile Agent Based Wireless Sensor Networks. Journal of Computers, 14–21 (2006)
6. Akyildiz, I.F., Su, W., Sankarasubramaniam, Y., Cayirci, E.: Wireless sensor networks: a survey. Computer Networks 38, 393–422 (2002)
7. Sardouk, A., Merghem-Boulahia, L., Gaiti, D.: Agents Cooperation for Power-Efficient Information Processing in Wireless Sensor Networks. In: Networking and Electronic Commerce Research Conference, Italy (2008)
8. Sardouk, A., Merghem-Boulahia, L., Gaiti, D.: Agent-Cooperation Based Communication Architecture for Wireless Sensor Network. In: IFIP Wireless Days/Ad-hoc and Wireless Sensor Networks, UAE, IFIP (2008)
9. Brooks, R.: A Robust Layered Control System for a Mobile Robot. IEEE Journal of Robotics and Automation 2(1), 14–23 (1985)
10. Rahim-Amoud, R., Merghem-Boulahia, L., Gaiti, D.: An Autonomic MPLS DiffServ-TE Domain. Whitestein Series in Software Agent Technologies and Autonomic Computing 9, 149–168 (2008)
11. Johnson, D., Hu, Y., Maltz, D.: The Dynamic Source Routing Protocol (DSR) for placeMobile Ad Hoc Networks for IPv4. Technical report (2007)
12. Laboratory, U., Laboratory, W.: Glomosim: A scalable simulation environment for wireless and wired network systems. In: The 3rd International Working Conference on Performance Modeling and Evaluation of Heterogeneous Networks (2005)
13. Sohraby, K., Minoli, D., Znati, T.: Wireless Sensor Networks, Technology, Protocols, and Applications. Willey, Chichester (2007)
14. Sun: Sun^{TM} Small Programmable Object Technology (Sun SPOT) Theory of Operation. Technical report, Sun Microsystem, Sun Labs (2008)

Considerations for
RFID-Based Indoor Simultaneous Tracking*

Apostolia Papapostolou and Hakima Chaouchi

Telecom-Sudparis, CNRS SAMOVAR, UMR 5157, LOR department
{apostolia.papapostolou,hakima.chaouchi}@it-sudparis.eu

Abstract. Context-aware applications is not just a vision. Advances in wireless communications and mobile capabilities have revolutionized the way services are brought to users, i.e. adapted to their context. Location is a key attribute of the term context and thus, an accurate location determination system is of paramount importance. RFID (Radio Frequency IDentification) is an emerging technology and recently has been explored for its applicability in location sensing systems. In this paper, we focus on an RFID-based localization approach in an indoor multi-user environment and model its most adverse implicating factors, that is collisions among its main components and interference from indoor characteristics. Extensive simulations are conducted to characterize and evaluate the performance behavior of the proposed scheme in environments with different levels of severity.

Keywords: RFID, Location, Simultaneous Trucking, Reader Collisions.

The proliferation of wireless technologies, mobile computing and Internet has spurred the development of innovative context-aware services, with location being a crucial attribute of the general term context. Additionally, network functionalities such as mobility management, network planning, load planning etc, can be improved if location information is available. However, for accomplishing these visions, a location determination system which provides an accurate, reliable and fast estimation of users or devices position is required.

WLAN-based indoor location systems have attracted the interest of the research and industry communities. The main benefits of such systems is their low cost and ease of deployment since they rely on the existing WLAN infrastructure. Moreover, the number of WiFi-enabled devices is rapidly increasing in the markets. However, the achieved accuracy is limited.

RFID is a rapidly developing short range technology which uses wireless communication for automatic identification of objects. An RFID system consists of two main components, the tag and the reader. The RFID reader can read data emitted from the RFID tags within its reading range by using a defined Radio Frequency (RF) and protocol. The RFID tags can be either passive or active.

---

* This paper is conducted in the French national research project ANR 2006 SUN: Situated and Ubiquitous Networks.

J. Wozniak et al. (Eds.): WMNC 2009, IFIP AICT 308, pp. 309–320, 2009.

Passive tags operate without battery, they just backscatter (far-field case) the RF signal transmitted to them from a reader. Active tags contain both a radio transceiver and a battery. Thus, passive tags compared to active tags are less expensive and with unlimited lifetime but with reduced read range capability. Due to its relatively low cost and its technical capabilities, RFID is considered as an attractive technology for location sensing. Correlating tag ids with specific locations can aid determining the location of a reader based on the location information retrieved from the tags within its range. However, RFID technology has some limitations which should be considered before employing it for localization. Being a short-range communication technology requires a dense deployment of tags so that readers can always detect some of them. Additionally, interference from specific materials affect the read range of a reader. Last but not least the reader collision problem prevents the colliding readers from communicating with the RFID tags in their respective reading zones and thus the simultaneous tracking of multiple users is more complicated than the single-user case.

In this work, we address, model and evaluate the problem of tracking multiple reader-enabled users in the presence of interference and collisions among their readers when they simultaneously attempt to communicate their common tags.

The remainder of this paper is organized as follows. Section 1 outlines work related to the localization problem. In section 2 an overview of the RFID technology is provided. In section 3 the concept of localization relying on RFID and its limitations are introduced. In section 4 we model the system, communication properties and positioning methods of an RFID-based localization system. Simulations results in section 6 evaluate its performance and finally in section 7 we conclude our study.

1   Related Work

The most well known positioning system is the Global Positioning System (GPS), [1], which is satellite-based and is successful for tracking users in outdoor environments. However, the inability of satellites signals to penetrate buildings cause the complete failure of GPS indoors. For indoor location sensing a number of wireless technologies have been proposed, such as infrared, [7] ultrasound, [8], UWB, WiFi, [3]-[6], Bluetooth, RFID, ZigBee and three are the main positioning techniques that are employed, namely triangulation, scene analysis and proximity. For triangulation methods, either the time of arrival (TOA), time difference of arrival (DTOA), angle of arrival (AOA) or strength (RSS) of the signals received from the surrounding transmitters are used for locating a user. Scene analysis methods require an offline phase for characterizing the radio behavior of the area under study. Proximity methods are based on the identification of objects with known location or topology constraints. [2] provides an interesting survey for positioning methods and schemes.

RFID-based schemes, such as [10]-[14], rely on the existence of RFID readers in the area and modify any of the three well-known positioning methods to locate users who are equipped with an RFID tag. Others, such as [12] and [15], rely on the deployment of tags in the area and try to locate a single user who is equipped with an RFID reader.

2 RFID Technology Overview

RFID (Radio Frequency IDentification) is an automatic identification system that consists of two basic hardware components, a tag and a reader. A tag has an identification (ID) stored in its memory that is represented by a bit string. The reader, which typically is a powerful device with memory and computational resources, is able to read the IDs of the tags in the neighborhood by running a simple link-layer protocol over the wireless channel. Various types of tags exist which differ significantly in their computational capabilities. They range from dump passive tags which operate without battery but they respond simply to reader's queries to smart active tags which contain both radio transceiver, memory and a power supply. Thus, passive tags compared to active tags are less expensive and with unlimited lifetime but with reduced read range capability. Due to their low cost, passive tags are anticipated to be a popular choice especially for large scale deployment.

The communication between a reader and a passive tag is done using either magnetic or electromagnetic coupling. Coupling is the transfer of energy from one medium to another medium, and tags use it to obtain power from the reader to transfer data. Two are the main types of coupling, inductive or backscatter, depending on whether the tags are operating in the near-field or far-field of the interrogator, respectively. A key difference between them is that far-field communication has a longer read range compared to near field communication.

RFID systems operate in the Industry, Scientific and Medical (ISM) frequency band that ranges from 100 KHz to 5.8 GHz but they are further subdivided into four categories according to their operating frequency: Low Frequency (LF), High Frequency (HF), Ultra-High Frequency (UHF) and Microwave. Tags operating at UHF and microwave frequencies use far-field and couple with the interrogator using backscatter. Recently, ultra-high frequency (UHF) band passive RFID systems have drawn a great deal of attention. In this paper, we consider RFID tags operating in the UHF (890-960 MHz) frequency band.

3 RFID-Based Localization

RFID is a rapidly developing short range technology which uses wireless communication for automatic identification of objects. Due to the low cost of passive tags, the non-LOS requirement, the simultaneous reading of multiple tags, and the reduced sensitivity regarding user orientation, the RFID is considered as an attractive technology for location sensing.

3.1 Methology Concept

RFID-based positioning schemes can be classified into two general categories depending on the type of the RFID component supported by the target's device. For the first type we assume that the user device is equipped with a tag and a number of readers are placed in the area. In this case, any of the general positioning techniques, i.e. triangulation, scene analysis or proximity can be employed to estimate the location of the user. In the second case, the user's terminal is equipped with an RFID reader and passive or active tags with static and known coordinates are deployed in the area. Its location can be calculated based on the location information retrieved from the detected tags.

We focus on the second case and we attempt to determine the location of multiple reader-enabled user terminals. More precisely, passive tags are deployed in the floor of the areas such that a grid is formed and they are termed as *reference* tags.

3.2 Limitations

Even though RFID technology has promising key characteristics for location sensing, it has also some limitations which become more intense in the case of simultaneous tracking in a multi-user environment and thus should be taken into account before employing an RFID system for localization.

Since RFID technology uses electromagnetic waves for information exchange between tags and readers, how radio waves behave under various conditions in the RFID interrogation zone (IZ) affects the performance of the RFID system. Radio waves propagate from their source and reach the receiver. During their travel, they pass through different materials, encounter interference from their own reflection and from other signals, and may be absorbed or blocked by various objects in their path. The material of the object to which the tag is attached may change the property of the tag, even to the point it is not detected by its reader.

However, the most harmful type of interference is the one among its components which is known as the RFID collision problem. Three are its main types: tag collision, multiple reader-to-tag collision and reader-to-reader collision.

- *Tag Collisions*
  This type of interference arises when multiple tags are simultaneously energized by the reader and reflect their respective signals back to the reader. Due to a mixture of scattered waves, the reader cannot differentiate individual IDs from the tags; therefore, anti-collision mechanisms such as those known as binary-tree and ALOHA are needed. For dealing with this type of interference the EPCglobal class-1 standard implements an algorithm based on a Query Tree protocol, [17]. Adopting this anti-collision algorithm a reader can potentially read 500 collocated tags per second.
- *Reader-to-Reader Collisions*
  Reader-to-reader collision is induced when a signal from one reader reaches other readers. This can happen even if there is no intersection among reader

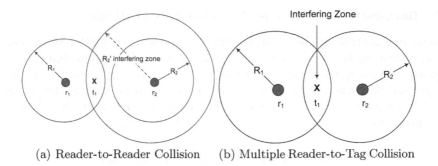

(a) Reader-to-Reader Collision   (b) Multiple Reader-to-Tag Collision

Fig. 1. The Collision Problem

interrogation ranges but because a neighbor reader's strong signal interferes with the weak reflected signal from a tag. Figure 1(a) demonstrates an example of collision from reader r_2 to reader r_1 when the latter tries to retrieve data from tag t_1. Generally, signal strength of a reader is superior to that of a tag and therefore if the frequency channel occupied by r_2 is the same as that between t_1 and r_1, r_1 is no longer able to listen to t_1s response.

- *Multiple Reader-to-Tag Collisions*
  Multiple reader-to-tag collision happens when a tag is located at the intersection of two or more reader interrogation ranges and the readers attempt to communicate with the tag simultaneously. As illustrated in figure 1(b), when readers r_1 and r_2 transmit query messages to a tag t_1, t_1 might not be able to read the query messages from r_1 and r_2 due to interference.

Reviewing the current literature several scheduling-based solutions can be found. However, applying any of the general Multiple Access (MA) mechanisms, i.e. Time Division MA (TDMA), Frequency Division MA (FDMA), Code Division MA (CDMA) and Carrier Sense MA (CSMA), directly in RFID is not trivial, mainly due to the limited capabilities of the RFID passive tags.

4 System and Communication Model

In this section we try to derive a realistic mathematical model for describing the communication properties among the component of an RFID by considering their specifications and the main sources of error related to the localization process.

4.1 Area Setup

We model an indoor environment as a 2-D area with L and W denoting its length and width respectively. A set, $\mathcal{T}$, of passive RFID tags with known coordinates (x_t, y_t), $\forall t \in \mathcal{T}$ are placed on the floor of this area such that a grid of reference tags is formed with inter-tag spacing δ. Within this area, a set, $\mathcal{N}$, of users with RFID reader-enabled terminals may be located and an accurate and fast estimation of their position, $(\widehat{x}_u, \widehat{y}_u), \forall u \in \mathcal{N}$, should be obtained.

4.2   Backscattered Path Loss Model and Read Range

The communication link between the RFID components is half duplex, reader to tag and then tag to reader. In forward link, reader sends a modulated carrier to tags and powers up the tags. In return link, the tag receives the carrier for power supply and backscatters by changing the reflection coefficients of the antenna. In such a way, its id is sent to reader. The path loss of this two way link may be expressed as:

$$PL(d) = PL_o + 10N \log\left(\frac{d}{d_o}\right) + X_\sigma(dB) \qquad (1)$$

where the variable X_σ, called the shadow fading, is a zero-mean Gaussian random variable in dB having a standard deviation of σ_{dB} and is used to model the random nature of indoor signal propagation due to the effect of various environmental factors such as multipath, obstruction, orientation, etc. N is the path loss exponent whose value depends on the frequency used.

$$PL_o = G_t G_r (g_t \Gamma g_r)\left(\frac{\lambda}{4\pi d_o}\right)^4 \qquad (2)$$

is the path loss at reference distance d_o, where G_t, g_t, and G_r, g_r are the gains of the reader and tag transmit and receive antennas, respectively. Γ is a reflection coefficient of the tag and λ the wavelength.

The combined forward and return path losses help calculate the maximum distance between the reader antenna and the tag at which the reader can decode the signals reflected from the tag, also called the read range and denoted as R_{max}. For a desired reader, the received backscattered power, $P_s(d)$, is given by

$$P_s(d) = \alpha_{BW} \Gamma P_t G_t G_r \times 10^{PL(d)/10}, \qquad (3)$$

where P_t denotes the total transmit power, and α_{BW} is the spectrum power of a used channel normalized by the total power. In the absence of interference, the maximum read range a reader receiver can decode the backscattered signal is such that

$$R_{max} = \arg\max_{d \geq 0} P_s(d) \geq TH, \qquad (4)$$

where a TH represents a threshold value.

4.3   Interference and Read Zone Reduction

In subsection 3.2 the interference problem was described. *Multiple reader-to-tag collisions, reader-to-tag collisions* and *interference from materials* are the most implicating factors and therefore modelling them is essential.

Regarding *multiple reader-to-tag collisions* we utilize the following model. Let R_i and R_j denote the read ranges of readers r_i and r_j and d_{ij} their distance. Apparently, if

$$R_i + R_j > d_{ij} \qquad (5)$$

and r_i and r_j communicate at the same time, they will collide and the tags in the common area will not be detected. For characterizing the probability of simultaneous communication we assume that each reader is in a scanning mode with probability p^{scan}. Thus, the probability of collision, P_{ij}^C, between r_i and r_j, if (5) is satisfied, depends on the probabilities r_i and r_j are in a scanning mode, p_i^{scan} and p_j^{scan}, respectively, i.e. $P_{ij}^C = p_i^{scan} \times p_j^{scan}$.

Reader-to-reader collisions affect the R_{max} parameter. In (4) this factor had been neglected. However, when interfering readers exist, the actual interrogation range of the desired reader decreases to a circular region with radius R_{max}^I, which can be represented by

$$R_{max}^I = \arg \max_{d \in [0, R_{max}]} SIR(d) \geq TH, \tag{6}$$

where

$$SIR(d) = \frac{P_s(d)}{\sum_i I_i} \tag{7}$$

and I_i, interference from reader r_i.

Finally, for incorporating in the model problem of *interference from materials*, each *reference* tag t is assigned a probability, p_t of not being detected. Obviously high values of p_t are assigned to tags which are mounted to such interfering materials.

5   Positioning Methology

5.1   Architecture

From architectural point of view, a location determination scheme can be either user-based or network-based. In the first case, each user is responsible for collecting and processing information necessary for determining his location, whereas, in the second case, a dedicated server is responsible for gathering all required data and finally providing the location estimates for all users. Processing capabilities, privacy and scalability issues are usually the main factors for selecting the appropriate approach. We propose a hybrid architecture as a compromise between them, i.e. both user and a central location server participate in the location decision process.

The reader of each user device queries for tags within its coverage in order to retrieve their *ids*. A list of these retrieved *ids* , called *Tag-List*, is forwarded to the *Location Server*. Based on these received *Tag-List* and a repository which correlates the *ids* of the *reference* tag with their location coordinates, the *Location Server* estimates the location of that user by employing an RFID-based positioning (see subsection 5.2) algorithm and finally returns this location estimate back to the corresponding user.

5.2   Positioning Algorithms

Let $\mathcal{D}_u$ denote the set of *reference* tags successfully detected from a user's reader r_u. Note that this happens if all factors described in 4.3 are considered.

- *Simple Average (SA)*
  This is the simplest scheme since it requires only *id* information of the detected *reference* tags. In this case, the user location is estimated as the simple average of the coordinates (x_t, y_t) of all tags $t \in \mathcal{D}_u$, i.e.

$$(\widehat{x}_u, \widehat{y}_u) = \left(\frac{\sum_{t \in \mathcal{D}_u} x_t}{|\mathcal{D}_u|}, \frac{\sum_{t \in \mathcal{D}_u} y_t}{|\mathcal{D}_u|} \right) \qquad (8)$$

- *Weighted Average (WA)*
  Since some of the detected tags may be closer than others, biasing the simple averaging method is proposed as an alternative approach. This can be achieved by assigning a weight, w_t to the coordinates of each tag $t \in \mathcal{D}_u$. These weights are based on their distance from the reader. Thus, (8) becomes

$$(\widehat{x}_u, \widehat{y}_u) = \left(\frac{\sum_{t \in \mathcal{D}_u} w_t \cdot x_t}{\sum_{t \in \mathcal{D}_u} w_t}, \frac{\sum_{t \in \mathcal{D}_u} w_t \cdot y_t}{\sum_{t \in \mathcal{D}_u} w_t} \right) \qquad (9)$$

where $w_t = 1/\widehat{d}_t$ and $\widehat{d}_t$ is the estimated distance from the tag t based on the level of the received signal strength.
- *Multi-Lateration (ML)*
  Finally, a multi-lateration based approach is also investigated, according to which $(\widehat{x}_u, \widehat{y}_u)$ can be obtained by solving the following system of $|\mathcal{D}_u|$ equations:

$$(x_t - x)^2 + (y_t - y)^2 = \widehat{d}_t, \qquad \forall t \in \mathcal{D}_u \qquad (10)$$

Since $\widehat{d}_t$ are not accurate, the above system of equations can be solved by a standard LS approach as

$$(\widehat{x}_u, \widehat{y}_u) = (A^T A)^{-1} A^T b \qquad (11)$$

where A and b are matrixes obtained after subtracting the last equation from the first $|\mathcal{D}_u| - 1$ equations in the system of equations (10) for linearizing it, such that:

$$A_t = \left[2(x_t - x_{|\mathcal{D}_u|}) \; 2(y_t - y_{|\mathcal{D}_u|}) \right], \qquad (12)$$

$$b_t = \left[x_t^2 - x_{|\mathcal{D}_u|}^2 + y_t^2 - y_{|\mathcal{D}_u|}^2 + \widehat{d}_t^2 - \widehat{d}_{|\mathcal{D}_u|}^2 \right], \qquad (13)$$

$$\forall t \in \mathcal{D}_u \backslash \{|\mathcal{D}_u|\}$$

6 Simulation

In this section we evaluate the performance of our approach through simulations, using Matlab, [18], as our simulation tool. As performance metric we use the *mean location error* (MLE), defined as the euclidean distance between the actual and the estimated position of a user. We provide and interpret results of the simulations we conducted for evaluating the influence of the parameters δ, α, and R_{max} in the system performance. In order to illustrate the performance

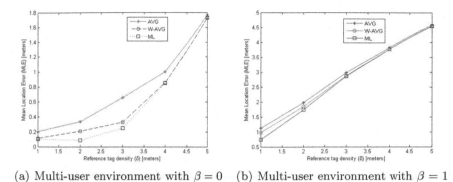

(a) Multi-user environment with $\beta = 0$ (b) Multi-user environment with $\beta = 1$

Fig. 2. Impact of tag density (δ)

degradation due to the collision problem and the essentiality of an anti-collision mechanism, we considered two multi-user environmental cases which differ in the level of the collision problem. Assuming that the probability user readers query their tags follows uniform distribution $U(\beta, 1)$, we set $\beta = 0$ for the first case and $\beta = 1$ for the second case. Apparently, for the second environment all users readers scan simultaneously for their tags and thus its performance is anticipated to be worse due to the collision problem among them.

6.1 Impact of Tag Density

Figures 2(a) and 2(b) illustrate the dependency of the *MLE* on the tag density, δ, when $\beta = 0$ and $\beta = 1$, respectively, and for the three rfid-based positioning methods described in subsection 5.2, i.e. *AVG*, *W-AVG* and *LS*. For all cases, increasing the inter-tag spacing reduces the accuracy. However, when the collision problem is severe, the achieved accuracy and performance reduction are worse and thus a dense tag deployment is required for providing robustness. Finally, comparing the behavior of the three positioning schemes, we note that there is a benefit from the added complexity but in highly colliding environments the achieved benefit is not significant.

6.2 Impact of Interference from Materials

Similarly, figures 3(a) and 3(b) show the impact of the level of interference (α) from environment materials for the two scenarios when $\delta = 2$ meters. We observe that this factor increases the MLE, however, its impact is less severe compared to the parameter δ.

6.3 Impact of Interrogation Range

Finally, in figures 4(a) and 4(b) the influence of the maximum read range, R_{max}, is depicted when $\delta = 2$. For both scenarios we observe that when $R_{max} = 1$,

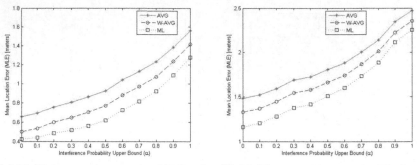

(a) Multi-user environment with $\beta = 0$ (b) Multi-user environment with $\beta = 1$

Fig. 3. Impact of material interference level (α)

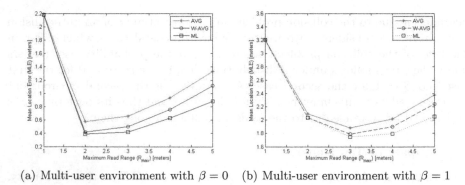

(a) Multi-user environment with $\beta = 0$ (b) Multi-user environment with $\beta = 1$

Fig. 4. Impact of maximum read range ($R_m ax$)

the MLE is increased and this is because tags are not detected. When $\beta = 0$, $R_{max} = 2$ gives the optimum performance for two main reasons; further than this collisions are more probable but also location information from far-away tags is included. For the second case, the optimum performance is achieved when $R_{max} = 3$ meters because of the collisions which prevents tags from being detected.

7   Conclusion

In this paper, we explore the applicability of the RFID technology in location sensing and the main design and environmental factors that should be considered before developing an RFID-based localization scheme. We focused on a scenario when the location of multiple reader-enabled terminals needs to be estimated based on the information retrieved from low cost passive tags, which are deployed in an area. We proposed a mathematical model for taking into account all implicating factors which affect the accuracy performance of the system, that is all types of collisions among its components, interference from materials, and temporal environmental changes. Extensive simulations were conducted to evaluate the impact of these parameters. More precisely, when reader collisions is not an issue, a low dense ($\delta \leq 4$ meters) deployment of passive tags can provide an accurate location information with error less than 1 meter. However, in a highly colliding environment, passive tags should be deployed with spacing of 1 meter in order to have similar location error resilience. Interesting remarks can be drawn regarding the communication range of readers. In the absence of collisions, short read range (2 meters) is beneficial. In contrast when readers attempt simultaneously access to the medium, a higher range (3-4 meters) results in better accuracy.

To summarize, RFID technology is suitable for positioning, but its performance degrades in highly populated environments and thus a denser tag deployment or/and a mechanism for controlling reader transmissions are required.

References

1. Kaplan, E.: Understanding GPS: Principles and Applications. Artech House (2005)
2. Hightower, J., Borrielo, G.: Location Systems for Ubiquitous Computing. IEEE Computer 34(8), 57–66 (2001)
3. Bahl, P., Padmanbhan, V.N.: RADAR: An In-Building RF-based User Location and Tracking System. In: IEEE Infocom, vol. 2, pp. 775–784 (2000)
4. Youssef, M., Agrawala, A.: The Horus Location Determination System. In: Mobisys, pp. 205–219 (2005)
5. King, T., Kopf, S., Haenselmann, T., Lubberger, C., Effelsberg, W.: COMPASS: a Probabilistic Indoor Positioning System Based on 802.11 and Digital Compasses. In: WinTeck, pp. 24–40 (2006)
6. Papapostolou, A., Chaouchi, H.: WIFE: Wireless Indoor positioning based on Fingerprint Evaluation. In: IFIP Networking (2009)
7. Want, R., Hopper, A., Falcao, V., Gibbons, J.: The Active Badge Location System. ACM Transactions on Information Systems 40(1), 91–102 (1992)
8. Priyyantha, N.B., Chakraborty, A., Balakrishnan, H.: The Cricket Location-Support System. In: 6th International ACM MOBICOM (2000)
9. Want, R.: An introduction to RFID technology. IEEE Pervasive Computing 5(1), 25–33 (2006)
10. Ni, L.M., Liu, Y.: LANDMARK: Indoor Location Sensing Using Active RFID. Wireless Networks, 701–710 (2004)
11. Hightower, J., Want, R., Borriello, G.: SpotON: An indoor 3D location sensing technology based on RF signal strength. University of Washington, Department of Computer Science and Engineering, Seattle, WA, UW CSE 00-02-02 (February 2000)
12. Wang, C., Wu, H., Tzeng, N.-F.: RFID-based 3-D positioning schemes. In: IEEE INFOCOM, pp. 1235–1243 (2007)
13. Bouet, M., Pujolle, G.: A Range-Free 3-D Localization Method for RFID Tags Based on Virtual Landmarks. In: IEEE PIMRC (2008)

14. Bekkali, A., Sanson, H., Matsumoto, M.: RFID Indoor Positioning Based on Probabilistic RFID Map and Kalman Filtering. In: Third IEEE International Conference on Wireless and Mobile Computing, Networking and Communications, WiMob 2007 (2007)
15. Yamanoi, K., Tanaka, K., Hirayma, M., Kondo, E., Kimuro, Y., Matsumototi, M.: Self-localization of Mobile Robots with WID System by using Support Vector Machine. In: Proceedings of 2004 IEEWRSI International Conference on Intelligent Robots and Systems, Sendai, Japan (2004)
16. Leong, K.S., Ng, M.L., Cole, P.H.: The reader collision problem in RFID systems. In: IEEE 2005 International Symposium on Microwave, Antenna, Propagation and EMC Technologies for Wireless Communications, Beijing, China, August 8-12 (2005)
17. http://www.epcglobalinc.org
18. http://www.mathworks.com

The Efficiency Performance on Handover's Scanning Process of IEEE802.16m

Ardian Ulvan and Robert Bestak

Czech Technical University in Prague, Faculty of Electrical Engineering,
Technicka 2, Prague, 16627, Czech Republic
{ulvana1,bestar1}@fel.cvut.cz

Abstract. This paper investigates the efficiency of scanning process on handover procedure in IEEE802.16m. The performance evaluation is based on current running standard of 802.16e since .16m is still under development. The evaluation is done at both parts of handover procedure, i.e. network topology acquisition and handover process. An efficient scanning process is proposed to reduce the scanning time and obtained a minimal handover interruption time.

Keywords: WiMAX network, handover, MAC efficiency.

1 Introduction

The IEEE 802.16m amends the IEEE802.16 WirelessMAN-OFDMA specification to provide an advanced air interface for operation in licensed frequency bands. It includes enhancements and extensions to the IEEE STD 802.16e to meet the cellular layer requirements of IMT-Advanced next generation mobile networks conducted by International Telecommunications Union – Radio Communications Sector (ITU-R) [1].

The handover is an integral part of all mobile wireless systems. Continuous connection during user movement among cells is allowed due to handover procedure, but on the other hand, the handover brings a significant increase of Medium Access Control (MAC) overhead and also causes an increase in delay of packet delivery to the destination user.

The handover WiMAX is introduced in the latest version of IEEE802.16e [2]. Based on the emerging 802.16j proposals, it is very probable that in the next versions of WiMAX recommendations (IEEE 802.16j, and IEEE 802.16m) will be defined the same types of handovers. Moreover, the general principles (with regards to the requirements of new standards) of handover will be adopted from 802.16e.

The handover issues on IEEE802.16 have been carried out in several literatures. The fast handover in IEEE802.16e is analyzed in [3]. The single neighbor BS scanning, fast ranging and pre-registration have been proposed to reduce the handover time. To decrease hard handover, the authors in [4] propose handover modification that makes possible to receive downlink data just after synchronization with downlink channel of the target Base Station (BS). This is achieved by introducing a new MAC management message – Fast DL_MAP_IE message, which describes the fast downlink access definition in system's information element. This message is used for

J. Wozniak et al. (Eds.): WMNC 2009, IFIP AICT 308, pp. 321–331, 2009.

transmission of emergent packet (packet with payload of delay sensitive services) by target BS to mobile station (MS). In [5], the collision of connection identifiers (CIDs) in the target BS is solved by employing transport CID mapping scheme that increases the handover performance. The authors also introduce Passport handover to decrease the hard handover delay.

The aim of this paper is to analyse scanning time during the network topology acquisition stage of handover procedure and to determine handover interruption time that occurs during the handover procedure. We consider several scan types and several handover strategies in our analysis to obtain the most efficient scanning time and the lowest interruption time.

The remaining of the paper is organized as follows. The description of handover procedure based on IEEE802.16e standard, the analysis of inefficient aspect of handover, scanning time analysis and handover interruption time in IEEE802.16m's system profile are described in Section 2. The result analyses by numerical calculations are given in Section 3. Section 4 concludes our work and highlights our future work.

2 Handover Procedure

2.1 IEEE802.16e Handover

According to [2], the handover procedure can be divided in two stages (Figure 1). Stage that is executed before handover, called **Network Topology Acquisition,** contains network topology advertisement and MS scanning. In this stage, the Mobile Station (MS) investigates and collects information about neighborhood base stations of its Serving BS. During the scanning phase, the MS seeks a suitable handover to the target BS or Relay Station (RS) that are suitable to be added to the Diversity Set. The Diversity Set is a list of the BSs/RSs, which are involved in the handover procedure in case of Macro Diversity Handover (MDHO) or Fast Base Station Switching (FBSS).

The scanning is realized in the "scanning intervals" which interleave the normal operation of MS. Once the scanning phase is completed, the MS sends results to its serving BS. The reported results can be delivered by two report types. The first one is "event trigger report", in which MS sends reports based on a defined trigger, such as carrier-to-interference-and-noise ration (CINR), receive signal strength indicator (RSSI), relative delay, or round trip delay (RTD). In this type of report, the measurement report is sent to the Serving BS after each measurement case. In the second type of report, "periodic report", the MS sends reports periodically.

The results of scanning are used in the next stage of handover procedure called **Handover Process**. The first step is *cell reselection*. In this step, possible target BS is selected based on signal quality and offered QoS. Then, *handover decision* and *initiation* process can be initialized if all conditions and requirements for handover are met follow. The first step of handover process is ended by performing the synchronization to the new Target BS. However, before the synchronization is done the connection(s) to the serving BS have to be closed first.

As soon as the downlink synchronization is done, the MS can start the next step of handover: *network re-entry* procedure. The network re-entry consists of three substeps i.e. *ranging*, *re-authorization* and *re-registration*. In the ranging process the MS obtains information about uplink channel via Uplink MAP (UL-MAP) and Uplink Channel Descriptor (UCD) messages. The Ranging is followed by authorization and registration of MS to the target BS. After successful authorization and registration to the target BS, the MS can start normal operation.

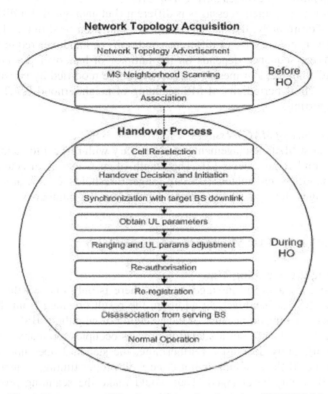

Fig. 1. Two stages of Handover Procedure based on IEEE802.16e

2.2 MAC Management Overhead in Handover

The total handover overhead is given by the summarization of length of all handover MAC management messages exchanged during the handover procedure.

Generally, the total handover overhead is affected by following parameters:

- Frequency of handovers (number of handovers per time interval)
- Sequence of exchanged messages
- Length and structure of MAC messages

Frequency of handovers
The number of handovers per a time interval depends on numbers of MSs and BSs/RSs, speed of MSs, the trajectory of MSs and the setting of cells' boundary. The

number of MSs and BSs/RSs in a network and speed of MSs are random and cannot be changed or influenced without impact on users QoS. In contrast, the boundary of cells (or range of BS' areas) can be effected by network parameters setting (threshold levels, relative or absolute thresholds, hard handover threshold hysteresis, etc).

Sequence of exchanged messages during handover
The sequence of management messages is different (but analogical) for different types of handovers. Additionally, the message sequence can even be different for the same types of handover; e.g. in case of different conditions or handover requirements. Besides, the message sequence depends on the initiator of handover procedure (BS or MS). The MAC management message sequence can be modified by network parameter setting up such as periodicity of MS scanning of neighborhood BSs/RSs or scanning results reporting.

Length and structure of MAC messages
The length of most MAC management messages vary with respect to handover conditions, or types and requirements [2]. The length of messages is affected by proper setting up of handover parameters (BS cell range, thresholds, etc.). For example, the length of messages during scanning phase depends on the number of recommended BSs to scan.

2.3 Minimization of MAC Management Overhead

2.3.1 Reducing the Scanning Time
In network topology acquisition stage, in fact, there is only one BS that can be selected as target BS for handover. In addition, the result obtained from the scanning process may become invalid because of the changing of neighbour BS's channel quality. Consequently, if the scan or association process occupies too many resources, the throughput significantly decreases. Furthermore, the standard does not clearly state the scanning time. If the scanning is not done with proper timing, channel condition of neighbour BSs may be changed. That would make the scanning process results useless. Figure 2 shows the exchanges of MAC management message in the handover procedure.

These two parts of handover procedure, fulfill up by the MAC layer, contribute to the overhead in 802.16m. Fortunately, there are several schemes that can be implemented to cope with these issues. The main wastes of resources are represented by redundant scanning and association processes of neighbour BSs. To resolved the issues we propose the single BS scanning scheme, target BS fast ranging and the pre-registration mechanism.

Single BS scanning scheme can be achieved by exploiting the target BS estimation algorithm, in which a MS only scans or associates to the neighbour BS with the best CINR. In the mean time, target BS uses a Fast_Ranging_IE to grant a dedicated uplink ranging opportunity to MS in its broadcasting UL-MAP message. So, the MS does not need to do contention-based ranging. Finally, the pre-registration is deployed, in which target BS obtains the service flow and authentication information of this MS through backbone network before handover.

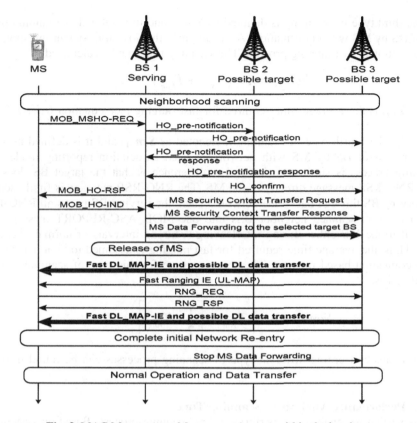

Fig. 2. MAC Management Message exchanges within the handover

To analyze the performance of efficient scanning scheme, we consider four types of scanning. The first type is denoted as $Scn_{\_1}$ and it is defined as scanning process of BSs by MS without association. In this scanning type, the MS is only requested to obtain downlink synchronization with the target BS in order to get information about quality of target BS's physical channel. The scanning time can be calculated as:

$$Scn_{\_1} = n \times T_{Sync} \qquad (1)$$

where n denotes as the number of neighbor BSs that need to be scanned and T_{Sync} is the average time required for downlink synchronization.

The second type is $Scn_{\_2}$ and it is defined as scanning process of BSs by MS without coordination. A MS requires the downlink synchronization as well as the execution of contention resolution-based ranging. The $Scn_{\_2}$ can be calculated as:

$$Scn_{\_2} = n \times \left(T_{Sync} + T_{Cont\_res}\right) \qquad (2)$$

where T_{Cont_res} is defined as the average time required for contention resolution-based ranging (see equation 5 for calculation).

The third type of scanning is denoted as Scn_3 and it is defined as scanning process of BS by MS with coordination. A MS requires the downlink synchronization and the execution of fast ranging process. The scanning time can be calculated as:

$$Scn_{\_3} = n \times \left(T_{Sync} + T_{Rng}\right) \tag{3}$$

where T_{Rng} is the average time required for fast ranging (see equation 6 for calculation).

Finaly, the fourth type of scanning is denoted as Scn_4 and it is defined as scanning process of BS by MS with network assisted association reporting mode. The scanning process is similar to Scn_3. The difference is that the target BS does not send RNG-RSP message directly to the MS. The RNG-RSP message is firstly sent to the serving BS through backbone network. Then, the serving BS packs all RNG-RSP messages from scanned neighbor BSs into one MOB_ASC-REPORT message and sends it to the MS either during interleaving scanning intervals or normal operation time. Thus, the average time required for fast ranging is assumed to be a half of that for a contention-based ranging. The scanning time for this type of scanning can be calculated as:

$$Scn_{\_4} = n \times \left(T_{Sync} + \left[\frac{1}{2} \times T_{Rng} \right] \right) \tag{4}$$

More details concerning different type of scanning processes can be found in reference [3].

2.3.2 Performance Analysis of Scanning Time

In this section we analyze the performance of scanning process. Based on system description document, we assumed a 20 ms super frame of IEEE802.16m is divided into 4 equally-size frames, where the frame length is assumed to be 5 ms [8]. The ratio of cell load (R_{Load}) is assumed to be $0 \le R_{Load} \le 100\%$. The T_{Cont_res} depends on the CINR and ratio of cell load. According to [9] the T_{Cont_res} is set to 75 ms and 150 ms when the ratio of cell load is 0% and 50% respectively. Since the ranging collision probability is higher as ratio of cell load increases, then T_{Cont_res} can be approximated as:

$$T_{Cont\_res} = 75 + \left(\left[150 \times R_{Load} \right] \times 2 \right) \tag{5}$$

The average time required for fast ranging (T_{Rng}) is also assumed to be directly proportional to the ratio of cell load. The T_{Rng} is set to 25 ms and 50 ms when the ration of cell load is 0% and 50% respectively [9]. It can be calculated as:

$$T_{Rng} = 25 + \left(\left[50 \times R_{Load} \right] \times 2 \right) \tag{6}$$

The average time required for authorization (T_{Auth}) is relatively longer than the other times since it is the time needed by target BS to obtain the authorization information of MS from the authorization server. T_{Auth} is assumed to be 150 ms [3]. Finally, the average time required for registration (T_{Reg}) is assumed to be 2 frames. The considered values of delays in our paper are summarized in Table 1.

Table 1. Time paramaters in IEEE802.16m scanning process

Time required	Notation	Typical value
Synchronization	T_{Sync}	2 frames
Contention resolution-based ranging	T_{Cont_res}	hundredth of ms
Fast ranging	T_{Rng}	tenth of ms
Authorization	T_{Auth}	150 ms
Registration	T_{Reg}	2 frames

The results of scanning time performance with respect to cell load and number of neighbour BSs are depicted in Figure 3 and Figure 4. The discussion of the results can be found on section 3.

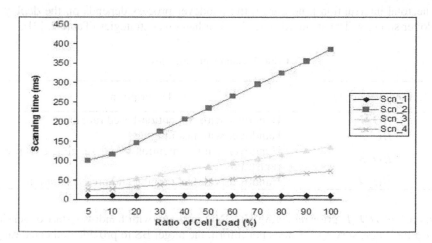

Fig. 3. The scanning performance in various ratio of cell loads

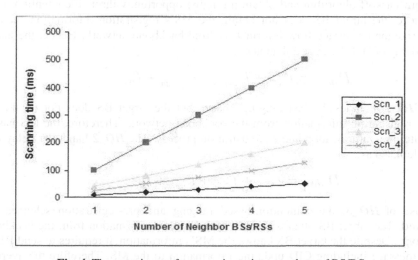

Fig. 4. The scanning performance in various numbers of BS/RS

2.3.3 Handover Interruption Time

The handover interruption time in 802.16e systems is caused by switching of MS form the serving BS to the target BS The same description is assumed for 802.16m. The system profile of 802.16m also designed the maximum interruption time for handover i.e. 30 ms for intra-frequency and 100 ms for inter-frequency [6].

When the MS crosses a boarder of cells between the serving BS and target BS, the connection with the serving BS is closed. After that, a new connection with the target BS is established. Notice that after closing the connection and before setting up the new one, the MS has no connection to the network for a short time. During the interruption, packets must be routed from the serving BS to the target BS. After establishing of the connection between the MS and the target BS, packets are again sent to MS. The packets are delayed due to re-connection (network re-entry) of MS to the target BS [7].

The total interruption time during the handover process depends on the deployed handover strategy. The standard specifies four handover strategies (Table 2, [2]).

Table 2. Handover strategies

Handover strategy	Description
HO_1	Handover with contention-based ranging
HO_2	Handover with fast ranging
HO_3	Handover with contention-based ranging and pre-registration
HO_4	Handover with fast ranging and pre-registration

In case of **HO_1**, prior receiving HO-IND message with handover start or service release indicator, serving BS does not inform the target BS to provide dedicated ranging opportunity for MS. During the network re-entry, if collision occurs, MS executes random backoff algorithm and obtain a ranging opportunity through contention. Furthermore, the target BS does not obtain the MS's registration information such as authorization or service flow information from backbone network. Hence, the handover delay of HO_1 can be calculated as:

$$D_{HO\_1} = T_{Sync} + T_{Cont\_res} + T_{Auth} + T_{Reg} \qquad (7)$$

The **HO_2** includes the fast ranging phase, ,but the target BS does not obtain the MS's registration information from the backbone network. Therefore, the MS has to execute a re-authorization and re-registration process. The *HO_2* handover delay can be calculated as:

$$D_{HO\_2} = T_{Sync} + T_{Rng} + T_{Auth} + T_{Reg} \qquad (8)$$

In case of **HO_3**, the contention-based ranging and pre-registration schemes are adopted. The target BS obtains the MS's registration information from the backbone network.. Despite the target BS knows the MS's information, it requires to send REQ-RSP message including CID updating information to the MS. Therefore the average time for pre-registration is assumed to be a half of the average time of whole registration process. The handover delay of *HO_3* is given as:

$$D_{HO\_3} = T_{Sync} + T_{Cont\_res} + \frac{T_{Reg}}{2} \qquad (9)$$

The last type of handover, **HO_4,** adopts the fast ranging and pre-registration schemes. The target BS provides a dedicated ranging opportunity to MS and obtaining the service flow and authorization information of the MS from backbone network. The target BS also requires to send REQ-RSP message including CID update information the MS. Thus, the average time required for registration is assumed to be half of complete registration. The handover delay of **HO-4** can be calculated as:

$$D_{HO\_4} = T_{Sync} + T_{Rng} + \frac{T_{Reg}}{2} \qquad (10)$$

The results of handover interruption time of four handover strategies are shown in Figure 5. The results are discussed in section 3.

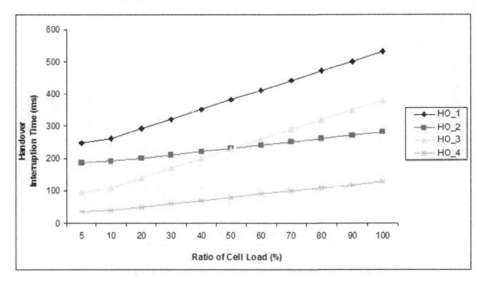

Fig. 5. Handover interruption time in various handover strategies

3 Analysis of Efficiency

Based on presented results in previous section, analysis of efficiency of scanning process and handover interruption time is discussed.

Figure 3 shows that the ratio of cell load has significant impact on the scanning time for some of scanning types. In case of the scanning without association *(Scn_1)* the scanning time remains constant for what different cell load increases. This is due to the fact that a MS only needs to synchronize with the target BS to learn the quality its physical channel. For other three scanning mechanisms, the scanning time increases as the cell load increases.

From Figure 4, it can be observed that the number of neighbour BSs affects the duration of scanning time. For the ratio of cell load of 5%, and 5 neighbor BSs, the increase of scanning time is almost linear to the increase of number of neighbour BSs. The scanning without association (*Scn_1*) also has a better performance than the ones in all cases.

Figure 5 indicates that the performance of handover interruption time depends on several conditions such as the cell load and frame duration. Handover with fast ranging and pre-registration (*HO_4*) seems to be the only handover strategy that can fulfilled the IEEE802.16m's requirements since it can reach the interruption time around 34 ms.

Impact of different frame duration on the scanning time is shown in Figure 6. As it can be observed, the scanning time linearly increase as the number of frame growing. Again, the scanning type *Scn_1* gives better results since it has the lowest scanning time (10 ms).

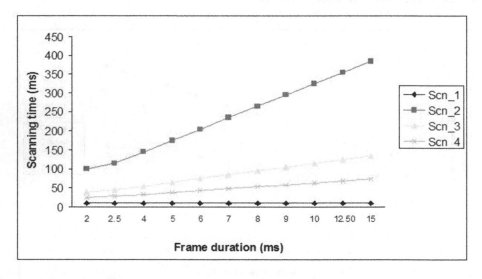

Fig. 6. The effect of frame duration on the scanning time

4   Conclusions

This paper analyses handover's scanning process of IEEE802.16m. Since the standard is still under development, the paper assumes similar handover procedure as it is described in IEEE802.16e. But, we are considering the 802.16m's system profiles in our analysis.

The analysis focuses on scanning time and handover interruption time. Based on our analysis, there are some inefficient issues in 802.16e's handover procedure which is critical for 802.16m. These are redundant scanning and association process of neighbour BSs. Additionally, the defined handover strategies also contribute the overhead in handover procedure. The results suggest using the scanning process without association since it has the lowest scanning time. In addition, the handover with the fast ranging and pre-registration is favorable since it has the lowest interruption time.

The IEEE802.16m standard is designed to support high speed mobility therefore the handovers procedures can be expected to very frequently occur. Thus, the scanning process plays an important role in the handover procedure. In our future work, we would like to investigate the single BS scanning and avoiding the inefficient handover by means of mobility prediction.

Acknowledgement. This work has been performed in the framework of FP7 project ROCKET ICT-215282, which is funded by European Commission. The research work was further supported by grant of Czech Ministry of Education, Youth and Sports No. MSM6840770014.

References

1. Rec. ITU-R M.1645: Framework and overall objectives of the future development of IMT-2000 and systems beyond IMT-2000 (January 2003)
2. IEEE P802.16e/D12, Air Interface for Fixed and Mobile Broadband Wireless Access Systems: Amendment for Physical and Medium Access Control Layers for Combined Fixed and Mobile Operation in Licensed Bands, New York (December 2005)
3. Wang, L., Liu, F., Ji, Y.: Perfomance Analysis of Fast Handover Schemes in IEEE802.16e Broadband Wireless Networks. In: Asia Pacific Advanced Network (2007)
4. Choi, S.: Fast handover scheme for real-time downlink services in IEEE 802.16e BWA systems. In: Proceeding of IEEE 61st Vehicular Technology Conference, vol. 3, pp. 2028–2032 (2005)
5. Jiao, W., Jiang, P., Ma, Y.: Fast Handover Scheme for Real-Time Applications in Mobile WiMAX. In: Proceeding of International Conference on Communication (ICC 2007), Glasgow, Scotland (June 2007)
6. IEEE 802.16m-07/002r4, IEEE 802.16m System Requirements (October 2007)
7. Lee, H., Wong, W.C., Sydir, J., Johnsson, K., Yang, S., Lee, M.: MS MAC handover Procedure in an MR Network – Network Topology Advertisement, Proposal paper on IEEE 802.16j, CTP 06/218 (November 2006)
8. IEEE 802.16m-08/003r7, IEEE 802.16m System Description Document (February 2009)
9. Lee, D.H., Kyamakya, K., Umondi, J.P.: Fast Handover Algorithm for IEEE802.16e Broadband Wireless Access System. In: Proceeding of the 1st International Symposium on Wireless Pervasive Computing (2006)

Optimized Handover Schemes over WiMAX

Zina Jerjees, H.S. Al-Raweshidy, and Zaineb Al-Banna

Brunel University, Wireless Networks and Communications Centre (WNCC),
London, UB8 3PH, UK
{Zina.Jerjees,Zaineb.Al-Banna}@brunel.ac.uk

Abstract. Voice Over Internet Protocol (VoIP) applications have received significant interests from the Mobile WiMAX standard in terms of capabilities and means of delivery multimedia services, by providing high bandwidth over long-range transmission. However, one of the main problems of IEEE 802.16 is that it covers multi BS with too many profiled layers, which can lead to potential interoperability problems. The multi BS mode requires multiple BSs to be scanned synchronously before initiating the transmission of broadcast data. In this paper, we first identify the key issues for VoIP over WiMAX. Then we present a MAC Layer solution to guarantee the demanded bandwidth and supporting a higher possible throughput between two WiMAX end points during the handover. Moreover, we propose a PHY and MAC layers scheme to maintain the required communication channel quality for VoIP during handover. Results show that our proposed schemes can significantly improve the network throughput up to 55%, reducing the data dropped to 70% while satisfying VoIP quality requirements.

Keywords: WiMAX, handover, cross-layer.

1 Introduction

The IEEE 802.16 standard has emerged as an important technology for delivering packet data service in a wide area cellular network [1] [2]. As multi-homed WiMAX will characterize future wireless systems investigating the development of intelligent and efficient handover management mechanisms that can provide seamless roaming capability to mobile users moving between several different access networks, providing a high bandwidth for high speed applications and reduce the overhead cost. However, a traditional problem may occur in the standard, is that it covers too many profiles and PHY layers [3], which can lead to potential interoperability problems as the mobile user traverse between large number of access devices. In ubiquitous networks, mobile nodes can traverse hotspots while maintaining Voice over Internet Protocol (VoIP) communication. Thus, a single VoIP communication can experience several handovers, which can lead to packet loss, resulting in the deterioration of VoIP communication quality. On the other hand, the VoIP different streams may follow the same path through between WiMAX two end points, the Mobile Station (MS) and the base station (BS). This means that the capacity (bandwidth) of that path (defined by its Link layer) severely restricts the amount of critical video streams that can be viewed in emergency situations considering Quality of Service (QoS) parameters. Therefore, preventing packet losses

J. Wozniak et al. (Eds.): WMNC 2009, IFIP AICT 308, pp. 332–346, 2009.

during handover is crucial for high-quality VoIP communication. Thus, enhance us to introduce our handover proposals in order to guarantee VoIP minimum requirements. Since the Mobile WiMAX also provides high performance due to relying upon the upgraded PHY layer *Orthogonal Frequency Division Multiplexing Access (OFDMA)* which, as multiplexing technique, supports multipath environments, thus resulting in the ability to generate higher throughput and improved network coverage. The MAC handover process introduces a large number of interactions between the MS and adjacent (*neighboring*) BSs for the purpose of scanning, ranging, parameter negotiation and information exchanging.

Our first proposed solution, to improve the mobile WiMAX handover and address the latency issue. The second solution is to adapt the channel congestion problem and how it severely restricts the amount of critical video streams, introducing an algorithm that based on the channel bandwidth information, which effectively reduce the mobile WiMAX handover (HO) latency.

In section 2 of this paper, background and related work with an overview of WiMAX MAC layer and handover process moving forward to discuss the handover performance requirements, showing different levels of services for providing Wi-MAX QoS. Then in section 3 we present our proposed schemes to improve the throughput and traffic loss issues and reduce the handover latency. Section 4 presents simulations network model, section 5 presents the performed results and discussion, and finally section 6 concludes the paper.

2 Background and Related Work

2.1 Related Work

A handover solution in [4] proposes a Link-layer handover algorithm that enhances the ranging of data streams between the serving BS and the neighboring BSs, thus enables the MS to receive the data downstream at the time it becomes synchronized with the neighboring BSs, the limitation of this scheme to downstream only from BS to MS may lead to the deterioration of VoIP communication quality as it requires a two directions streams (down and upstream) handover enhanced solution. The proposal in [5] introduce a cross-layer approach for fast handover, the concept relies on the use of an extended MAC-layer information being sent to every channel between the MS and the BSs, by including the IP address information to the MAC probing-messages, as this reduces the time to scan all available channels. The drawbacks of this scheme, the availability of the MS IP address in this open way may grant the chance to an unsecure access device to obtain that address just simply. The other drawback is the buffering of the probe message by an access device for a certain fixed time before it sends it out may cause an additional delay for the MS to connect to its communicating channel. [6] Through the backbone message, a BS sends information to neighbor BSs. The MS selects candidate target BS based on the signal strength and response time of that BS.

Our proposals present a fast handover scheme which reduces HO latency in mobile WiMAX scenarios. The MS will follow the scanning/ranging steps with eliminated number of BSs. As the VoIP quality will be degraded for a reason or another, our schemes will try to solve some channel issues such the traffic congestion and traffic routing without affecting the HO prospects as in the other proposed schemes.

2.2   IEEE 802.16 MAC Layer Handover Support

In IEEE 802.16e standard, three kinds of handover types are supported [10]:

1. Hard handover (HHO)
2. Fast BS switching (FBSS) and
3. Macro Diversity Handover (MDHO)

Among these handover types, HHO is the simplest one while the other two types are more complicated and optional. Up till now, only HHO and FBSS are adequately defined in the standard for practical use. However, both had the physical radio link broken before it is re-established at the target access point that results in handshake latency. In this paper, when handover is mentioned, it refers to HHO. The handover process in WiMAX is concluded in the movement of the MS from a BS to another BS with a connection to a different air interface to provide the desired communication requirements.

The WiMAX handover procedure is consisting of several scenarios as shown in the following section for its all-over acquaintance [7] [11].

2.2.1   Network Topology Advertisement
The BSs periodically broadcast Mobile Neighbor Advertisement (MOB_NBR_ADV) control messages which contain both physical layer (i.e., radio channel) and link layer (e.g., MAC address) information, Bandwidth and QoS. By means of such broadcasts, the MS becomes aware of the neighboring BSs. The MS then triggers the scan-phase.

2.2.2   Scanning/Ranging Procedure
In the scan phase of HO, the MS scans and synchronizes with the neighboring BSs based on channel information from the neighbor advertisement. In order to find an appropriate BS target, the serving BS scans the available channels of the down-link (DL) frequency band. The MS is looking for a well-known DL preamble frame [11]. Since the preamble is transmitted periodically, the BS should gather the frame duration and having synchronized it on time. As a result of receiving preamble frame; the channel estimation, initialization and equalization procedures are taken place. The Frame Control Header (FCH) or the Down Link Frame Prefix (DLFP) which are following the DL preamble can be decoded to the DL burst, which contains the DL mapping message (DL MAP) and the Downlink Channel Descriptor (DCD). These burst messages contain all the required parameters for enabling the DL transmission. When successfully the DL MAP and the DCD are being received, the MAC and the PHY synchronization of the BS is started. If the synchronization successes, it then starts the ranging procedure. The scanning and ranging processes are shown in Figure 1. the BS decodes the Uplink Channel Descriptor (UCD) that contains the uplink centre frequency and the size of the initial ranging contention slots in order to obtain the required uplink parameters, as the BS may need to correct its offset frequency timers and adjusts its transmission power. The MS regularly receives channel information about the neighboring BSs through the MOB_NBR-ADV message from

the current serving BS. If the signal strength weakens, The MS will send a Mobile Scanning Request (MOB_SCN_REQ) message to the neighboring BS with a potential target BS list (selected in the previous phase). The serving BS replies a Mobile Scanning Response (MOB_SCN-RSP) message to the MS to allocate a scanning duration. The MS selects candidate target BSs based on the signal strength and response time of each BS, acquired from scanning. The serving BS may negotiate directly with the listed BSs the allocation of a unicast ranging opportunity. If successful, the ranging procedure can be non-contention-based.

Else, the MS starts a contention-based CDMA procedure to be allocated a ranging slot by the neighboring BS. For the sake of simplicity, in analysis we always assume non-contention-based-ranging. Then the MS starts a hand-shake ranging procedure with the neighboring BS for the OFDMA uplink synchronization and parameter (e.g., transmission power) adjustment. The MS starts choosing an initial ranging contention slot to send its ranging messages (Ranging Request (RNG-REQ)). These messages are addressed to the reserved CID. Then the neighboring BS responds with Ranging Response (RNG-RSP). This response message includes the primary management CID, transmission power information and the frequency timing offset adjustments. This procedure ends after the MS has completed ranging with all its neighbors. In the ranging phase, a MS may switch to a new channel, thus temporally loosing connection with the serving BS.

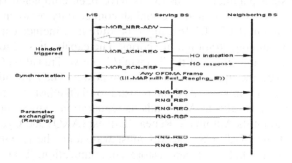

Fig. 1. Scanning and Ranging procedure

From the physical view; Synchronization of a MS across multi-BS is hard to achieve. The synchronization here means for an example; same video content is transmitted in the same OFDMA frame, the same OFDMA data region by the same channel coding scheme [11]. MS synchronization is critical not only for achieving macro-diversity and reducing interference, but also for smooth handover.

2.2.3 HO Decision and Initiation
The HO trigger decision and initiation can be originated by both the MS and the BS using a MS HO Request message (MOB_MSHO-REQ) or a BS HO Request message (MOB_BSHO-REQ) respectively. Here we use the HO started by the MS as an example as illustrated in Figure 2. The MS makes a decision about which BS(s) is (are) its target(s). A HO begins with when the MS sends a MOB_MSHO-REQ message to its serving BS indicating one or more possible target BSs. The serving BS may obtain directly from potential target BSs the expected MS performance at the target BSs through the

exchange of HO indication and response messages. After receiving a response from a target BS (MOB_BSHORSP), the MS notifies the serving BS about its decision to perform a HO by means of a HO Indication (MOB_HO-IND) message. The MS can also ask the serving BS to negotiate with the target BS the allocation of a ranging opportunity. If necessary, the MS may start ranging after HO initiation. The HO decision and initiate process does not provoke connectivity break-up nor does it add latency. However, the possible ranging procedure after does introduce additional latency.

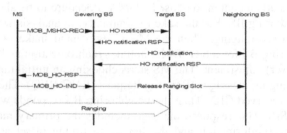

Fig. 2. Handover Decision and Initiation

2.2.4 Network Re-entry
After all the physical parameter adjustments have been completed successfully, the network re-entry process is initiated to establish connectivity between the MS and the target BS. As defined in IEEE 802.16e, this procedure may include capability negotiation, authentication and registration transactions. The capabilities negotiation includes information such like; the length of the Transmit/ Receive Transition Gaps (TTG) and (RTG) and how much the maximum transmission power. The authentication includes the exchange of encryption keys. And the registration is the last step before the MS is allowed to enter the network; the MS sends its registration request message that contains information about its Automatic Repeat Request (ARQ) and Cyclic Redundancy Check (CRC) and the MAC CS. the MS must wait until the re-entry procedure is completed successfully before it can restore communication. Then the MS has to manage its IP connectivity establishment to download the configuration file from its server. When this download is done, the network re-entry is completed. The duration of this phase should be taken account into the entire HO latency.

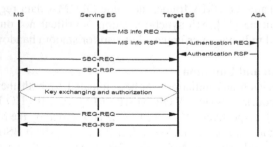

Fig. 3. Network Re-entry

2.3 Criterion Causes Handover

All above procedures may affect the video broadcast performance over WiMAX. This motivates us to highlight some issues which may lead to handover decision in ubiquitous WiMAX:

1. Synchronization of MS across multi-BS is hard to achieve since the same application i.e., video stream content has to be transmitted in the same OFDMA frame and by the same channel usage. The MS synchronization is critical to achieve macro-diversity, reducing interference and reducing the handover. The reason why it is hard to achieve synchronization is as follows: different levels of delay to transmit video data packets copy from MS to BSs, as this video packet copy could also be lost during transmission on the channel between the MS and the BSs. Therefore the same OFDMA will not transmit the same video data packets.

2. There is a possibility of large-size video frame to be transmitted and share the medium. This will arise a congestion issue since the same channel will carry the same video data packets. Thus the video quality will degrade significantly when MS experience temporal fading or interference. In addition, the burst transmission mechanism that allows multiple MAC PDUs belonging to the same video channel to be exchanged in an aggregate way [9]. And if the burst size is big, this may cause an additional delay rate to the channel access operation.

Therefore, to achieve a seamless handover the following three requirements should be considered:

1. Prompt and reliable detection of degradation in wireless link (channel) quality between end-points (MS and BS).

2. Elimination of handover processing elements on Network (if possible), Link and Physical layers.

3 Proposed Handover Solutions over WIMAX

3.1 Fast Channel Scanning (FCS) Based on Bandwidth Information

The first proposed scheme to show that the scanning time is reduced in accordance with the number of eliminated BSs based on the negotiations procedure. Therefore the handover latency will be reduced.

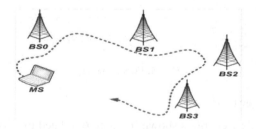

Fig. 4. FCS Scanning Scenario

The MS scans the neighboring BSs based on the channel information provided by the serving BS. Once scanning is complete, the serving BS performs negotiation with the target BSs through a backbone message, and selects a single target BS that meets the bandwidth requirements requested by the MS. While the MS scans, data cannot be transmitted, causing data buffering and undermining the system's performance. Also we have to take care of the MS mobility factor in account. So, adaptive channel scanning [13] technique also is taken into consideration. The proposed scheme involves eliminating unnecessary BSs by performing negotiations with neighboring BSs prior to scanning. Through the backbone message, a BS sends information to neighboring BSs regarding the bandwidth required by the MS before the MS scans. A response from neighboring BSs allows the protocol to assess the MS's expected performance if the service is provided by the corresponding BS. Then the MS scans only the BSs that satisfy its requirements. This approach reduces the number of BSs to be scanned and the time required for scanning, ultimately reducing system interruptions. BS0 is the serving BS, and BS1, BS2 and BS3 are neighboring BSs. Whereas BS1 and BS2 satisfy the MS' bandwidth requirements; BS3 cannot provide the necessary bandwidth. Target BS selection takes place in the order of negotiation scanning; BS3 is eliminated during the negotiation process, and therefore is no longer subjected to scanning. Scanning is performed only on BS1 and BS2. Upon receipt of the MOB_SCN-REQ message, BS0 transmits an HO pre-notification to BS1, BS2 and BS3 to notify them of the MS' ID, as well as its required bandwidth. The HO pre-notification response informs BS0 that although BS1 and BS2 can provide the bandwidth that the MS requires, BS3 can only offer a lower bandwidth. The MS then receives the MOB_SCN-RSP message containing this data, and scans only BS1 and BS2, but not BS3. Based on the values obtained from scanning, BS1 or BS2 is selected as the target BS. The known procedure involves scanning every neighboring BS, which decreases the system's performance due to suspended data transmission during the scanning process. On the other hand, the negotiation process takes place through the backbone network, and does not interrupt data transmission. Therefore, negotiation is performed prior to scanning, to identify the BSs that satisfy the MS requirements. Only the selected BSs are scanned, decreasing the number of scans. This ultimately reduces the suspension of data transmission.

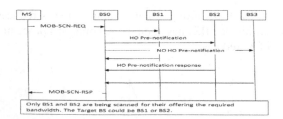

Fig. 5. FCS Scheme

3.2 Traffic Manager (TM)

Our second proposal's scenario is shown in figure 6, related to critical region applications that require QoS to ensure the bandwidth and delay requirements in case of high network loads.

The main idea is that when a MS is transmitting a VoIP application through the assigned channel to its serving BS, and this channel capacity is over loaded, therefore, this MS will try to find an alternative path to forward its additional application streams. This dilemma use case led to the development of our Traffic Manager (TM). Traffic Manager receives bandwidth requests from the applications (supporting different services that require QoS), and uses a real-time view of the network topology and link capacities to calculate the best path for that connection. On this path, the demanded bandwidth is reserved for the lifetime of the connection. The MS must have two interfaces at least. Two data streams pass by each other; they may occupy the same space and bandwidth. To avoid flow interference each data stream should use different channel, here will require channel assignment issue or use one of the interfaces for one flow and the second interface for the other flow with a condition that the two interfaces are using different channels not a single one. This means that the first connections between the MS and the serving BS will be reserved over this path, without disconnecting it, and when this is congested, other request will be routed along other paths through the network. As a result, the network is able to adjust to dynamically changing traffic demands of the applications, guaranteeing the demanded bandwidth, and supporting a higher possible throughput between two Wi-MAX end points.

In addition to WiMAX PHY/MAC layers, there are three more functional entities essential to support Traffic Manager Server over WiMAX. The first one is a correct central real-time view of the network topology, and an accurate estimation of the bandwidth capacity of every channel between end points. The second one is a central software module that can calculate the path and possible bandwidth for every available neighbor BS for a bandwidth request within the network coverage. The third one is a mechanism to ensure that a connection does not consume more bandwidth then it has been assigned.

MS is transmitting a VoIP application through the first interface channel to its serving BS, and this channel capacity is facing traffic congestion issue resulted in the application degradation. Therefore, MS will try to find an alternative path to forward its additional application streams. MS will start an association context with Traffic Manager.

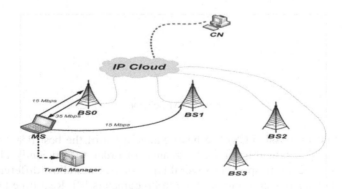

Fig. 6. TM Scenario

The MS first interface (IF1) informs the TM of the bandwidth rate whenever an ACK frame is received or the bandwidth rate reaches the Bandwidth-limit. After recording the bandwidth rate in BAN1, the TM compares BAN1 with the Bandwidth-limit which is the threshold for switching to another network interface. In the case that BAN1 less than the Bandwidth-limit, the TM detects the deterioration of the condition of the wireless link and switches to the Two-path transmission in order to prevent packet loss and transmit the data packets on both interfaces.

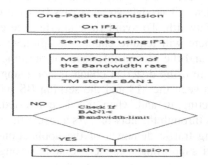

Fig. 7. Transmission to another Interface

In the Two-path transmission, since the MS sends data packets to the CN through both WiMAX interfaces, the networks load increases. When switching to the Two-path transmission, the bandwidth rate that a packet experiences is used as a switching criterion. The MS second WiMAX interface is assigned to communicate with the TM. In this case, there is a need for network load balancing in order not to decrease the connection capability, which is not considered in this study.

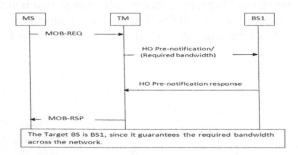

Fig. 8. TM Scheme

From the first proposal (FCS), we have gained a list of the best BSs estimated to offer the required bandwidth for data streams. In order to efficiently allocate UL bandwidth, UL scheduling specifies several types of services allow different levels of flexibility and efficiency carrying out the QoS parameters [9], Real-time Polling Service (rtPS) is selected in the parameter set. rtPS supports real-time data streams that generate variable size SDUs on a periodic basis, such as MPEG video. According to the UL bursts size that meets the flow's real-time needs, it allows the MS to specify the size of the desired grant. Therefore during the granted UL allocation, the MS sends PDUs containing the bandwidth request information. This service type requires more signalling overhead among the others, but supports flexible sizes for efficient data transport.Since the TM is aware of this list, it will forward the traffic to an alternative BS within the list. Here BS1 has been chosen to be the Target BS to communicate with the CN.

TM server consists of an important entity (its structure is built in OPNET network modeller [12]) that is responsible for mapping video channel ID to Connection ID (CID), transmitting TM-MAC-PDU over WiMAX PHY/MAC, burst scheduling and allocating OFDMA data region for each PDU. Mapping video channels to multicast CIDs shall be known to all BSs belonging to the same TM server geographical zone.

WiMAX PHY/MAC at BS side applies PHY channel coding (specified by TM server) to each TM-MAC-PDU and maps each TM-MAC-PDU to the corresponding OFDMA data region slots (determined by TM server) for transmission. One TM-DATA-IE for each TM-MAC-PDU will be transmitted in TM-MAP (at the beginning of TM zone) to indicate the CID and OFDMA data region.

At the MS side, according to video channel selected by a video channel selector [11] indicates WiMAX PHY/MAC to decodes only the TM-MAC-PDUs associated with the corresponding video channel (which translates into CID), thus saving power and bandwidth consumption. Communication entity between TM server and BS(s) is provided for transporting video packet from TM server to multiple BSs, which are physically separated in Multi-BS TM system.

When the MS intends to switch to another video channel as the available channel is congested, its TM server first gets the corresponding new Multicast CID via mapping from video channel to multicast CID. Then WiMAX PHY/MAC checks the TM-MAP in the new OFDMA frame locates the TM-DATA-IE containing the new multicast CID and starts decoding the corresponding TM-MAC-PDU. In the meantime, WiMAX PHY/MAC may stop decoding the TM-MAC-PDUs belonging to the congested video channel. Via TM-MAP redirection, WiMAX PHY/MAC knows the frame number of next frame containing interested TM-MAC-PDUs. Since TM server can locate and decode only the packets on the video channel currently being watched, no power will be wasted for decoding unwanted video packets. This may optimize the power consumption issue if required.

This proposal has the following features. First, we provide broadcast synchronization among multi-BSs through the cooperation between TM server and BSs. In other words, same video content will be transmitted at the same time and in same bandwidth across multi-BSs. In this way, a seamless handoff can be achieved from one cell to another in a feasible way.

4 Simulation Model

The OPNET v14.0 simulation tool [12]; is used in order to implement the scenarios shown in figure 4 and 6. OPNET provides flexibility through enabling the design of new communication protocols and devices, which may be used along with a readily available library that has been developed. For the first scheme, we build our simulations on a simplified but typical scenario; for the physical layer specification, we assume OFDMA/TDD is used, the OFDMA frame duration (*Ft*) is 5ms, and the minimum OF-DMA slot rate is 122kps. We rely on a network topology which contains a serving BS and three neighboring BSs. The MS is aware of the neighboring BS list from neighbor advertisements before the HO. During scanning, the MS gets synchronized with 2 neighbors and finally chooses one after several ranging (according to the best bandwidth information) transactions.

Our second scenario; the mobile station has two WiMAX interfaces. MS is communicating with the correspondent node (CN) through IF1 to BS1, connected the wireless network to IP cloud which is used here to represent the internet backbone connectivity. The MS' second interface is connected to the Traffic Manager server. Scalable Video coding (SVC) [14] is the codec used in the server which is an extended version of MPEG-4 H.264/AVC (for Advanced Video Coding) where the video quality degradation is smooth when users move within and across cells. SVC is a video coding technology that encodes the video at the highest resolution and allows the data bit-stream to be adapted to provide various lower resolutions streams, results in several types of encoded data streams with a consideration to their bandwidth requirement.

The one-way delay to the CN from each WiMAX BS network is varied: that from BS0 is set to 35ms and that from BS1 is set to 10ms. Note the hidden Nodes issue is not considered in this study.

Two video streams are requested between MS and the serving BS in each direction, for UL 30 and DL 35 Mb/s respectively. Because the direct link between MS and BS has a capacity of 50 Mb/s, the first stream is forwarded via this link. But the second stream needs another 15 Mb/s, meaning that if the stream was forwarded over the same path, the link would be loaded with 65 Mb/s, which is too much. Therefore, the Traffic Manager ensures that the second stream will be forwarded from MS to BS1at this time, to the corresponding node (CN), enabling the streaming of both video sources from MS to CN. Since the SVC coding mechanism is being used as this will give the priority to the most important data streams to pass through the best channel. Therefore, the first stream is assigned to provide basic video quality with low bit-rate that is required by most users, and the second stream will pass the higher resolution data which is less important and can be removed when the available bandwidth is not sufficient.

As Synchronization of MS across multi-BS is hard to achieve since the same application i.e., video stream content has to be transmitted in the same OFDMA frame and by the same channel usage, the MS must try at least more than 1 OFDMA frame duration (*Ft*) in *T* slot to get synchronized before ranging. Therefore assuming it take the MS (μ (0, *Tsync_Max*) and *Tsync_Max* is (2T*Ft*) according to the initial synchronization and downlink quality estimations. If the synchronization is not successful after *Tsync_Max*, the neighboring BS is assumed invalid within the BS obtained list and no further actions are taken. So the Synchronization latency will be reduced:

$$Tsync(proposed) = Tsyncs * \frac{N-S}{N} \ . \tag{1}$$

As the *Tsyncs* is the standard Synchronization latency.

N: number of BSs,   *S*: number of invalid BSs.

The second part is the ranging transaction duration based on the bandwidth information: the MS must estimate an initial ranging request (B). We assume the rate of this estimation follows a normal threshold limit N $(Bmin, \delta)$. δ is the variance which is assumed as one bandwidth rate adjustment step. $Bmin$ is the lower limit for the successful ranging bandwidth request. In this model, if and only if $Bmin > B$, do the neighboring BS receive ranging requests. The MS must increase B by 1 step and retry until $B \geq Bmin$ as it's done in the two path algorithm. In the optimized HO, the MS directly ranges with valid BSs that provide the required bandwidth. The total ranging transaction latency can be calculated as:

$$\sum_{i}^{N}\sum_{j}^{M} Trang_{(i,j)} \, . \tag{2}$$

While N here is the number of valid neighboring BSs and M represents the number of ranging retrials for the i neighboring BS. In the seamless HO, the MS need not wait for ranging responses from the neighboring BS. Therefore, the ranging latency is $M*Trang\_REQ$, while $Trang\_REQ$ is the ranging request transmission delay (Ft). The RNG_RSP transporting latency through the backbone does not affect the ranging transaction latency since the MS is still served by the serving BS during ranging transactions. So the Scanning latency will be:

$$\sum_{i}^{N} Tscan_{(i)} = N * (T_{sync(proposed)} + \sum_{j}^{M} Trang_{(j)}) \, . \tag{3}$$

The network re-entry delay in the optimized scheme Tent is the capability negotiation, authentication and registration plus a random handling duration is assumed to be 110ms in heavy load. The assumed length of the Transmit Transition Gaps (TTG) is 5ms which is small according to the forwarding algorithm in TM. Transmit Transition Gap (TTG), is the gap between the last sample of the downlink burst and the first sample of the subsequent uplink burst in a time division duplex (TDD) transceiver. This gap allows time for the base stations (BS) to switch from transmit to receive mode. During this gap, the BS is not transmitting modulated data but simply allowing the BS transmitter carrier to ramp down.This latency is eliminated in the optimized scheme since the MS still keeps the communication with the severing BS during the network re-entry process. As the result, HO latency of the optimised schemes can be calculated as:

$$T_{handover} = \sum_{i}^{N} Tscan + (Tsync\_Max * N * Tent)) \, . \tag{4}$$

5   Results and Discussion

We run the OPNET simulation to analysis effects of different parameters on the HO latency for both scenarios respectively.

The delay is a function of handover algorithm and traffic characteristics. FCS resulted in less delay as shown in figure below.

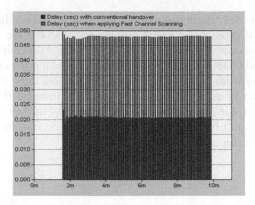

Fig. 9. The Handover Delay

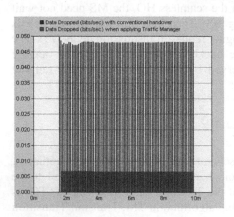

Fig. 10. Data Dropped During Handover

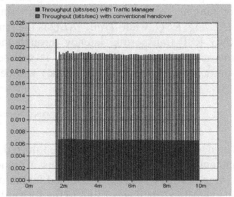

Fig. 11. Throughput performance

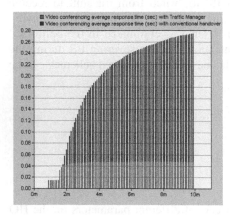

Fig. 12. Video Conferencing Average Response Time

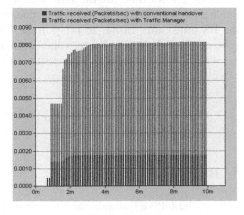

Fig. 13. Traffic Received

The analysis results show that the FCS scheme as compared to the known conventional scheme can reduce the scan time, therefore handover latency up to 50% in the network topology acquisition process especially in association modes and successfully keeps handover delay under 30 ms. The reduction comes from elimination of network unnecessary BSs by performing negotiations with neighboring BSs prior to scanning. Only selected BSs will be scanned and ranged within the network. This ultimately reduces number of scans, thus the suspension of data transmission

From figures 10, 11, 12 and 13, it can be seen that the TM scheme can improve the video heavy load application's throughput in the network to more than 55% (This statistic represents the average number of packets successfully received or transmitted by the receiver or transmitter channel per second), reduces the data dropped to 70%, with assurance of traffic received to more than 60% by the target node (This statistic represents the total number of data packets received per second by target node from the network, across all interfaces).

It can be seen that the proposed handover algorithm improved the system efficiency and enhanced the overall network performance enormously in terms of fast application response time, of Video Conferencing applications in response time are 65% (20.766sec) in 10 ms respectively as compared to conventional handover technique for WiMAX integrated networks.

The advantage of our cross-layer scheme is that the Performance does not degrade when the load increases, as the load will be distributed across different channels. In a word, these fast handover schemes can successfully reduce the waste of wireless resources and improve the performance of IEEE 802.16 broadband wireless networks and its mobility requirement during the handover process.

6 Conclusion

In this paper, we have proposed solutions to select a better base station and initiating handover to that station. The aim of the (FCS) Fast Channel Scanning is to reduce the scanning time pre- handover by elimination the number of BSs to be scanned according to the required bandwidth support, providing a clear evidence of how it successfully improves the mobile WiMAX handover and addresses the latency issue. The second proposal Traffic Manager (TM), when a mobile station detects that the current channel quality of the primary path becoming degraded as a result of traffic congestion on that channel, the mode is switched to another transmission interface according to the bandwidth condition. Through simulations, we have shown several performances at the time the MS switches to different network to adapt the channel congestion problem and how it severely restricts the amount of critical video streams, introducing an algorithm that based on the channel bandwidth information, which effectively reduce the mobile WiMAX handover (HO) latency. Future work, will aim to convert the TM server scheme to a protocol design that will add more applications in addition to video traffic in large mesh networks, to improve the system scalability and other capacity issues of the network. Since the proposed traffic management scheme had MS equipped with two WiMAX interfaces that is impractical as it will increase the hardware cost of MSs If an intention to bring this product/service to market.

References

1. IEEE Std 802.16-2004, IEEE Standard for Local and Metropolitan Area Networks, Part 16: Air Interface for Fixed Broadband Wireless Access Systems
2. IEEE Std 802.16e-2005, IEEE Standard for Local and Metropolitan Area Networks, Part 16: Air Interface for Fixed Broadband Wireless Access Systems, Amendment 2: Physical and Medium Access Control Layers for Combined Fixed and Mobile Operation in Licensed Bands, Corrigendum 1
3. Etemad, K.: Overview of mobile WiMAX technology and evolution. IEEE Communications Magazine (October 2008)
4. Choi, S., Hwang, G.-H., Kwon, T., Lim, A.-R., Cho, D.-H.: Fast Handover Scheme for Real-Time Downlink Services in IEEE 802.16e BWA System. In: Vehicular Technology Conference (2005 IEEE 61st), vol. 3, pp. 2028–2032 (2005)
5. Speicher, S., Bünnig, C.: Fast MAC-Layer Scanning in IEEE 802.11 Fixed Relay Radio Access Networks. In: ICN/ICONS/MCL, April 23-29 (2006)
6. Kyung-ah, K., Chong-Kwon, K., Tongsok, K.: A seamless handover Mechanism for IEEE 802.16e Broadband Wireless Access. In: ISPC 2004 (2004)
7. Chen, L., Cai, X., Sofia, R., Huang, Z.: A Cross layer Fast Handover Scheme For Mobile WiMAX. IEEE Press, Los Alamitos (2007)
8. Li, B.: Hong Kong University of Science and Technology Yang Qin and Chor Ping Low, Nanyang Technological University Choon Lim Gwee, Republic Polytechnic: A Survey on Mobile WiMAX. IEEE Communications Magazine (December 2007)
9. Walke, B.H., Mangold, S., Berlemann, L.: IEEE 802 Wireless Systems. John Wiley and Sons Ltd., Chichester (2006)
10. Hu, R.Q., Paranchych, D., Fong, M.-H., Wu, G.: On the Evolution of Handoff Management and Network Architecture in WiMAX. IEEE, Los Alamitos (2007)
11. Optimized Network Engineering Tools, OPNET Technologies, session 1827 (2007), http://www.opnet.com
12. Optimized Network Engineering Tools (2008), http://www.opnet.com
13. Rouil, R., Golmie, N.: Adaptive Channel Scanning for IEEE 802.16e. In: Proceedings of 25th Annual Military Communications Conference (MILCOM 2006), Washington, D.C., October 23-25 (2006)
14. Schwarz, H., Marpe, D. (Member, IEEE), Wiegand, T., (Member, IEEE): Overview of the Scalable Video Coding Extension of the H.264/AVC Standard. IEEE transactions on circuits and systems for video technology 17(9) (September 2007)

Data Collection for Heterogeneous Handover Decisions in beyond 3G Networks

Marc Fouquet[1], Christian Hoene[2], Morten Schläger[3], and Georg Carle[1]

[1] Technical University of Munich
[2] University of Tübingen
[3] Nokia Siemens Networks, Berlin

Abstract. Future mobile networks will be increasingly heterogeneous. Already today, wireless LAN is used by many mobile network operators as an addition to traditional technologies like GSM and UMTS; WiMax and 3GPP Long Term Evolution (LTE) will be added. Having heterogeneous wireless networks, one challenging research question needs to be answered: Which user should be served by which access network at which time and when to conduct a handover? For such decisions, information on the state of networks and terminals is required.

In this publication we simulate mobile networks in which a central entity called Network Resource Management (N-RM) gives handover recommendations to mobile terminals. Based on these recommendations and local knowledge on link qualities, the terminals choose the cell to switch to.

The N-RM should have a global view on the networks to give best recommendations. We designed the Generic Metering Infrastructure (GMI), a publish/subscribe system to collect information about access networks and terminals efficiently.

We investigate the tradeoff between the signalling overhead caused by data collection and the quality of the handover decisions and show, how smart monitoring can reduce the amount of measurement data while ensuring the efficient use of heterogeneous networks.

1 Introduction

Future mobile networks will be heterogeneous, i.e. consisting of GSM, UMTS, WLAN, WiMax [1], LTE [2], and other radio access technologies. 3GPP is currently standardizing IP-based mobility solutions that will allow seamless handovers between such technologies [3].

Having a heterogeneous network, we need to decide, which user should be served by which access technology at which time. As different user and operator preferences must be considered, a policy-driven decision engine has to make the handover decisions [4]. It is probably impossible to develop one "perfect" algorithm that fits all needs.

Deciding handovers locally in the mobile terminal is sub-optimal, as the global situation in the network, i.e. the load in the different cells, can not be taken into account. It would be advantageous to have a "wizard of oz" view on the access networks and make network-side handover decisions with perfect information, but, of course, it is impossible to collect all data and to make decisions on time. In addition flat hierarchies

J. Wozniak et al. (Eds.): WMNC 2009, IFIP AICT 308, pp. 347–358, 2009.

in future mobile networks make the process of data collection increasingly difficult. In LTE networks there will no longer be a node like the UMTS RNC, which already has load- and radio data of hundreds of cells, but only evolved nodeBs located much closer to the antennas. With wireless LAN access points, the situation is similar. Collecting management data is expensive as the base stations are spread in the countryside, with costly rented or wireless "backhaul"-links connecting them to the operator's core network. Thus, any central view on the network will be inaccurate and delayed.

To get an efficient view on the state of the network, we have designed the **Generic Metering Infrastructure (GMI)**[5], a publish/subscribe system for collecting and distributing measurement data in an operator's network. It uses various compression techniques for efficient data transport. The GMI is intended for all kind of management applications, i.e. for fault management, security management, and resource management. In this publication we concentrate on the usage of GMI for making handover decision.

We developed a network-side decision engine called **Network Resource Management (N-RM)**, which resides in the core network (Figure 1). It implements algorithms, which give the terminals handover recommendations. The mobile terminals make the final handover decisions based on their local knowledge on the radio conditions and the N-RM recommendations. We base this exemplary system on research results by Fan et al. [6] achieved in the BmBF project "ScaleNet".

In this work we do not aim at increasing the performance of the handover system because it could follow diverse goals and any arbitrary policies. Instead, we take a few algorithms as examples and study, on how the accuracy of information influences the quality of central decisions. More precisely, we primarily focus on the effect of different strategies for data collection and the resulting quality of mobility decisions.

We describe the GMI in Section 2 and the setup for our experiments in Section 3. Results are presented in Section 4 and interpreted in terms of data volume in Section 5. We discuss related work in Section 6 and finally conclude the paper in Section 7.

2 The Generic Metering Infrastructure

In this section we describe the Generic Metering Infrastructure (GMI), our publish/-subscribe system for future mobile networks. It allows for efficient data distribution by enabling the clients to selectively subscribe for information they need, by distributing the data to interested parties in a multicast-like fashion and by compressing data.

Like all publish/subscribe systems, the GMI offers an event service. Clients can subscribe for "types" of information they are interested in. Whenever new information is created at a producer, it is sent to the event service which then takes care of distributing the new data to all interested clients.

2.1 GMI Design

The GMI's event brokers are called **Metering Management and Collection Entities (MMCEs)**. These nodes manage the subscriptions and create distribution trees if multiple receivers are interested in the same data.

Our generic metering infrastructure is a subject-based publish/subscribe system. This means that the "types" of information as described above are organized in a tree, an individual leaf of the tree is identified by a DNS-like address.

The GMI supports 3 different types of requests:

Periodic measurements: A metering client can subscribe to periodic metering tasks. In this case the subscriber specifies a desired report period and the subject it wants to stay informed about.

Triggers: It is also possible to set triggers for measurements. For such a subscription the metering client specifies one or multiple thresholds for a metered value. If the value rises above or falls below the given threshold the client is informed immediately.

Request/Reply: The last type of reporting is an immediate response to a request of a metering client for a certain value of data. This notification is not an event in the classical sense of a p/s-system. In this case the metering client simply sends a request for a subject (which is a message similar to a subscription) and receives a reply containing the value (which is handled like a notification).

In each case the GMI uses appropriate methods to optimize the data transfer, i.e. by merging similar tasks and building distribution trees.

2.2 Granularity Periods and Report Periods

Monitoring in 3GPP networks usually does not directly deal with raw data from the network elements, but with derived "key performance indicators" (KPI).

A network element collects raw data for an interval called **granularity period (GP)**. After the GP is finished, this data is used to calculate the KPIs [7]. With today's network elements like RNCs, the minimum GP is 5 minutes, while 30 minutes are a far more typical value. Of course monitoring with such a granularity barely helps when building a heterogeneous resource management system.

For the GMI we expect the meters to work the same way as they do today, but with shorter granularity periods. Each meter has a specific minimum GP. New data for the GMI's publish/subscribe system is only available when a GP has ended.

This means that with *periodic measurements*, the minimum report interval is one GP. With *triggers*, the system can only evaluate at the end of each GP, whether or not the trigger has fired. *Request/reply-style queries* are answered after the current GP has ended, so they may get delayed for one GP.

For periodic measurement jobs, the client specifies the reporting interval in multiples of the granularity period. We call this property of measurement jobs the **report period (RP)**. So a report period of 10 GPs means that each 10th metered value is actually reported to the client.

2.3 GMI in Future Mobile Networks

Figure 1 shows, how the GMI could be deployed in a 3G-beyond network. For UMTS, meters would be placed at the Radio Network Controller (RNC), a central node that controls hundreds of cells. However in other radio access networks, the necessary data has to be collected at much more distributed locations.

The links between the cells (that may be located somewhere in the countryside) and the packet core network are a scarce and expensive resource for mobile network operators. So any management task has to be careful to save bandwidth here.

The interface between MMCEs and meters depends on the type of meter and the link between meter and MMCE. Legacy meters could use well-known protocols like SNMP here, while GMI-enabled meters would use protocols that are optimized to save data.

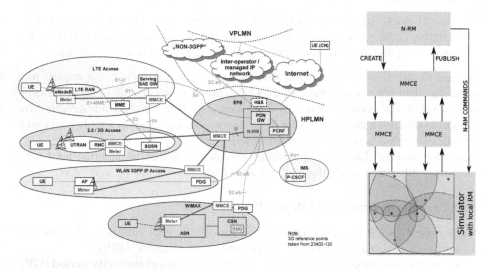

Fig. 1. Mapping of the GMI to the SAE network architecture

Fig. 2. Overview of the experiment setup

3   Simulation Setup

In this section we describe the setup of our experiments. All components of our simulations can be seen in Figure 2. They were run on a single machine, but as separate applications using TCP/IP communications.

3.1   The User- and Radio Network Simulator

The simulator is an application that simulates the mobile networks and their users. Several cells belonging to different **radio access networks (RANs)** are placed on a map. The users move around on this map and start and stop sessions. We use an "accelerated real-time" simulation, events in the simulator happen roughly 15 times faster than they would in reality.

The map is also shown in Figure 2. It contains one LTE cell, five UMTS cells and three WLAN hotspots. Signal quality is based on the distance between the user and the center of the cell, so our UMTS, LTE and WLAN cells only differ in bandwidth and range.

The movement model of the users in our simulations is a modified random waypoint pattern. The users select a new waypoint somewhere on the map with a probability of

40%. With 10% probability they will move to one of the three interesting locations that are covered by the WLAN hotspots. In 50% of the cases the user will stay at his current position for an exponentially distributed random time. Session duration is exponentially distributed as well, between the sessions there is an also exponentially distributed waiting time with same expectation, therefore each user has 0.5 sessions on average.

Our users produce actual load on the air interface by opening sessions, while the GMI produces signalling load which only occurs on the backhaul links between the base stations and the core network. Our assumption here is that there is no direct interaction between these two kinds of traffic, even though they have to share the backhaul link. The rationale for this assumption is the idea that the end-users' traffic will primarily be limited by the available bandwidth on the air interface. The backhaul-link should be dimensioned as small as possible as it produces costs for the operator, but this will be done according to estimates for the sum of user and signalling load.

On startup, users will connect to the strongest RAN. There is a network-internal resource management build into the simulator, which allows handovers between cells of the same RAN, the handover logic for this case is given in Algorithm 1 which is executed for each user periodically.

There are no handovers between different RANs as long as no global resource management application (N-RM) is running. When users leave the coverage area of the current RAN they will lose their sessions and scan for a new RAN as soon as they notice that the old one is gone.

The simulator is able to accept GMI subscriptions for a large number of parameters regarding cells and users. The minimum granularity period of all meters is one second, which would be 15 seconds in the real system.

3.2 N-RM

The network resource management (N-RM) is our global application for handover management. It gets data from the simulator via GMI, its decisions are passed back to the simulator and influence the mobile terminal's cell selection (see Figure 2).

We assume that N-RM does not know the exact location of the user — it only knows the user's location on a "per cell" granularity. Therefore it can not tell exactly, which cells are available at the user's current position, it only has a rough idea, which cells overlap.

The mobile terminal initially also does not know about cells of other RANs. In real life it would have to scan different frequencies to find alternative cells — and this constant scanning would waste battery power. So in our case, N-RM will give the mobile terminal hints, which cells to search for. In reality these hints would also contain radio parameters. The mobile terminal will only scan for cells of different RANs after receiving such a hint.

We are testing five different setups:

– *Experiments without N-RM*
   In these runs, the users will stay in their RAN until they lose connectivity.

Algorithm 1. Mobile terminal- and RAN-internal RM decisions

---

input: users, cells
foreach $u \in users$ **do**

```
// mobile terminal-side logic. If N-RM recommended cells,
   the mobile terminal tries to connect there.
```
  foreach $c \in cells\_recommended\_by\_global\_rm(u)$ **do**
```
    // check signal quality
```
    if $sq(c, u) > th_{minimum\_sq}$ **then**
      $u.handoverTo(c)$;
      $continue\_with\_next\_user$;

```
// This is the "local RM" logic. It only makes handovers
   within cells of the same RAN.
```
  foreach $c \in nearby\_cells\_of\_current\_ran$ **do**
```
    // check minimum requirements for load and signal
```
    if $sq(c, u) > th_{minimum\_sq}$ **and** $l(c) < th_{critical\_load}$ **then**
```
      // find best available cell
```
      $score(c) \leftarrow calculate\_score(l(c), sq(c, u))$;
      if $score(c) > score(former\_best\_cell)$ **then**
        $new\_best\_cell \leftarrow c$;

  if new_best_cell $found$ **then**
    $u.handoverTo(new\_best\_cell)$;

Algorithm 2. N-RM decision logic on bad signal quality

---

precondition: u $\in$ users, sq(u) $< th_{minimum\_sq}$
input   : users, cells
$cells_{recommended} \leftarrow \emptyset$;
foreach $n \in neighbour\_cells(cell(u))$ **do**
  if $l(n) < th_{acceptable\_load}$ **then**
    $cells_{recommended} \leftarrow cells_{recommended} + \{n\}$;

$sort\_by\_load(cells_{recommended})$;
$send\_recommendation(u, cells_{recommended})$;

Algorithm 3. N-RM decision logic on cell overload

---

precondition: c $\in$ cells, l(c) $> th_{critical\_load}$
input   : users, cells, user_list(c)
$cells_{recommended} \leftarrow \emptyset$;
foreach $n \in neighbour\_cells(c)$ **do**
  if $l(n) < th_{acceptable\_load}$ **then**
    $cells_{recommended} \leftarrow cells_{recommended} + \{n\}$;

$sort\_by\_load(cells_{recommended})$;
$receivers = random\_subset(users(c))$;
foreach $r \in receivers$ **do**
  $send\_recommendation(r, cells_{recommended})$;

- *Experiments with a purely trigger-based N-RM*
The N-RM subscribes for the load in each cell and the signal quality of each user. The subscriptions are trigger-based, which means that N-RM is notified whenever a threshold-value is crossed.

  In case of a user with bad signal quality, N-RM will request load information from surrounding cells using ONCE requests. Based on the results it will recommend cells to the user (see Algorithm 2). This recommendation will have an impact on the next run of the local RM (Algorithm 1) inside the simulator.

  In case of an overloaded cell, N-RM will again request load information from the neighbour cells. It will choose a subset of the current users in the overloaded cell and send its recommendations to this subset (see Algorithm 3).

  With this basic triggered N-RM, the resource management will only be informed when a critical situation occurs, there is no feedback whether the situation persists despite the countermeasures. As our triggers are defined with hysteresis, N-RM will only take action again when i.e. the load in a cell crosses the upper threshold again after it has crossed the lower threshold.
- *Experiments with an N-RM based on periodic reports*
N-RM subscribes to load information of each cell and signal quality information of each user on a periodic basis. This means that there is a constant flow of reports which does not change during the simulation. N-RM does not request current information when it needs specific data, but it uses the last reported value.

  The basic decision algorithms are still Algorithm 2 and Algorithm 3. These are run periodically whenever new data has arrived.
- *Combination of triggers and periodic reports*
This variant works with triggers again, but after a trigger has fired and N-RM has taken action, it enters a success control loop. In this state the trigger will be turned off and replaced by a periodic measurement job which continuously monitors the critical value.

  With each arrival of a current value, N-RM will check if the system still is in a critical state. If this is the case, it will again request additional data which is needed for the decision and take action accordingly. There is a different (much lower) threshold for cancelling the periodic measurement job and returning to the triggers when the situation has been resolved.
- *N-RM with full information*
For comparison all simulations have also been run using a N-RM which subscribes for periodic reports of all values in the simulator with a report period of 1 GP. This can be seen as the theoretical maximum amount of data that N-RM could possibly subscribe to. The decision algorithms are still the same.

4 Results

Figure 3 shows the amount of traffic which is produced by GMI reports. On the horizontal axis we have increased the number of users in the system and therefore the offered load.

One can see that the curve for periodic reports with a report period of 10 GP is linear in the number of users as expected. The curve for a report period of 1 GP is also linear but comparably high.

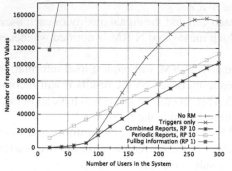

Fig. 3. Number of transmitted values

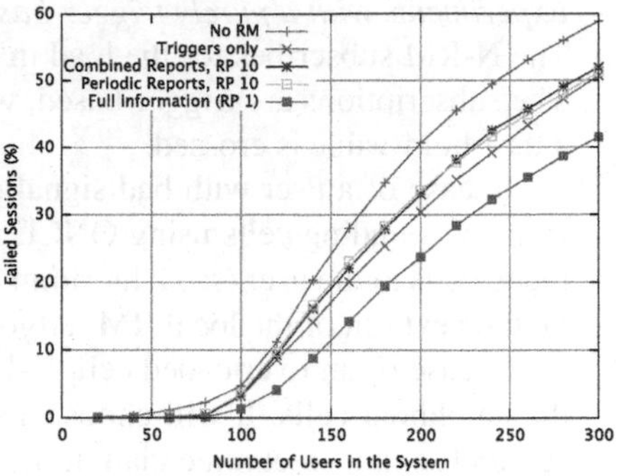

Fig. 4. Failed sessions

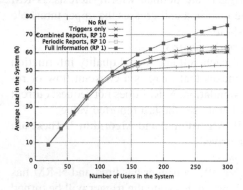

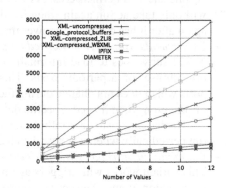

Fig. 5. Average load

Fig. 6. Data volume when transporting measurement data using different protocols

When the system load is low, triggers produce only very few messages as almost no critical events happen. However, when the load increases, the number of trigger-messages explodes. As one can see, the number of messages decreases again when the load is extremely high, as the parameters always stay above their thresholds.

The "combined reports" curve shows the result of an N-RM algorithm which switches to periodic reports after receiving a trigger for a subject. As one can see this curve always produces less messages than the triggered or periodic curves.

The quality of the decisions is evaluated using Figure 4, which shows the percentage of failed sessions. One astonishing fact is that the trigger curve is better than periodic reports and combined reports. This was not expected as the used trigger algorithm is quite primitive, it only acts when it gets an ascending trigger and does not control the success of its actions. However the variance in the measured parameters is high, so the resource management is still triggered very often — which causes N-RM to send more recommendations to the users.

Figure 5 shows the average load in the system. Here the number of users can be seen as a measure of the offered load, while the vertical axis is the actual load which could be handled by the system. The main room for improvement through a network resource management are the WLAN cells here. Without resource management, the users will never switch to WLAN but stay in UMTS or LTE even when passing by a WLAN cell. However with RM, some users can be moved to WLAN, therefore the available bandwidth of the system can be used better.

5   Data Volume

So far, we have only been talking about the number of values that were sent through the GMI. In this section we will discuss the volume of management data which has to be sent over the backhaul links for heterogeneous resource management.

In our implementation, the GMI uses an uncompressed XML data format. However in a production environment it would be favourable to use more bandwidth-efficient protocols. Basically any protocol which transmits key-value-pairs is suitable for transporting the GMI messages.

Figure 6 shows the performance of six different candidate protocols when transmitting a representatively chosen GMI dataset. Note that this is not a hard comparison of these protocols which is simply not possible in such a short analysis. This section is rather meant to give a hint on how to estimate the practically required data volume. With each of the candidates there are almost unlimited possibilities to encode the data differently which of course affects the volume. It should also be noted that for Figure 6, only application layer data volume has been considered, there are no IP-headers and also no headers of the layer 4 protocols that might be used in combination with our six candidates (TCP, UDP, SCTP).

- Uncompressed XML has advantages in terms of transparency as the data is transmitted in plain text. However it wastes bandwidth on the expensive backhaul links.
- WBXML [8] is a standard by the Open Mobile Alliance, which replaces XML tags by shorter binary strings, but leaves the actual content of the tags unchanged. In absence of a Document Type Definition (DTD), it builds a string table from the tag names and uses references to this table afterwards, which was the case here. In our example the data has been compressed to 69% of the size of the original XML.
- Google Protocol Buffers [9] is another representation which preserves the hierarchical structure of XML. In the example the data was compressed to 45% of the original size. The direct comparison with WBXML may be a bit unfair, as the Protocol Buffers encoder was able to use meta-information about the structure of the document. With a DTD, we would expect WBXML to perform roughly equivalent to Google Protocol Buffers.
- Diameter [10] is shown in this comparison as it is a common accounting protocol in 3GPP networks. It requires the GMI messages to be mapped to flat key-value pairs. With our example data, the data volume used by Diameter was 25% of the original XML.
- IPFIX [11] is based on Cisco's Netflow v9 protocol. It is basically meant to export traffic flow data, but it is flexible enough for our purpose. IPFIX also works on flat attribute-value-pairs, but on the link it separates the attributes from the values. The attributes are sent once in so-called template records in the beginning of the transmission, while the values are sent separately in data records. Therefore IPFIX

can save bandwidth compared to Diameter, our example-data was compressed to 12% of the original size.

– The last candidate protocol is a simple LZ77-zipped [12] version of the original XML data. The messages have not been compressed one by one, but the state on both sides is held during the transmission of multiple messages. In the long term, after 50 messages, the data volume could be reduced to 7% of the original XML size. However this advantage comes with increased costs in terms of memory- and CPU-usage at sender and receiver.

Table 1. Data volume vs. success

Resource Management	Number of Users	Failed Sessions	Data Volume per Cell
None	80	2.28%	0.00 kBit/s
Combined Reports (RP 10)	80	0.40%	0.01 kBit/s
Maximum Information (RP 1)	80	0.09%	0.64 kBit/s
None	160	28.01%	0.00 kBit/s
Combined Reports (RP 10)	160	22.09%	0.09 kBit/s
Maximum Information (RP 1)	160	14.16%	1.18 kBit/s

For Table 1 the results from Section 4 have been combined with knowledge about message sizes. Here we assume data transport by IPFIX, which was the second-best solution in the comparison above, as we want to avoid the computational effort of compressing the data using LZ77. We also assume that each value is sent in a separate packet — which is a worst-case scenario — and add TCP and IP headers. As mentioned in Section 3.1, our granularity period is 15 seconds.

As we can see, heterogeneous access management gives a high benefit while using very little bandwidth. With 80 users in the system the result with our combined periodic and triggered reporting is almost as good as the result obtained with the maximum available information.

With 160 users the radio network is already overloaded as we can see from the success rate of the sessions. However the number of messages we need for N-RM still remains negligible compared to the amount of user data transferred through HSDPA or LTE cells. Here one could even consider to send all available data (GP 1) to the N-RM, which produces 14 times the data volume of the combined reports method, but still stays around 1 kBit/s per cell.

6   Related Work

In recent years, there has been a lot of research on handovers in heterogeneous networks. The approaches in [13], [4] and [14] leave the decision which network to choose to the mobile terminal. This is a reasonable design concept since the information about signal quality of the surrounding base stations is available there, while with network-centric decision engines this information must be transported to the core network. Transporting this data also introduces undesired delay that leads to potentially imprecise values.

On the other hand a management facility inside the network is able to take global information, i.e. on the load situation into account and therefore is able to make better decisions. Additionally, it can help the mobile terminal to find adjacent networks without forcing it to scan for available access points which would deplete its battery power.

The authors of [15] use a network-centric approach which basically integrates WLAN into an UMTS UTRAN network. Our approach attempts to be more general by abstracting the handover-logic from details of the RANs. [16] adjusts load triggers on the network-side to optimize handover performance. This work is about the actual handover decision process, i.e. in a combined GERAN/UTRAN network, while we primarily focus on data collection and intentionally keep the decision process as simple as possible.

The authors of [6] use a network-assisted policy-based approach. They're using two decision engines, one of which is located in the core network while the other resides on the mobile terminal. This scenario is also the base of our work, the GMI could be used here to provide the network- and RAN-related information which is required by the decision engine on the network side.

A related standard is IEEE 802.21 [17], which specifies an information service, an event service and a command service to support heterogeneous access decisions and therefore consists of several building blocks that were similarly realized for our simulations. However IEEE 802.21 leaves the actual question of data transport open.

7 Conclusion

Today, users of mobile networks increasingly demand data services and voice-calls for flat prices. This makes business difficult for network operators, there is hard competition on the market and revenues are shrinking. Operators have to cut costs and one way to do so is increasing network efficiency. This requires heterogeneous networks with smart management, as different access technologies have different strengths and weaknesses, which leads to a need for new methods of data collection.

We have shown that the flexibility that our GMI provides gives advantages when making heterogeneous handover decisions. It is possible to make good decisions with fewer data by switching between periodic reporting and setting triggers. Our simulation results show that customer experience could be enhanced significantly, while the cost in terms of produced overhead was negligible.

Our concept of the GMI is not only suitable for heterogeneous access management, but also for general management tasks or for security (i.e. distributed intrusion detection). If the GMI would be used for those purposes as well, the gains could be even bigger because of the late duplication property of the GMI's publish/subscribe system.

Acknowledgements

The authors would like to thank Mark Schmidt and Vladimir Maldybaev for their help regarding the evaluation and Andreas Monger for his conceptual GMI work and the initial implementation of the MMCEs.

References

1. IEEE802.16, Air Interface for Fixed Broadband Wireless Access Systems, IEEE Standard for Local and Metropolitan Area Networks (October 2004)
2. 3GPP, TS 23.401 v8.4.1: General Packet Radio Service (GPRS) enhancements for Evolved Universal Terrestrial Radio Access Network (E-UTRAN) access (2008)
3. 3GPP, TS 23.402 v8.4.1: Architecture enhancements for non-3GPP accesses (2008)
4. Wang, H.J., Katz, R.H., Giese, J.: Policy-Enabled Handoffs Across Heterogeneous Wireless Networks. In: WMCSA 1999: Proceedings of the Second IEEE Workshop on Mobile Computer Systems and Applications, p. 51. IEEE Computer Society, Washington (1999)
5. Monger, A., Fouquet, M., Hoene, C., Carle, G., Schläger, M.: A metering infrastructure for heterogeneous mobile networks. In: First International Conference on Communication Systems and Networks (COMSNETS), Bangalore, India (January 2009)
6. Fan, C., Schläger, M., Udugama, A., Pangboonyanon, V., Toker, A.C., Coskun, G.: Managing Heterogeneous Access Networks. Coordinated policy based decision engines for mobility management. In: LCN 2007: Proceedings of the 32nd IEEE Conference on Local Computer Networks, pp. 651–660. IEEE Computer Society, Washington (2007)
7. 3GPP, TS 32.401 V8.0.0: Telecommunication management; Performance Management (PM); Concept and requirements (Release 8) (2008)
8. Martin, B., Jano, B.: WAP Binary XML Content Format (1999),
   http://www.w3.org/TR/wbxml/
9. Google Inc., Protocol Buffers (2008),
   http://code.google.com/apis/protocolbuffers/
10. Calhoun, P., Loughney, J., Guttman, E., Zorn, G., Arkko, J.: Diameter Base Protocol, RFC 3588 (Proposed Standard) (September 2003),
    http://www.ietf.org/rfc/rfc3588.txt
11. Claise, B.: Specification of the IP Flow Information Export (IPFIX) Protocol for the Exchange of IP Traffic Flow Information, RFC 5101 (Proposed Standard) (January 2008),
    http://www.ietf.org/rfc/rfc5101.txt
12. Ziv, J., Lempel, A.: A universal algorithm for sequential data compression. IEEE Transactions on Information Theory 23, 337–343 (1977)
13. Gazis, V., Alonistioti, N., Merakos, L.: Toward a generic "always best connected" capability in integrated WLAN/UMTS cellular mobile networks (and beyond). IEEE Wireless Communications 12, 20–29 (2005)
14. Stevens-Navarro, E., Wong, V.: Comparison between vertical handoff decision algorithms for heterogeneous wireless networks. In: IEEE 63rd Vehicular Technology Conference, 2006. VTC 2006-Spring , May 2006, vol. 2, pp. 947–951 (2006)
15. Pries, R., Mäder, A., Staehle, D.: A Network Architecture for a Policy-Based Handover Across Heterogeneous Networks. In: OPNETWORK 2006, Washington D.C., USA (August 2006)
16. Tölli, A., Hakalin, P.: Adaptive load balancing between multiple cell layers. In: 2002 IEEE 56th Vehicular Technology Conference, 2002. Proceedings. VTC 2002-Fall, vol. 3, pp. 1691–1695 (2002)
17. IEEE802.21, Media Independent Handover Services (2007),
    http://www.ieee802.org/21/

A Cross-Layer User Centric Vertical Handover Decision Approach Based on MIH Local Triggers

Maaz Rehan[1], Muhammad Yousaf[2], Amir Qayyum[1], and Shahzad Malik[3]

[1] CoReNeT, M. A. Jinnah University, Islamabad
[2] Center for Advanced Studies in Engineering (C@SE), Islamabad
[3] COMSATS Institute of Information Technology (CIIT), Islamabad
maazrehan@yahoo.com, myousaf@ymail.com, aqayyum@ieee.org,
smalik@comsats.edu.pk

Abstract. Vertical handover decision algorithm that is based on user preferences and coupled with Media Independent Handover (MIH) local triggers have not been explored much in the literature. We have developed a comprehensive cross-layer solution, called Vertical Handover Decision (VHOD) approach, which consists of three parts viz. mechanism for collecting and storing user preferences, Vertical Handover Decision (VHOD) algorithm and the MIH Function (MIHF). MIHF triggers the VHOD algorithm which operates on user preferences to issue handover commands to mobility management protocol. VHOD algorithm is an MIH User and therefore needs to subscribe events and configure thresholds for receiving triggers from MIHF. In this regard, we have performed experiments in WLAN to suggest thresholds for Link Going Down trigger. We have also critically evaluated the handover decision process, proposed Just-in-time interface activation technique, compared our proposed approach with prominent user centric approaches and analyzed our approach from different aspects.

Keywords: MIH, Vertical Handover Decision, Cross-layer, User Centric Approach.

1 Introduction

The process of migration of connection from one type of network to another involves decisions and multiple information as input to the handover decision phase. This information pertains to different layers. For example information like received signal strength, packet error rate, missed beacons, link speed, etc. can only be taken from MAC Layer; connection characteristics like achieved throughput, delay, jitter, etc. can only be taken from Transport Layer; and commercial networks subscription details and user preferences regarding network selection are application layer features. While moving, whenever a Mobile Node (MN) is in overlapping region, effective network selection becomes a task that requires intelligent decision making based upon selected information from multiple layers and therefore it openly speaks off the necessity of a cross-layer design [5].

J. Wozniak et al. (Eds.): WMNC 2009, IFIP AICT 308, pp. 359–369, 2009.
© IFIP International Federation for Information Processing 2009

Users would like to avoid inappropriate handover decisions due to varying cost of different wireless access networks. The user may require from the system to choose either a cost effective network, or a best performance network even if it is costly because the connectivity is more important, or the user needs a dynamic hybrid approach which exhibits different behavior in different situations.

802 family of IEEE includes a variety of wireless technologies like 802.11, 802.15, 802.16 that help to establish Local, Personal and Metropolitan area networks respectively. Similarly, cellular networks like GPRS, EDGE, UMTS provide IP support and allow devices to be connected to Wide Area Network. This builds an overall picture in which networks with wide coverage encompass networks with small coverage thus creating overlapping regions and the need of handover. When a Mobile Node (MN) leaves its current network and enters into a new network, a Handover (HO) process is required so that the current end-to-end services of MN may continue. Horizontal Handover (HHO) happens when MN moves into same network technology. Otherwise it is Vertical Handover (VHO).

Link Layer (L2) notifications help to speed up the process of HO. Abstract or Unified L2 notifications [19, 21, 22, 24] facilitate upper layers to receive these notification in an implementation/link technology independent way. Some of these abstractions have been specifically designed for L3 handover [21, 22], while others are for L3 & above in general [19, 24]. Test-bed implementations [21, 23] are also available that use Link Up and Link Down triggers to facilitate L2/L3 handover. Our cross-layer solution focuses on handover decision making in a user centric way that intelligently selects a target network among the candidates. After HO decision phase, any mobility management protocol (e.g. MIPv6[21, 22, 23], EMF[25], TCP-migrates[26] etc.) can be used for handover.

Media Independent Handover (MIH) [19] is a proposed framework of IEEE 802.21 WG which provides a generic interface between 'L3 & above' and 'L2 & below' for different network technologies, e.g. 802 family, 3GPP and 3GPP2. MIH divides the handover into Initiation, Preparation and Execution phases [17]. Handover execution is the phase in which mobility management protocols execute and MIH has nothing to do with it. Handover Initiation and Preparation are the phases where MIH is involved. Handover is initiated when observed link layer parameter, e.g. RSS, missed beacons, packet error rate, etc. degrade enough to indicate either a connection breakage or network load. As a result, handover preparation phase starts in which information about the neighboring networks is accumulated through the already active interface with the help of Point of Service (PoS) entity of current network, as proposed by MIH. MIH provides the aforementioned services to the MIH User through MIH_SAP and MIH_LINK_SAP. The MIH_LINK_SAP is replaced by media specific SAPs of the underlying interface. Fig. 1 shows MIH communication of MIH User with WLAN through MIH_SAP and 802.11u MSGCF [20].

Rest of the paper is organized as follows. Section 2 arguments on the existing approaches by discussing different aspects, Section 3 presents proposed idea, Section 4 gives analysis of the proposed idea and Section 5 is conclusion.

2 Service Continuity, Network Availability and Subscriptions

For handover, [11] lists vertical handover decision strategies including Traditional (RSS-based), Decision Function-based (DF), User-Centric (UC), Multiple Attribute Decision Making (MADM), Fuzzy Logic and Neural Network (FL/NN) and Context Aware (CA) based. Normally, decision strategies work as follow: network selection module or score function first calculates weights of parameter(s) like RSS (Received Signal Strength), QoS parameters like delay, packet error rate, jitter, bandwidth, throughput, etc. and contextual parameters like battery status, etc. either through Analytic Hierarchy Process (AHP) [2] or take them from configuration files as input by user or take fixed values. Secondly it computes weighted sum of the selected parameters to obtain network score and compare the cumulative value obtained against each network to select the best network. In the following subsections we discuss various aspects of handover to make the whole process transparent with the examples from state-of-the-art techniques.

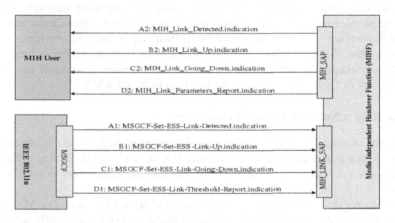

Fig. 1. IEEE 802.21 & IEEE 802.11u interworking

2.1 Assignment of Weights, Score Function and Network Selection

The weighted sum approach [1,2,3,6,7,8,12,13,14] plays a key role in network selection, therefore, inappropriate weight fixing may not bring desired results [13,14]. Weights can either be input from user directly [12] or as relative important values (AHP) of QoS parameters. AHP for weight calculation [1,2,15] and ranking is indeed a trusted mechanism but it can prove to be less useful when incorporated in handover scenarios. A mobile user may not have enough knowledge and experience to relate meaningful QoS parameters like jitter, packet error rate, bit error rate, etc in the way they should be related. Since user has to work with relative numbers when adding, deleting or updating a QoS parameter in AHP, the user can undesirably select costly network or pay less cost but with degraded service, because of selecting a network that provides poor services.

Network selection among candidates, based on end-to-end QoS parameters [1,13,14,15] like achieved bandwidth, delay, jitter, throughput, etc., requires a transport connection on each interface in order to get access network's QoS parameters. The parameters are then calculated using multi-criteria input function called Score function. The cumulative value is compared with the currently selected network's cumulative QoS value to assess handover requirement. Another mechanism for comparing QoS parameters can be to compare currently selected network's transport parameters with the transport parameters of each candidate network, obtained in the last N sessions [10], but this approach may not be effective if the network undergoes through frequent changes in a short span of time.

2.2 Free Access Zone v/s Subscriptions

When a user moves from his home to office, many small coverage networks (WLANs) may come in his way. MN can connect to any of them, if *i) the network is unprotected*, or *ii) the MN is an authorized user*. It means that MN can only connect to at most a few of them if above condition is satisfied. For the rest of the time, on the way, it has to connect to broadband or cellular network for service continuity, and therefore, MN needs to be a subscriber of the corresponding wide area network. Practically speaking, most of the subscribers may have a cellular, a braodband and a wireless LAN interface that can adequately fulfill their requirement of service continuity. We also believe that a network's differentiated service to its subscriber is based on the subscribed package, e.g. pre-pay, post-pay etc., and therefore, techniques like network status inquiry or QoS negotiation my be less beneficial for a MN [1].

2.3 Need for a Network Entity in Handover

Philosophy of using network entity, e.g. Information Server (IS) is discussed in [1,8,14]. When a user carrying multi-homed MN advances to a vicinity where, there is either a single network or multiple overlapped networks, two possibilities can emerge.

MN does not take 'beforehand QoS information' from IS. In this case MN senses a network when it enters the vicinity. In case of single known network MN will attempt to associate the AP/BS after performing RSS stability check. In case of multiple known networks, it will select the best network based on network pre-subscription information, then perform RSS stability check on the selected network and will keep moving down the list if a network does not pass this check. Finally, it will attempt to associate to the chosen AP/BS.

MN takes 'beforehand QoS information' from IS. In this case, MN first performs IS discovery and then requests for network map i.e. APs/BSs in its way. Now MN can compute best network, well before sensing the beacons, based on network pre-subscription information. When user enters the vicinity, MN can sensing the signals and perform RSS stability check on the selected network. Rest of the procedure is same as stated above.

This shows that IS benefits in pre-deciding the best network based on subscribed theoretical QoS promises. The disadvantage is its cost to install it in the network which includes IS discovery protocol, a protocol to gather & maintain network map and a request/response protocol to retrieve network map from IS.

3 Proposed VHOD Approach

Although our proposed approach collects cross-layer context i.e. user preferences and link layer parameters, yet we call it user centric, for, handover is purely based on user preferences. The proposed handover decision policy, Fig. 2, has two major input providers- user preferences and MIHF. As per [19], VHOD algorithm is an MIH User so, it configures link thresholds for receiving link state triggers from MIHF. VHOD algorithm strictly obeys user preferences and executes them when appropriate link state event occurs i.e. when MIHF informs VHOD algorithm about changes in link state through well defined triggers like MIH_Link Detected/Up/Down/Going_Down etc. If handover is required as per user preferences, VHOD issues handover command for mobility management protocol being used in the MN. The proposed handover philosophy is explained below.

3.1 Cross-Layer Information Gathering

Cross-layer information gathering comprises of two parts, one is User Preferences gathering while other is Link Layer Information gathering.

Gathering User Preferences. We take network subscription information from user and store them in our cross-layer module. For the user, some networks e.g. Ethernet and WiFi, may be free as he is an authorized user of some local area network (LAN) in a hotel, on the airport, in the office, in the campus or in a conference. Therefore, these

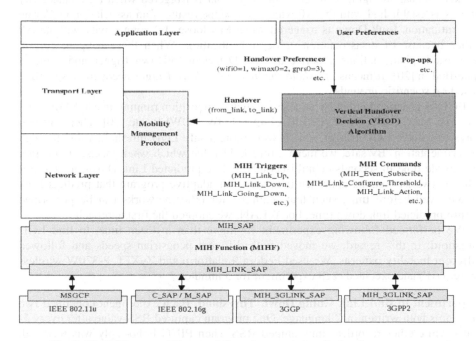

Fig. 2. Block Diagram of Proposed Cross-layer solution for handover management

networks may always be at the first preference of the user. Finally user is asked to provide his network preferences in descending order for different type of network traffic like Video, Voice and File Download, etc. VHOD will subscribe triggers against those access technologies which are selected by the user as his handover preferences.

User may specify more than one interfaces at same preference e.g. two GPRS connections from two different service providers. In case of links at the same priority we perform scaling of units based on cost. In this way we decide which preference is the best for the user in terms of cost and link utilization.

1. Cost and data rates of both networks (Network-A and Network-B) are in same units, so

$$DataRate_{Scaled} = (DataRate_{Network-A} / Cost_{Network-A}) * Cost_{Network-B}$$

2. Data rates are specified in different units, e.g. Net_A in per Mb while Net_B in per Hour, often the case in cellular network. First we have to find megabits per hour for Link_B, and then to perform step 1, described above, to scale the two costs. $DataRate_{Mb/h} = DataRate_{b/s} * 3600 / (8*1024*1024)$

Gathering Link Layer Information. At L2, when a threshold set by MIH User is crossed, or an adaptive algorithm running in that layer detects link state change, it notifies MIHF and MIHF in turn notifies MIH User.

Link Detected event is triggered when MN senses beacons from nearby APs and informs upper layers about the details of that AP like MAC address, frequency, networkID, signal strength, SNR, etc. Link Up event is triggered when L2 connectivity of MN is established with the AP which can be the result of an association and/or an authentication. Link Down is triggered when MN looses L2 connectivity with an AP either because of disassociation or de-authentication. When these happen, current wireless implementations generate a Link Detected/Up/Down trigger and same is specified in [20]. It means *threshold for link going down trigger* needs to be specified based on scientific grounds.

LGD is generated either by an RSS predictive algorithm running inside SME or by 802.11u MAC due to KEY_EXPIRATION, LOW_POWER, etc [20]. There are two ways to receive LGD trigger. One is to explore a safe LGD threshold value through experimentation. By safe, we mean a threshold value which when meets, application has ample time in seconds or milliseconds, before predicted Link Down, to scan for other networks. Second approach is to write an adaptive program that predicts Link Down, well before time, such that scanning for other networks can be performed within predicted link down time. For WLAN, we adopted the first approach.

We performed following experiment for more than a dozen time, to find LGD threshold. In this regard, we moved with different pedestrian speeds and followed different mobility patterns. We used Fedora 9 platform and ZyXEL P-320W wireless router (AP) and executed each experiment for 5 minutes.

Experiment to find Link Going Down Threshold. Each experiment contained two programs both written in C language. One program captured RSS value after every 5-7 ms while other recorded timestamped RSS when PING Echo-reply was received. Features of recording time-stamp and RSS were added in source code of PING [18] program available in fedora 9. PING packets were sent after every 0 milliseconds.

RSS behaviour was analyzed with different speeds and mobility patterns inside building, 'outside building within AP boundary' and very close to AP boundary (inside/outside). Results of experiments are summarized in Table 1. We conclude that,

1. Safe and adequate RSS capturing interval is every 200th millisecond
2. Near AP boundary, packets drop frequently between -89dbm and -93dbm
3. As soon as RSS lowers to -94dbm or -95dbm, MN looses L2 connectivity with AP, very often, i.e. MN frequently disassociates with AP.

Table 1. Summary of Link Going Down Thresholds

Application Type	LGD Threshold (RSS in dbm)	Link Down Predict Time (milliseconds)	L2 Connectivity / Packet Drop
Sensitive to Data Loss	> = -88	~ 1000	Present / Occasional
Other	-89 to -93	~1000 to ~0	Present / Frequent
	-94 & -95	< = 0	Rare / Frequent

3.2 Network Selection Procedure: VHOD Algorithm

VHOD is based on cross-layer information i.e. the information taken from application layer in the form of user preferences and information taken from MIH in the form of triggers. When MIHF trigger is received, VHOD algorithm immediately runs strictly following user preferences. Network selection algorithm is depicted in Fig. 3., where we can notice that 'No' procedure is specified against MIH_Link_Detected trigger because the functionality against this trigger is understood, which is, that, MN will only connect to an APs whose ESSID or any other network ID is specified in configuration file.

4 Analysis of Proposed Technique

The parameters like throughput, handover signaling overhead, handover latency, etc. are related to mobility management protocol and out of the scope of this work.

4.1 Complexity Analysis of VHOD

Time complexity of our VHOD algorithms is $O(n)$ for single traffic type. We use linked list (LL) for storing user preferences thus space complexity is CONSTANT. The reason for using LL is that user preferences may change over the time frequently or infrequently and therefore, storing data in 2D static or dynamic array would be more costly than LL in terms of addition, deletion or changing a preference priority. The space complexity of LL in this particular case is CONSTANT because the number of interfaces on MN will be limited, as discussed above. VHOD algorithm performs look-up in the nodes and selects the most suitable from the ordered list after simple comparison.

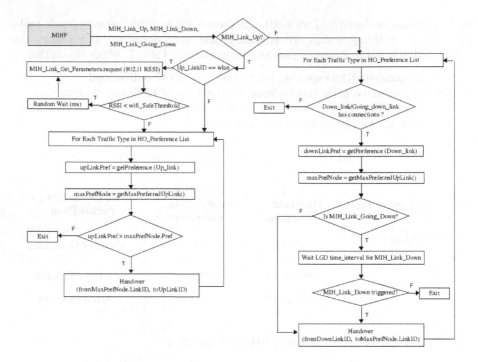

Fig. 3. Proposed Vertical Handover Decision (VHOD) Algorithm

4.2 Delay in Link Selection

In our handover approach, link selection delay depends on two factors, viz. i) *the time to take handover decision* and, ii) *the handover decision process*. We have shown that complexity of handover decision process is $O(n)$, for *single traffic type* (audio, video or download). Time to take handover decision is tied with two triggers, i) *link down* and, ii) *link going down*. In our case LGD is related to threshold value that we obtained through experimentation. Out of 7 experiments, 5 experiments generated in time HO command which proves lower delay.

4.3 Interface Activations for Scanning Candidate Links

Interface activations within a network depends on the dwell time of MN which in turn depends on the speed of user. We have compared our *'Just-in-time'* interface activation technique with *'always on'* and *'periodic'* interface activation techniques. If each interface activation consumes One Unit of battery, then *Just-in-time* consumes almost CONSTANT units of battery from 2-20km/h speed while other techniques consume more battery, even at 20km/h speed (see Fig. 4). *Just-in-time* only suffers when MN's movement is near cell boundary. Here, MN can also suffer from handover ping-pong effect, but we have remedied it (see Fig. 3).

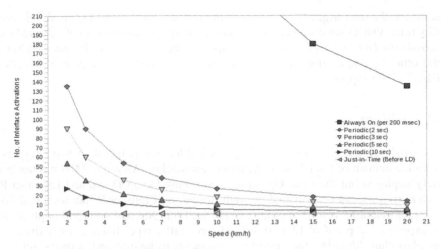

Fig. 4. Comparison of interface activations for different intervals

4.4 Handover Ping-Pong Effect

If MN keeps moving near cell boundary, the threshold will be crossed time and again which can lead to "HO to WiMAX" followed by "HO to WLAN", and so on. Once WiMAX is joined then coming back to WLAN, in case of same or different AP, just based on MIH_Link_Up trigger may be short sighted decision, so a stability check is performed before handing traffic over to WLAN from broadband or cellular network. When MN receives MIH_Link_Up event, it requests MIHF for RSSI of 802.11 through MIH_Link_Get_Parameter. If captured RSS is greater than *wifiSafeThreshold value (-86dbm)*, normal handover process is activated otherwise this process is repeated after short random interval until MIH_Link_Down is received for WLAN (see Fig. 3).

4.5 Comparison with Other User Centric Techniques

Table 2 highlights that, the weaknesses of other prominent User Centric techniques are the strengths of VHOD approach e.g. RSS threshold obtained through experiments,

Table 2. Table showing comparison of User Centric HO algorithms with VHOD algorithm

Criteria	[1]	[2]	[4]	[8]	[9]	[12]	[15]	**VHOD**
Provides experimental RSS threshold value	N	N	N	N	N	N	N	Y
Generates 'Scan' trigger before Handover	N	N	Y	N	Y	N	N	Y
Requires Transport Connection for getting Candidate Networks' QoS parameters	Y	Y	N	N	N	Y	N	N
Requires Network Entity, e.g. IS to get Candidate Networks' QoS Parameters	N	N	Y	Y	Y	N	Y	N
Dwell Timer before handover	Y	N	N	Y	Y	N	Y	Y
'MIH or L2' Trigger Support	N	N	N	N	N	N	N	Y
Defines expected link availability time	N	N	N	N	N	N	N	Y

generation of a scan trigger well before Handover and mentioning expected link availability time. VHOD selects a network based on the agreed QoS instead of achieved QoS of *Candidate Links* by establishing a Transport connection or using Information Server while other techniques use one of them. Similarly VHOD takes additional benefit of MIH/L2 trigger support.

5 Conclusion

Our cross-layer approach for performing vertical handovers based on MIH triggers is new in the domain of User Centric handover approaches. Although MIH has not been widely deployed but the Link Up, Link Down, Link Going Down and Link Get Parameters triggers can be provided through the current Linux or Windows based drivers implementation. VHOD approach works well for QoS and non-QoS applications. Our experimental proof of LGD threshold, per 'traffic type' linear complexity handover algorithm, "Just-in-time" interface activation technique and network selection method based on subscribed QoS makes VHOD approach prominent among the prevailing User Centric approaches.

References

1. Qingyang, S., Abbas, J.: A quality of service negotiation-based vertical handoff decision scheme in heterogeneous wireless systems. European Journal of Operational Research 191, 1059–1074 (2008)
2. Qingyang, S., Abbas, J.: Network Selection in an Integrated Wireless LAN and UMTS Environment using Mathematical Modeling and Computing Techniques. IEEE Wireless Communications, 42–48 (2005)
3. Anita, S., Dr. Nupur, P.: A Review of Vertical Handoff Decision Algorithm in Heterogeneous Networks. In: Mobility 2007, Singapore (2007)
4. Andrea, C., Giuseppe Di, M.: A User-Centric Analysis of Vertical Handovers. In: 2nd ACM International Workshop on Wireless Mobile Applications and Services on WLAN Hotspots, pp. 137–146 (2004)
5. Vineet, S., Mehul, M.: Cross-layer design: A survey and the road ahead. IEEE Communications Magazine 43(12), 112–119 (2005)
6. Wei, S., Qing-An, Z.: A Novel Decision Strategy of Vertical Handoff in Overlay Wireless Networks. In: Fifth IEEE International Symposium on Network Computing and Applications, NCA 2006 (2006)
7. Paramad, G., Saxena, S.K.: A Dynamic Decision Model for Vertical Handoffs across heterogeneous Wireless Networks. In: PWASET, July 2008, vol. 31 (2008)
8. Christian, M., Samuel, P.: Adaptive handoff scheme for heterogeneous IP wireless networks. Computer Communications 31, 2094–2108 (2008)
9. Wen-Tsuen, C., Yen-Yuan, S.: Active Application Oriented Vertical Handoff in Next-Generation Wireless Networks. In: IEEE Communications Society/WCNC (2005)
10. Olga, O., John, M., Gabriel-Miro, M.: Utility-based intelligent network selection in beyond 3G systems. In: IEEE International Conference on Communications (ICC 2006), vol. 4, pp. 1831–1836 (2006)
11. Meriem, K., Brigitte, K., Guy, P.: An overview of vertical handover decision strategies in heterogeneous wireless networks. Computer Communications 31, 2607–2620 (2008)

12. Jiping, L., Yuanchen, M., Satoshi, Y.: Intelligent Seamless Vertical Handoff Algorithm for the next generation wireless networks. In: Mobilware 2008, Innsbruck, Austria, vol. 278, Article No. 23 (2008)
13. Rami, T., Jacques, D., Guy, P.: A Trusted Handoff Decision Algorithm for the Next Generation Wireless Networks. IJCSNS 8(6) (June 2008)
14. Rami, T., Salazar, O., Guy, P.: Vertical Handoff Decision Scheme Using MADM for Wireless Networks. In: WCNC (2008)
15. Shun-Fang, Y., Jung-Shyr, W., Hsu-Hung, H.: A Vertical Media-Independent Handover Decision Algorithm across Wi-Fi and WiMAX Networks. In: 5th IFIP International Conference on Wireless and Optical Communication Networks, Indonesia (2008)
16. IEEE Standard for Local and Metropolitan Area Networks: Part16: Air Interface for Fixed and Mobile Broadband Wireless Access Systems, Amendment 3: Management Plane Procedures and Services, IEEE Std 802.16g^{TM}-2007 (2007)
17. IEEE 802.21 official website- 802.21 Tutorial, p. 18 (2007),
    http://www.ieee802.org/21/Tutorials/
    802%2021-IEEE-Tutorial.ppt
    (accessed March 09, 2009)
18. Code of PING Command for Fedora 9 platform,
    http://www.skbuff.net/iputils/iputils-current.tar.bz2
    (accessed March 09, 2009)
19. IEEE Standard for Local and Metropolitan Area Networks: Media Independent Handover Services, IEEE 802.21-2008 (January 21, 2009)
20. IEEE Standard for Local and Metropolitan Area Networks: Part11: Wireless LAN Medium Access Control (MAC) and Physical Layer (PHY) specifications
21. Amendment 7: Interworking with External Networks, IEEE p802.11u TM/D5.0 (February 2009)
22. Gogo, K., Shibui, R., Teraoka, F.: An L3-Driven Fast Handover Mechanism in IPv6 Mobility. In: SAINTW 2006, USA, January 23-27 (2006)
23. Teraoka, F., et al.: Unified Layer 2 (L2) Abstractions for Layer 3 (L3)-Driven Fast Handover. RFC 5184 (2008)
24. Montavont, N., Noel, T.: Stronger Interaction between link layer and network layer for an optimized mobility management in heterogeneous IPv6 Networks. Pervasive and Mobile Computing 2(3), 233–261 (2006)
25. Wellens, M.: Enabling Seamless Vertical Handovers Using Unified Link-Layer API. In: Mobility 2006, Bangkok, Thailand, October 25–27 (2006)
26. Yousaf, M., Qayyum, A.: On End-to-End Mobility Management in 4G Heterogeneous Wireless Networks. This paper appears in INCC 2008, Pakistan, May 1-3 (2008)
27. Snoeren, A.C., Balakrishnan, H.: An End-to-End Approach to Host Mobility. In: ACM MOBICOM Conference Boston, USA, pp. 155–166 (2000)

Author Index